This is a new and greatly expanded edition of what has become one of the best-known introductions to the principles, techniques, and applications of optical holography. Where necessary, existing sections have been updated, and two new chapters, on holographic optical elements and advanced techniques in holographic interferometry, have been added.

The book begins by presenting the theory of holographic imaging, the characteristics of the reconstructed image, and the various types of holograms. Practical aspects of holography are then covered (including optical systems, light sources, and recording media), as are the production of holograms for display, colour holography, and computer-generated holograms. A variety of the applications of holography are then discussed in detail, such as high-resolution imaging, holographic optical elements, information storage and processing, and holographic interferometry, including vibration analysis and holographic interferometry with photorefractive crystals.

This book assumes only undergraduate training in science or engineering and contains more than 1,000 selected references; anyone wishing to learn more about optical holography, as well as established researchers and engineers in this field, will find it invaluable.

CAMBRIDGE STUDIES IN MODERN OPTICS

*Series Editors*

P. L. KNIGHT

*Department of Physics,*
*Imperial College of Science, Technology and Medicine*

A. MILLER

*Department of Physics and Astronomy, University of St. Andrews*

Optical Holography

# TITLES IN PRINT IN THIS SERIES

# Optical Holography

## Principles, techniques, and applications

SECOND EDITION

## P. HARIHARAN

*CSIRO Division of Applied Physics, Sydney, Australia*

CAMBRIDGE
UNIVERSITY PRESS

Published by the Press Syndicate of University of Cambridge
The Pitt Building, Trumpington Street, Cambridge CB2 1RP
40 West 20th Street, New York, NY 10011-4211, USA
10 Stamford Road, Oakleigh, Melbourne 3166, Australia

First edition published 1983
Second edition published 1996


*Library of Congress Cataloging-in-Publication Data*
Hariharan, P.
Optical holography : principles, techniques, and applications / P.
Hariharan. – 2nd ed.
p.  cm. – (Cambridge studies in modern optics)
Includes index.
ISBN 0-521-43348-7 (hardcover). – ISBN 0-521-43965-5 (pbk.)
1. Holography.  I. Title.  II. Series.
TA1540.H37  1996                                        95-23194
621.36′75 – dc20                                           CIP


A catalog record for this book is available from the British Library.

ISBN 0-521-43348-7 Hardback
ISBN 0-521-43965-5 Paperback


Transferred to digital printing 2004

## To Raj

but for your encouragement and support,
this book would never have been completed.

# Contents

*Contents*

# Preface

The ten years since the first edition of this book appeared have seen significant progress in the techniques of optical holography as well as several new and interesting applications. As always, progress has been uneven; some areas have matured, while others have experienced explosive growth.

This revised and expanded edition attempts to describe some of these new developments and to fit them into the context of earlier work. This task has not been easy because of limitations of space and the increasing level of activity in the field, as can be seen from the fact that it has been necessary to add two more chapters covering new techniques and applications as well as more than 300 additional references to selected original papers.

I am grateful to many of my colleagues for their assistance and, in particular, to Dianne Douglass for the typescript, Stuart Morris and Dick Rattle for the figures, and Gil Webster for guiding me through the pitfalls of LaTeX.

*P. Hariharan*
Sydney, January 1995

# Preface to the first edition

The past ten years have seen an upsurge of interest in optical holography because of several major advances in its technology. Holography is now firmly established as a display medium as well as a tool for scientific and engineering studies, and it has found a remarkably wide range of applications for which it is uniquely suited.

My aim in writing this book is to present a self-contained treatment of the principles, techniques, and applications of optical holography, with particular emphasis on recent developments. After a brief historical introduction, three chapters outline the theory of holographic imaging, the characteristics of the reconstructed image, and the different types of holograms. Five chapters then deal with the practical aspects of holography – optical systems, light sources and recording media – as well as the production of holograms for displays and colour holography. The next two chapters discuss computer-generated holograms and some specialized techniques such as polarization recording, holography with incoherent light, and hologram copying. These are followed by four chapters describing the more important applications of holography. Particle-size analysis, high-resolution imaging, multiple imaging, holographic optical elements, and information storage and processing are covered in two of these, and the other two are devoted to holographic interferometry and its use in stress analysis, vibration studies, and contouring.

To make the best use of the available space, the scope of the book has been limited to optical holography. No attempt has been made to cover related techniques such as acoustical and microwave holography, which have not made such rapid progress. In addition, much of the material on basic concepts of optics found in earlier books on holography now forms part of many introductory science and engineering courses. This has therefore been summarized in a set of appendices, where it is available for reference.

This book is intended for people who would like to learn more about optical holography as well as those who would like to use it. Students will find the book useful as a supplementary text, whereas researchers can use it as a reference work. The initial presentation of each topic is at a level that is accessible to anyone with a working knowledge of physical optics. This is then followed by a more detailed treatment for the serious worker. References to about 700 selected original papers identify sources of additional information and will, it is hoped, guide the reader through the voluminous literature in this field.

I am grateful to many of my colleagues for their assistance. A special debt is due to Dr W. H. Steel, who has been a continuous source of encouragement and helpful criticism.

*P. Hariharan*
Sydney, January 1983

# 1
# Introduction

When confronted with a hologram for the first time, most people react with disbelief. They look through an almost clear piece of film to see what looks like a solid object floating in space. Sometimes, they even reach out to touch it and find their fingers meet only thin air.

A hologram is a two-dimensional recording but produces a three-dimensional image. In addition, making a hologram does not involve recording an image in the usual sense. To resolve these apparent contradictions and understand how a hologram works, we have to start from first principles.

## 1.1 The concept of holographic imaging

In all conventional imaging techniques, such as photography, a picture of a three-dimensional scene is recorded on a light-sensitive surface by a lens or, more simply, by a pinhole in an opaque screen. What is recorded is merely the intensity distribution in the original scene. As a result, all information on the relative phases of the light waves from different points or, in other words, information about the relative optical paths to different parts of the scene is lost.

The unique characteristic of holography is the idea of recording the complete wave field, that is to say, both the phase and the amplitude of the light waves scattered by an object. Since all recording media respond only to the intensity, it is necessary to convert the phase information into variations of intensity. This is done by using coherent illumination and, as shown in fig. 1.1, having a reference plane, or spherical, wave incident on the recording medium (a photographic plate) along with the wave scattered by the object.

Without going into the detailed theory, it is apparent that what is recorded on the photographic plate is the interference pattern produced by the two waves. The intensity at any point in this pattern depends on the phase as well as the amplitude of the original object wave. Accordingly, the processed photographic

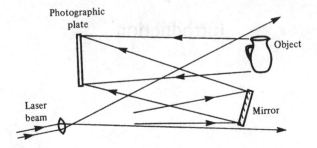

Fig. 1.1. Recording a hologram. The photographic plate records the interference pattern produced by the light waves scattered from the object and a reference wave reflected to it by the mirror.

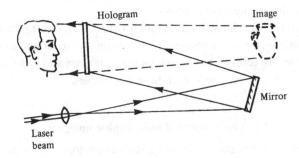

Fig. 1.2. Reconstruction of the image. The hologram, after processing, is illuminated with the reference wave from the laser. Light diffracted by the hologram appears to come from the original object.

plate, which is called a hologram (or whole record), contains information on both the phase and the amplitude of the object wave. However, this information is in a coded form, and the hologram itself bears no resemblance to the object.

The reason for the success of holography is that the object wave can be regenerated from the hologram merely by illuminating it once again with the reference wave, as shown in fig. 1.2.

To an observer, the reconstructed wave is indistinguishable from the original object wave; the observer sees a three-dimensional image that, as shown in figs. 1.3 and 1.4, exhibits all the normal effects of perspective and depth of focus that the object would exhibit if it were still there.

Fig. 1.3. Views from different angles of the image reconstructed by a hologram, showing changes in perspective.

Fig. 1.4. Picture of the reconstructed image taken with the camera lens at full aperture (f/1.8), showing the effect of limited depth of focus.

## 1.2 Early studies

The roots of holographic imaging can be traced back to work by Wolfke [1920] and by Bragg [1939,1942] in x-ray crystallography, which led to the development of the Bragg x-ray microscope. When a crystal is illuminated with a beam of x-rays, the diffraction pattern obtained corresponds to the square of the modulus of the Fourier transform of the amplitude scattered by the crystal lattice. As a result, all information on the phase is lost, creating a major problem in the determination of crystal structures. However, for a centrosymmetric unit cell, the Fourier transform is real, and only the sign of the diffracted amplitude can change. In fact, in the case of a crystal with such a structure and with a heavy atom at the centre, the scattered amplitude from the heavy atom can be large enough that the weaker scattered amplitudes from the other atoms can add to it, or subtract from it, without altering the sign of the resultant. If then a mask is made whose transmittance at any point is proportional to the measured amplitude at the corresponding point in the x-ray diffraction pattern, and this mask is illuminated with a collimated beam of monochromatic light, the Fraunhofer diffraction pattern obtained, which corresponds to a second Fourier transform, yields an image of the structure of the crystal.

Gabor's aim when he proposed the idea of holographic imaging [Gabor, 1948, 1949, 1951] was to obtain increased resolution in electron microscopy. Since it was difficult to correct the spherical aberration of magnetic electron lenses, he proposed to record the scattered field of the object when it was irradiated with electrons and then to reconstruct the image from this record with visible light. To demonstrate the feasibility of his proposal, he devised an arrangement in which light waves were used both to record the hologram and to reconstruct the image from it.

In Gabor's experiment, the object was a transparency consisting of a clear background with a few fine opaque lines on it. This transparency was illuminated with a collimated beam of monochromatic light, and its Fresnel diffraction pattern was recorded on a photographic plate. The complex amplitude in the diffraction pattern could be considered then as the sum of two terms, a constant term due to the directly transmitted beam, which constituted the reference wave, and a spatially varying term due to the light scattered by the details in the object. Since the reference wave was much stronger than the scattered wave, variations in the phase of the scattered wave resulted mainly in variations in the amplitude in the diffraction pattern, the phase being always very nearly that of the reference wave.

The actual hologram was a positive transparency made from this negative, the exposure and processing conditions being adjusted so that the amplitude transmittance of the hologram at any point was proportional to the intensity in the original diffraction pattern. When this hologram was illuminated once again with a collimated beam of monochromatic light, the transmitted wave had a uniform phase but exhibited the same variations in amplitude as those in the diffraction pattern. These spatial variations in amplitude gave rise to two diffracted waves, one corresponding to the original scattered wave from the object, and another, with the same amplitude but with variations in phase of the opposite sign, which formed another image, *the conjugate image*.

The similarities of Gabor's experiment to Bragg's x-ray microscope are evident, but the differences are also extremely significant. In the latter, because of the symmetry of the object, there was no phase information to be lost, and an exact reconstructed image was obtained; in the former, a much wider range of objects could be handled, but the loss of phase information, though tolerable, led to the formation of an additional (conjugate) image.

Following Gabor's work, several attempts were made to produce holograms with an electron microscope. However, the application of his technique to this field has been limited by several practical problems.

Optical holography was also not very successful initially, even though the validity of Gabor's ideas was confirmed by a number of workers, and some of the later developments in holography were anticipated by them [Rogers, 1952; El-Sum & Kirkpatrick, 1952; Lohmann, 1956]. The main reason for the lack of progress was the poor quality of holographic images. This was because the reconstructed image was superposed on a background caused by the conjugate image and the direct beam. Several techniques were proposed to eliminate the conjugate image, but none was really successful. As a result, interest in optical holography declined after a few years.

The breakthrough that effectively solved the twin-image problem and opened the way to the large-scale development of optical holography was the off-axis reference beam technique invented by Leith and Upatnieks [1962, 1963]. This technique had its origin in their earlier work on the optical processing of synthetic-aperture radar data. They argued that the conjugate image was essentially due to aliasing, and they introduced a spatial carrier frequency by using a separate reference beam, which was incident on the photographic plate at an appreciable angle with respect to the object beam. Such a hologram, when illuminated with the original reference beam, produced a pair of images that were separated by a large enough angle from the directly transmitted beam, and from each other, to ensure that they did not overlap.

The invention of the off-axis reference beam technique was followed by the development of the laser, which made available, for the first time, a powerful source of coherent light with which it was possible to record holograms of diffusely reflecting objects with appreciable depth [Leith & Upatnieks, 1964].

At almost the same time, another major advance in holography was reported by Denisyuk [1962, 1963, 1965]. In his technique, which has some similarities to Lippmann's technique of colour photography, the object and reference beams are incident on the photographic emulsion from opposite sides. As a result, the interference fringes recorded are actually layers, almost parallel to the surface of the emulsion and about half a wavelength apart. Such holograms, when illuminated with white light from a point source, reflect a sufficiently narrow wavelength band to reconstruct an image of acceptable quality, similar to that normally obtained with monochromatic illumination.

## 1.3 The development of optical holography

These advances set off an explosive growth of activity, and optical holography soon found a very large number of scientific applications. They included high-resolution imaging of aerosols [Thompson, Ward & Zinky, 1967], imaging through diffusing and aberrating media [Kogelnik, 1965; Leith & Upatnieks,

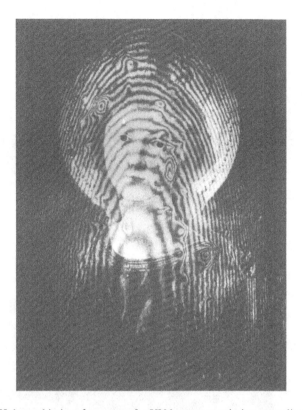

Fig. 1.5. Holographic interferogram of a XVth-century painting, revealing separation of the surface layers [Amadesi, Gori, Grella & Guattari, 1974].

1966], multiple imaging [Lu, 1968; Groh, 1968], computer-generated holograms [Lohmann & Paris, 1967], and the production and correction of optical elements [Upatnieks, Vander Lugt & Leith, 1966]. Other applications were related to information storage and information processing and included image deblurring [Stroke, Restrick, Funkhouser & Brumm, 1965] and pattern recognition [Vander Lugt, Rotz & Klooster, 1965].

Perhaps the most significant of these applications was holographic interferometry, which was discovered almost simultaneously by several groups [Brooks, Heflinger & Wuerker, 1965; Burch, 1965; Collier, Doherty & Pennington, 1965; Haines & Hildebrand, 1965; Powell & Stetson, 1965]. The technical advance it represented was astonishing. It became possible, for the first time, as shown in fig. 1.5, to map the displacements of a rough surface with an accuracy of a fraction of a micrometre; it was even possible to make interferometric comparisons of stored wavefronts that existed at different times.

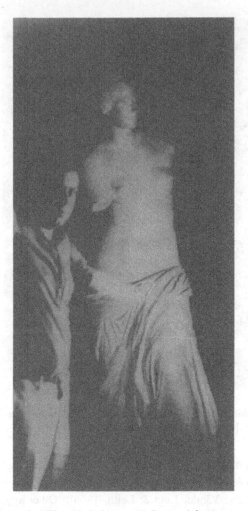

Fig. 1.6. The Venus de Milo; this hologram (1.5 m × 1.0 m) was produced by J. M. Fournier and G. Tribillon at the Laboratoire de Physique Général et Optique, Université de Besançon, France, in 1976.

Heterodyne techniques [Dändliker, 1980] and digital phase stepping techniques [Hariharan, Oreb & Brown, 1983] now permit direct measurements of the optical phase difference at a uniformly spaced network of points covering the interference pattern with even higher accuracy, and have opened up new applications for holographic interferometry.

In the field of three-dimensional displays, quite spectacular developments followed, including life-size holograms, and portraits with pulsed lasers (see figs. 1.6 and 1.7), as well as multicolour images. However, these holograms

Fig. 1.7. Professor Gabor with his holographic portrait; this hologram was produced by R. Rinehart at the McDonnell Douglas Electronics Company in 1971, using a pulsed laser.

represented a scientific *tour de force* rather than a viable technology, and little further progress was made until Benton [1969] invented the rainbow hologram. This was a transmission hologram in which, by sacrificing vertical parallax, two major advantages were gained – the hologram could be illuminated with white light and it reconstructed a very bright monochromatic image.

While it took some time for the practical advantages of the rainbow hologram to be appreciated, it ultimately resulted in developments in two areas. One was a series of new techniques for multicolour and achromatic imaging [Hariharan, Steel & Hegedus,1977; Tamura, 1978a; Benton, 1978]; the other was the white-light holographic stereogram of Cross [see Benton, 1975]. The latter was based on earlier work in which a three-dimensional image was built up from holograms of a number of views of an object from different angles in the horizontal plane, but which could be viewed in white light, since it made use of rainbow holograms.

Another significant area of progress has been in recording materials. Fine-grain silver halide photographic emulsions are still the most widely used recording medium, and considerable work has been done to improve their characteristics as well as to optimize processing techniques. In addition, other materials such as dichromated gelatin, photopolymers, photoresists, photothermoplastics, and photorefractive crystals are now being used to an increasing extent for specific applications for which they offer definite advantages. Techniques have also been developed for the large-scale replication of holograms on metallized foil by embossing.

The striking realism of holographic images has always fascinated scientists as well as laymen. However, as a result of developments in the last three decades, holography has ceased to be a novelty and has become a well-established technique with many valuable applications including art and advertising, high-resolution imaging, information processing, security coding, holographic optical elements, and nondestructive testing and strain analysis. As always in scientific research, some of these advances were not anticipated; we can expect more surprises in the next few years.

# 2

# Wavefront reconstruction

The concepts of holography outlined in Chapter 1 can now be formulated and discussed in more specific terms.

## 2.1 The in-line (Gabor) hologram

Consider the optical system shown in fig. 2.1, which is essentially that used by Gabor [1948] to demonstrate holographic imaging. In this arrangement, the object (a transparency containing small opaque details on a clear background) and the light source are located along an axis normal to the photographic plate.

When the object is illuminated with a monochromatic collimated beam, the light incident on the photographic plate can be regarded as consisting of two parts. The first is a uniform plane wave, corresponding to the directly transmitted light, which constitutes the reference wave. Since its amplitude and phase do not vary across the photographic plate, its complex amplitude (see Appendix 1) can be written as a real constant $r$. The second is a weak scattered wave caused by the transmittance variations in the object. The complex amplitude of this wave, at the photographic plate, can be written as $o(x, y)$, where $|o(x, y)| \ll r$.

The resultant complex amplitude at any point on the photographic plate is the sum of these two complex amplitudes, so that the intensity at this point is

$$
\begin{aligned}
I(x, y) &= |r + o(x, y)|^2, \\
&= r^2 + |o(x, y)|^2 + ro(x, y) + ro^*(x, y),
\end{aligned} \qquad (2.1)
$$

where $o^*(x, y)$ is the complex conjugate of $o(x, y)$.

A positive transparency is made from this recording. For simplicity, we shall assume that this transparency has been processed so that its amplitude transmittance (the ratio of the transmitted amplitude to that incident on it) is a

11

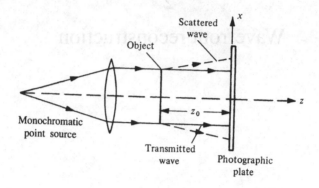

Fig. 2.1. Optical system used to record an in-line hologram.

linear function of the intensity and can be written as

$$\mathbf{t} = \mathbf{t}_0 + \beta T I, \tag{2.2}$$

where $\mathbf{t}_0$ is a constant background transmittance, $T$ is the exposure time, and $\beta$ is a parameter determined by the photographic material used and the processing conditions. The amplitude transmittance of this transparency (the hologram) is, accordingly,

$$\mathbf{t}(x, y) = \mathbf{t}_0 + \beta T [r^2 |o(x, y)|^2 + ro(x, y) + ro^*(x, y)]. \tag{2.3}$$

To view the reconstructed image, the hologram is replaced in the same position as the original photographic plate and, as shown in fig. 2.2, is illuminated with the same collimated beam of monochromatic light used to make the original recording. Since the complex amplitude at any point in this beam is, apart from a constant factor, the same as that in the reference beam, the complex amplitude transmitted by the hologram can be written as

$$\begin{aligned}
u(x, y) &= r\mathbf{t}(x, y), \\
&= r(\mathbf{t}_0 + \beta T r^2) + \beta T r |o(x, y)|^2 \\
&\quad + \beta T r^2 o(x, y) + \beta T r^2 o^*(x, y). \tag{2.4}
\end{aligned}$$

The expression for the complex amplitude of the transmitted wave contains four terms. The first of these terms, $r(\mathbf{t}_0 + \beta T r^2)$, which represents a uniformly attenuated plane wave, corresponds to the directly transmitted beam.

The second term, $\beta T r |o(x, y)|^2$, is extremely small in comparison to the other terms, since it has been assumed initially that $|o(x, y)| \ll r$. Accordingly, it can be neglected.

The third term, $\beta T r^2 o(x, y)$, is, except for a constant factor, identical with

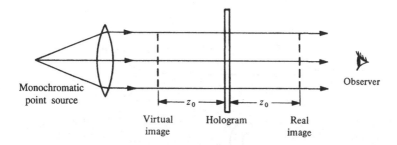

Fig. 2.2. Optical system used to reconstruct the image from an in-line hologram, showing the formation of the twin images.

the complex amplitude of the scattered wave from the object that was originally incident on the photographic plate. This wave reconstructs an image of the object in its original position. Since this image is located behind the transparency at a distance $z_0$ from it, and the reconstructed wave appears to diverge from it, it is a virtual image.

Similarly, the fourth term corresponds to a wavefront that resembles the original object wavefront, except that it has the opposite curvature. This wave converges to form a real image, the conjugate image, at the same distance $z_0$ in front of the hologram.

It is apparent that, with an in-line hologram, an observer viewing one image sees it superposed on the out-of-focus twin image as well as a strong coherent background. The presence of this unwanted image constitutes the most serious limitation of the in-line hologram.

Another major limitation is the need for the object to have a high average transmittance, if the second term on the right-hand side of (2.4), which has been assumed to be negligible, is not to interfere with the reconstructed image. Typically, it is possible to form images of fine opaque lines on a transparent background, but not *vice versa*.

Finally, it should be noted that the hologram must be a positive transparency because the image-forming waves interfere with the background during reconstruction. If the photographic plate that has been exposed in the recording arrangement is used directly (in which case, $\beta$ in (2.2) is negative), the reconstructed image resembles a photographic negative of the object.

## 2.2 The off-axis (Leith–Upatnieks) hologram

The first successful method for separating the twin images was developed by Leith and Upatnieks [1962, 1963, 1964]. They used a separate reference beam

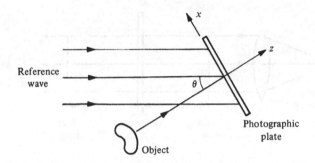

Fig. 2.3. Hologram recording with an off-axis reference beam.

derived from the same source to record the hologram. As shown in fig. 2.3, this reference beam was incident on the photographic plate at an offset angle $\theta$ to the beam from the object.

For simplicity, we assume this reference beam to be a collimated beam of uniform intensity. The complex amplitude due to the object beam at any point $(x, y)$ on the photographic plate can then be written as

$$o(x, y) = |o(x, y)| \exp[-i\phi(x, y)], \qquad (2.5)$$

while that due to the reference beam is

$$r(x, y) = r \exp(i2\pi\xi_r x), \qquad (2.6)$$

where $\xi_r = (\sin\theta)/\lambda$, since only the phase of the reference beam varies across the photographic plate.

The resultant intensity at the photographic plate is

$$
\begin{aligned}
I(x, y) &= |r(x, y) + o(x, y)|^2, \\
&= |r(x, y)|^2 + |o(x, y)|^2 \\
&\quad + r\,|o(x, y)| \exp[-i\phi(x, y)] \exp(-i2\pi\xi_r x) \\
&\quad + r\,|o(x, y)| \exp[i\phi(x, y)] \exp(i2\pi\xi_r x), \\
&= r^2 + |o(x, y)|^2 + 2r\,|o(x, y)| \cos[2\pi\xi_r x + \phi(x, y)]. \quad (2.7)
\end{aligned}
$$

The amplitude and phase of the object wave are therefore encoded as amplitude and phase modulation, respectively, of a set of interference fringes equivalent to a carrier with a spatial frequency of $\xi_r$.

If, as in (2.2), we assume that the resultant amplitude transmittance of the photographic plate is linearly related to the intensity in the interference pattern,

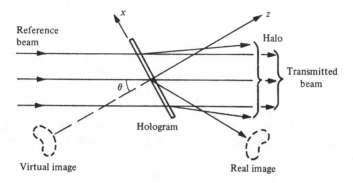

Fig. 2.4. Image reconstruction by a hologram recorded with an off-axis reference beam.

the amplitude transmittance of the hologram can be written as

$$\mathbf{t}(x, y) = \mathbf{t}_0 + \beta T\{|o(x, y)|^2$$
$$+ r\,|o(x, y)|\exp[-i\phi(x, y)]\exp(-i2\pi\xi_r x)$$
$$+ r\,|o(x, y)|\exp[i\phi(x, y)]\exp(i2\pi\xi_r x)\},\qquad(2.8)$$

where $\beta$ is the slope (in this case, negative) of the amplitude transmittance *versus* exposure characteristic of the photographic material, $T$ is the exposure time, and $\mathbf{t}_0$ is a constant background transmittance.

To reconstruct the image, the hologram is illuminated once again, as shown in fig. 2.4, with the same reference beam used to record it. The complex amplitude $u(x, y)$ of the transmitted wave is, in this case also, the sum of four terms, each corresponding to one of the terms of (2.8), and can be written as

$$u(x, y) = r(x, y)\mathbf{t}(x, y),$$
$$= u_1(x, y) + u_2(x, y) + u_3(x, y) + u_4(x, y),\qquad(2.9)$$

where

$$u_1(x, y) = \mathbf{t}_0 r\,\exp(i2\pi\xi_r x),\qquad(2.10)$$
$$u_2(x, y) = \beta T r\,|o(x, y)|^2\exp(i2\pi\xi_r x),\qquad(2.11)$$
$$u_3(x, y) = \beta T r^2 o(x, y),\qquad(2.12)$$
$$u_4(x, y) = \beta T r^2 o^*(x, y)\exp(i4\pi\xi_r x).\qquad(2.13)$$

The first term on the right-hand side of (2.9), $u_1(x, y)$, is, as before, merely the attenuated reference beam, which is a plane wave directly transmitted through the hologram. This directly transmitted beam is surrounded by a halo due to the second term, $u_2(x, y)$, whose angular spread is determined by the extent of the object.

The third term, $u_3(x, y)$, is identical with the original object wave, except for a constant factor, and produces a virtual image of the object in its original position; this wave makes an angle $\theta$ with the directly transmitted wave. Similarly, the fourth term, $u_4(x, y)$, gives rise to the conjugate image. However, in this case, the fourth term includes a factor $\exp(\mathrm{i}4\pi\xi_r x)$, which indicates that the conjugate wave is deflected from the $z$ axis at an angle approximately twice that which the reference wave makes with it.

Accordingly, even though two images – one real and the other virtual – are reconstructed in this arrangement, they are formed at different angles from the directly transmitted beam and from each other, and, if the offset angle $\theta$ of the reference beam is made large enough, the three will not overlap. This method therefore eliminates all the major drawbacks of Gabor's original in-line arrangement. In addition, the sign of $\beta$ affects only the phase of the reconstructed image, and a "positive" image is obtained even if the hologram is a photographic negative.

The minimum value of the offset angle $\theta$ required to ensure that each of the images can be observed without any interference from its twin, as well as from the directly transmitted beam and the halo of scattered light surrounding it, is determined by the minimum spatial carrier frequency $\xi_r$ for which there is no overlap between the angular spectra of the third and fourth terms, and those of the first and second terms.

These angular spectra are the Fourier transforms (see Appendix 2) of the four terms and can be written as follows:

$$U_1(\xi, \eta) = \mathcal{F}\{\mathbf{t}_0 r \exp(\mathrm{i}2\pi\xi_r x)\},$$
$$= \mathbf{t}_0 r \delta(\xi + \xi_r, \eta), \tag{2.14}$$
$$U_2(\xi, \eta) = \mathcal{F}\{\beta T r |o(x, y)|^2 \exp(\mathrm{i}2\pi\xi_r x)\},$$
$$= \beta T r [O(\xi, \eta) \star O(\xi, \eta) * \delta(\xi + \xi_r, \eta)], \tag{2.15}$$
$$U_3(\xi, \eta) = \mathcal{F}\{\beta T r^2 o(x, y)\},$$
$$= \beta T r^2 O(\xi, \eta), \tag{2.16}$$
$$U_4(\xi, \eta) = \mathcal{F}\left\{\beta T r^2 o^*(x, y) \exp(\mathrm{i}4\pi\xi_r x)\right\},$$
$$= \beta T r^2 O^*(\xi, \eta) * \delta(\xi + 2\xi_r, \eta), \tag{2.17}$$

where $O(\xi, \eta) = \mathcal{F}\{o(x, y)\}$ is the spatial frequency spectrum of the object beam, and the symbols $\star$ and $*$ denote, respectively, the operations of correlation and convolution. These spectra are shown schematically in fig. 2.5.

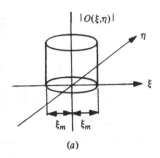

(a)

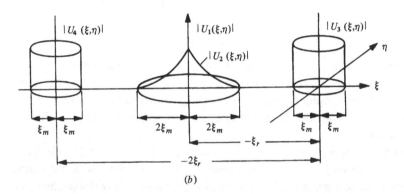

(b)

Fig. 2.5. Spatial frequency spectra of (a) the object beam and (b) a hologram recorded with an off-axis reference beam.

As can be seen, the term $|U_3(\xi, \eta)|$ is merely the object-beam spectrum $|O(\xi, \eta)|$ multiplied by a constant and is centred at the origin of the spatial frequency plane. The term $|U_1(\xi, \eta)|$ corresponds to the spatial frequency of the carrier fringes and is a $\delta$ function located at $(-\xi_r, 0)$, while the term $|U_2(\xi, \eta)|$ is centred on this $\delta$ function and, being proportional to the auto-correlation function of $O(\xi, \eta)$, has twice the extent of the object-beam spectrum. Finally, $|U_4(\xi, \eta)|$ is similar to $|U_3(\xi, \eta)|$ but is displaced to a centre frequency $(-2\xi_r, 0)$. Evidently, $|U_3(\xi, \eta)|$ and $|U_4(\xi, \eta)|$ will not overlap $|U_1(\xi, \eta)|$ and $|U_2(\xi, \eta)|$ if the offset angle $\theta$ is chosen so that the spatial carrier frequency $\xi_r$ satisfies the condition

$$\xi_r \geq 3\xi_{max}, \tag{2.18}$$

where $\xi_{max}$ is the highest frequency in the spatial frequency spectrum of the object beam.

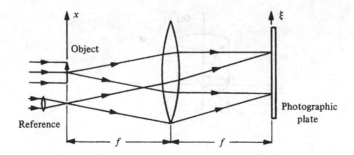

Fig. 2.6. Optical system used to record a Fourier hologram.

## 2.3 Fourier holograms

Another interesting hologram recording configuration is one in which the complex amplitudes of the waves that interfere at the hologram are the Fourier transforms of the complex amplitudes of the original object and reference waves. Normally, this implies an object that lies in a single plane or is of limited thickness.

A typical optical arrangement for recording such a hologram [Vander Lugt, 1964] is shown in fig. 2.6. The object is a transparency located in the front focal plane of a lens and is illuminated by a collimated beam of monochromatic light. If the complex amplitude of the wave leaving the object plane is $o(x, y)$, its complex amplitude at the photographic plate located in the back focal plane of the lens is

$$O(\xi, \eta) = \mathcal{F}\{o(x, y)\}. \tag{2.19}$$

The reference beam is derived from a point source also located in the front focal plane of the lens. If $\delta(x + b, y)$ is the complex amplitude of the wave leaving this point source, the complex amplitude of the reference wave at the hologram plane can be written as

$$R(\xi, \eta) = \exp(-i2\pi\xi b). \tag{2.20}$$

The intensity in the interference pattern produced by these two waves is, therefore,

$$I(\xi, \eta) = 1 + |O(\xi, \eta)|^2 + O(\xi, \eta)\exp(i2\pi\xi b)$$
$$+ O^*(\xi, \eta)\exp(-i2\pi\xi b). \tag{2.21}$$

To reconstruct the image, the processed hologram is placed in the front focal plane of the lens and illuminated with a collimated beam of monochromatic

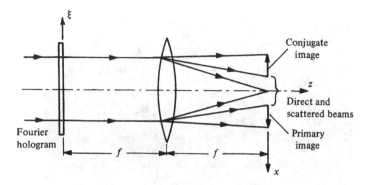

Fig. 2.7. Image reconstruction by a Fourier hologram.

light, as shown in fig. 2.7. If we assume that this wave has unit amplitude and that, as before, the amplitude transmittance of the processed hologram is a linear function of $I(\xi, \eta)$, the intensity in the interference pattern, the complex amplitude of the wave transmitted by the hologram is

$$U(\xi, \eta) = \mathbf{t}_0 + \beta T I(\xi, \eta). \tag{2.22}$$

The complex amplitude in the back focal plane of the lens is then the Fourier transform of $U(\xi, \eta)$,

$$
\begin{aligned}
u(x, y) &= \mathcal{F}\{U(\xi, \eta)\}, \\
&= (t_0 + \beta T)\delta(x, y) + \beta T o(x, y) \star o(x, y) \\
&\quad + \beta T o(x - b, y) + \beta T o^*(-x + b, -y).
\end{aligned} \tag{2.23}
$$

As shown in figs. 2.7 and 2.8, the wave corresponding to the first term on the right-hand side of (2.23) comes to a focus on the axis, while that corresponding to the second term forms a halo around it. The third term produces an image of the original object, shifted downwards by a distance $b$, while the fourth term gives rise to a conjugate image, inverted and shifted upwards by the same distance $b$. Both the images are real and can be recorded on a photographic film placed in the back focal plane of the lens. Since the film records the intensity distribution in the image, the conjugate image can be identified, in this case, only by the fact that it is inverted.

Fourier holograms have the useful property that the reconstructed image does not move when the hologram is translated in its own plane. This is because the only effect of a shift of a function in the spatial domain is to multiply its Fourier transform by a phase factor that is a linear function of the spatial frequency (see

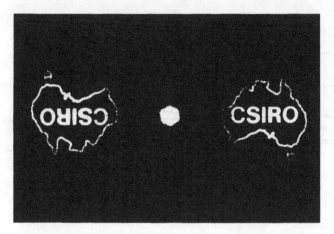

Fig. 2.8. Images reconstructed by a Fourier hologram.

Appendix 2). This phase factor has no effect on the intensity distribution in the image.

## 2.4 The lensless Fourier hologram

A hologram with the same properties as a Fourier hologram can be obtained without using a lens to produce the Fourier transform of the object wave, if the reference wave comes from a point source in the plane of the object [Stroke, 1965; Stroke, Brumm & Funkhouser, 1965]. Consider the system shown in fig. 2.9 in which the object is, as before, a transparency illuminated with a plane wave, and let the complex amplitude of the wave leaving the object plane be $o(x_1, y_1)$. It can be shown, using the Fresnel–Kirchhoff integral (see Appendix 3), that the complex amplitude at a point $(x_2, y_2)$ on the photographic plate due to this wave can be written as

$$o(x_2, y_2) = (i/\lambda z_0) \exp[-(i\pi/\lambda z_0)(x_2^2 + y_2^2)]O(\xi, \eta), \qquad (2.24)$$

where $z_0$ is the distance from the object plane to the hologram plane, $\xi = x_2/\lambda z_0$, $\eta = y_2/\lambda z_0$, and

$$O(\xi, \eta) = \mathcal{F}\{o(x_1, y_1) \exp[-(i\pi/\lambda z_0)(x_1^2 + y_1^2)]\}. \qquad (2.25)$$

The expression on the right-hand side of (2.25) is the Fourier transform of the object wave modified by a spherical phase factor that depends on the distance from the object to the hologram. Similarly, it can be shown that the complex

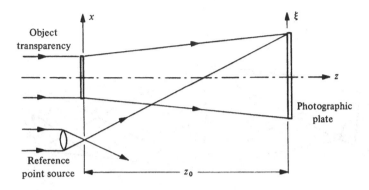

Fig. 2.9. Optical system used to record a lensless Fourier hologram.

amplitude at the photographic plate due to the reference wave is

$$r(x_2, y_2) = r \exp[-(\mathrm{i}\pi/\lambda z_0)(x_2^2 + y_2^2)] \exp(-\mathrm{i}2\pi\xi b), \qquad (2.26)$$

where $b$ is the distance of the reference point source from the $z$ axis.

The intensity in the resulting interference pattern is then

$$\begin{aligned} I(x_2, y_2) = r^2 + |o(x_2, y_2|^2 \\ + (\mathrm{i}/\lambda z_0)O(\xi, \eta) \exp(\mathrm{i}2\pi\xi b) \\ + (\mathrm{i}/\lambda z_0)O^*(\xi, \eta) \exp(-\mathrm{i}2\pi\xi b). \end{aligned} \qquad (2.27)$$

A comparison with (2.21) shows that this intensity distribution is very similar to that obtained with the arrangement used earlier to produce a Fourier hologram, and the resulting hologram has essentially the same properties.

In this recording configuration, the effect of the spherical phase factor associated with the near-field (or Fresnel) diffraction pattern of the object transparency is eliminated by the use of a spherical reference wave with the same average curvature.

## 2.5 Image holograms

It is often advantageous to record a hologram of a real image of the object, instead of the object itself [Rosen, 1966; Stroke, 1966]. A typical optical arrangement for this purpose, in which a large lens is used to project a real image of the object, is shown in fig. 2.10.

With such an arrangement, it is possible to position the projected image of the object, so that it straddles the photographic plate used to record the hologram.

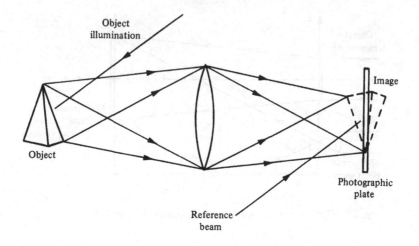

Fig. 2.10. Optical system used to record an image hologram.

The reconstructed image is then formed in the same position with respect to the hologram, so that part of the image appears to be in front of the hologram, and the remainder is behind it.

As will be shown later (see section 3.5.3), such an image hologram has the very useful property that it can be illuminated with a source of appreciable size and spectral bandwidth and will still produce an acceptably sharp image. Another advantage of such a hologram (see section 3.6.2) is increased image luminance, but this is offset partially by the fact that the range of angles over which the image can be viewed is limited by the aperture of the imaging lens.

### 2.5.1 Pinhole holograms

A pinhole placed between the object and the plate in the recording system, as shown in fig. 2.11(*a*), plays the role of a lens and produces an image hologram. When this hologram is illuminated by the conjugate of the reference beam, as shown in fig. 2.11(*b*), it reconstructs in addition to a real image of the object, a real image of the pinhole, and all the image-forming light passes through this image of the pinhole [Xu, Mendes, Hart & Dainty, 1989].

If, then, holograms of a series of objects are recorded on the same plate, with the pinhole moved to a different position for each object, individual images can be recalled by moving the pinhole to the appropriate position. Nondiffusing objects, such as amplitude masks, can be illuminated by a converging beam

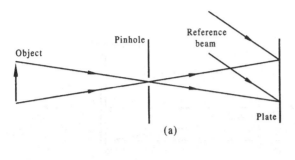

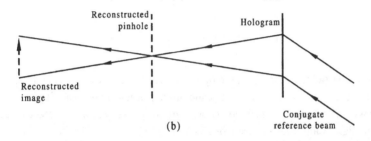

Fig. 2.11. (*a*) Recording, and (*b*) reconstructing, a pinhole hologram [Xu *et al.*, 1989].

from a lens; images can then be selected by placing a spatial light modulator in the focal plane and programming it to generate an appropriate array of pinholes.

## 2.6 Fraunhofer holograms

The Fraunhofer hologram exploits a unique situation in which an in-line hologram can be used without the usual problems associated with the presence of the conjugate image. This situation arises when, as shown in fig. 2.12, the object is small enough for its Fraunhofer diffraction pattern to be formed at the photographic plate. For this to happen, $z_0$, the distance of the object from the photographic plate, must satisfy the far-field condition (see Appendix 3),

$$z_0 \gg (x_0^2 + y_0^2)/\lambda, \tag{2.28}$$

where $x_0$ and $y_0$ define the lateral dimensions of the object.

One of the earliest successful applications of Fraunhofer holograms was in recording images of a three-dimensional distribution of aerosol particles [Thompson, 1963]. The particles are so small that, even at distances of a few millimetres, the hologram can be considered to lie in the far field of any

*Wavefront reconstruction*

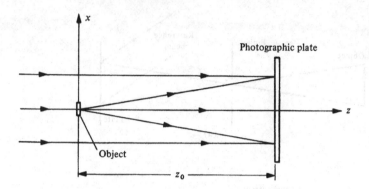

Fig. 2.12. Formation of a Fraunhofer hologram.

individual particle. In this case, the light contributing to the conjugate image is spread over such a large area in the plane of the primary image that it produces only a weak, uniform background. As a result, the primary image can be viewed without significant interference from its conjugate.

# 3

# The reconstructed image

This chapter discusses the characteristics of the reconstructed image and the dependence of these characteristics on various parameters of the optical system used for recording and reconstruction.

## 3.1 Images of a point

To simplify the analysis, it is convenient to consider the hologram of a point object $O$, whose coordinates are $(x_O, y_O, z_O)$, recorded with a reference wave from a point source $R$ located at $(x_R, y_R, z_R)$, as shown in fig. 3.1($a$) [Meier, 1965].

The complex amplitude of the object wave at a point $H(x_H, y_H, z_H)$ in the hologram plane can be written as $a_O = |a_O| \exp(-i\phi_O)$, where $\phi_O$ is the phase of the wave at this point relative to that at $O$. The phase $\phi_O$ can be computed from the optical path (see Appendix 3) and is given, to a first-order approximation, by the expression

$$\phi_O = (\pi/\lambda_1)[(1/z_O)(x_H^2 + y_H^2 - 2x_H x_O - 2y_H y_O)], \quad (3.1)$$

where $\lambda_1$ is the wavelength of the light used to record the hologram.

Similarly, the complex amplitude of the reference wave at the point $H(x_H, y_H, z_H)$ can be written as $a_R = |a_R| \exp(-i\phi_R)$, where

$$\phi_R = (\pi/\lambda_1)[(1/z_R)(x_H^2 + y_H^2 - 2x_H x_R - 2y_H y_R)]. \quad (3.2)$$

From (2.7), it follows that the position and spacing of the interference fringes in the hologram produced by these two waves are determined by the phase difference $(\phi_R - \phi_O)$.

We shall assume that the processed hologram is illuminated with monochromatic light of wavelength $\lambda_2$ from a point source $P$ located at $(x_P, y_P, z_P)$, as shown in fig. 3.1($b$). In the same manner, the complex amplitude of this

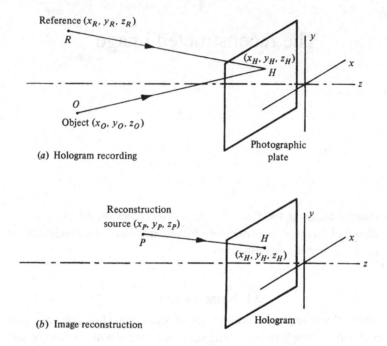

Fig. 3.1. Coordinate system used to study image formation by a hologram.

wave at the point $H(x_H, y_H, z_H)$ in the hologram plane can be written as $a_P = |a_P| \exp(-i\phi_P)$, where

$$\phi_P = (\pi/\lambda_2)[(1/z_P)(x_H^2 + y_H^2 - 2x_H x_P - 2y_H y_P)]. \qquad (3.3)$$

If we assume linear recording, as defined by (2.2) and (2.3), the terms in the expression for the complex amplitude of the transmitted wavefront that give rise to the two reconstructed images, as given by (2.12) and (2.13), are, apart from a constant factor,

$$u_3 = a_P a_R^* a_O,$$
$$= |a_P| |a_R| |a_O| \exp[-i(\phi_P - \phi_R + \phi_O)], \qquad (3.4)$$

and

$$u_4 = a_P a_R a_O^*,$$
$$= |a_P| |a_R| |a_O| \exp[-i(\phi_P + \phi_R - \phi_O)]. \qquad (3.5)$$

If $u_3$ is written in the form $u_3 = |u_3| \exp(-i\phi_3)$, it follows that the phase $\phi_3$

of this reconstructed wave at $H(x_H, y_H, z_H)$ is

$$\phi_3 = \phi_P - \phi_R + \phi_O. \tag{3.6}$$

When we substitute the values of $\phi_O$, $\phi_R$, and $\phi_P$ from (3.1), (3.2), and (3.3) in (3.6), and set $(\lambda_2/\lambda_1) = \mu$, we have

$$
\begin{aligned}
\phi_3 = (\pi/\lambda_2) \Bigg[ & (x_H^2 + y_H^2)\left(\frac{1}{z_P} + \frac{\mu}{z_O} - \frac{\mu}{z_R}\right) \\
& - 2x_H\left(\frac{x_P}{z_P} + \frac{\mu x_O}{z_O} - \frac{\mu x_R}{z_R}\right) \\
& - 2y_H\left(\frac{y_P}{z_P} + \frac{\mu y_O}{z_O} - \frac{\mu y_R}{z_R}\right)\Bigg].
\end{aligned}
\tag{3.7}
$$

Since the wave represented by $u_3$ is to produce a point image, it must be a spherical wave. Accordingly, it should be possible to express its phase at $H(x_H, y_H, z_H)$ in the form

$$\phi_3 = (\pi/\lambda_2)[(1/z_3)(x_H^2 + y_H^2 - 2x_H x_3 - 2y_H y_3)], \tag{3.8}$$

where $(x_3, y_3, z_3)$ are the coordinates of this point image.

If, therefore, we equate the coefficients of similar terms in (3.7) and (3.8), the coordinates of the image formed by the wave represented by $u_3$ can be written as

$$x_3 = \frac{x_P z_O z_R + \mu x_O z_P z_R - \mu x_R z_P z_O}{z_O z_R + \mu z_P z_R - \mu z_P z_O}, \tag{3.9}$$

$$y_3 = \frac{y_P z_O z_R + \mu y_O z_P z_R - \mu y_R z_P z_O}{z_O z_R + \mu z_P z_R - \mu z_P z_O}, \tag{3.10}$$

$$z_3 = \frac{z_P z_O z_R}{z_O z_R + \mu z_P z_R - \mu z_P z_O}. \tag{3.11}$$

Similarly, the coordinates of the image formed by the conjugate wave represented by $u_4$ can be written as

$$x_4 = \frac{x_P z_O z_R - \mu x_O z_P z_R + \mu x_R z_P z_O}{z_O z_R - \mu z_P z_R + \mu z_P z_O}, \tag{3.12}$$

$$y_4 = \frac{y_P z_O z_R - \mu y_O z_P z_R + \mu y_R z_P z_O}{z_O z_R - \mu z_P z_R + \mu z_P z_O}, \tag{3.13}$$

$$z_4 = \frac{z_P z_O z_R}{z_O z_R - \mu z_P z_R + \mu z_P z_O}. \tag{3.14}$$

## 3.2 Image magnification

We can consider an extended object to be made up of a number of point objects and apply the preceding analysis to evaluate the characteristics of the reconstructed image.

### 3.2.1 Lateral magnification

The lateral magnification of the primary image can be defined either as

$$M_{\text{lat},3} = (dx_3/dx_O), \tag{3.15}$$

or as

$$M_{\text{lat},3} = (dy_3/dy_O). \tag{3.16}$$

Both these definitions lead to the same result, since, from (3.9) and (3.10),

$$M_{\text{lat},3} = \left[ 1 + z_O \left( \frac{1}{\mu z_P} - \frac{1}{z_R} \right) \right]^{-1}. \tag{3.17}$$

Similarly, for the conjugate image, we have

$$M_{\text{lat},4} = (dx_4/dx_O), \tag{3.18}$$

or

$$M_{\text{lat},4} = (dy_4/dy_O), \tag{3.19}$$

which, from (3.12) and (3.13), yield

$$M_{\text{lat},4} = \left[ 1 - z_O \left( \frac{1}{\mu z_P} + \frac{1}{z_R} \right) \right]^{-1}. \tag{3.20}$$

### 3.2.2 Angular magnification

If we assume that the observer's eye is located in the hologram plane, the angular magnification of the primary image can be defined as

$$M_{\text{ang}} = \frac{d(x_3/z_3)}{d(x_O/z_O)}, \tag{3.21}$$

which reduces to

$$\left| M_{\text{ang}} \right| = \mu, \tag{3.22}$$

and is the same as that for the conjugate image.

### 3.2.3 Longitudinal magnification

The longitudinal magnification of the primary image can be calculated from the relation

$$M_{\text{long},3} = (\mathrm{d}z_3/\mathrm{d}z_O),$$

$$= \frac{1}{\mu}\frac{\mathrm{d}}{\mathrm{d}z_O}\left\{\frac{z_O}{1 + z_O[(1/\mu z_P) + (1/z_R)]}\right\},$$

$$= \frac{1}{\mu}\left\{\frac{1}{1 + z_O[(1/\mu z_P) + (1/z_R)]}\right\}^2,$$

$$= \frac{1}{\mu}M_{\text{lat},3}^2. \tag{3.23}$$

Similarly, the longitudinal magnification of the conjugate image is

$$M_{\text{long},4} = (\mathrm{d}z_4/\mathrm{d}z_O),$$

$$= -\frac{1}{\mu}\frac{\mathrm{d}}{\mathrm{d}z_O}\left\{\frac{z_O}{1 - z_O[(1/\mu z_P) + (1/z_R)]}\right\},$$

$$= -\frac{1}{\mu}\left\{\frac{1}{1 - z_O[(1/\mu z_P) + (1/z_R)]}\right\}^2,$$

$$= -\frac{1}{\mu}M_{\text{lat},4}^2. \tag{3.24}$$

Note that $M_{\text{long},3}$ and $M_{\text{long},4}$ are of the same magnitude but have opposite signs. The consequences of this are discussed in the next section.

## 3.3 Orthoscopic and pseudoscopic images

To understand the implications of the opposite signs of $M_{\text{long},3}$ and $M_{\text{long},4}$, consider an off-axis hologram recorded with a collimated reference beam incident normally to the photographic plate, as shown in fig. 3.2(*a*). When this hologram is illuminated once again with the same collimated reference beam, as shown in fig. 3.2(*b*), it reconstructs two images, one virtual and the other real, both of which, at first sight, are exact replicas of the object. However, the two images differ in one very important respect.

As we have seen earlier, while the virtual image is located in the same position as the object and exhibits the same parallax properties, the real image is formed

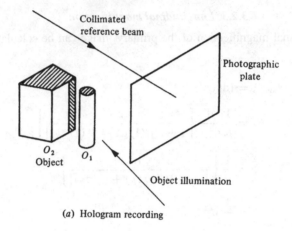

(*a*) Hologram recording

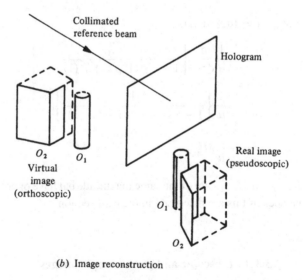

(*b*) Image reconstruction

Fig. 3.2. Formation of orthoscopic and pseudoscopic images by a hologram.

at the same distance from the hologram, but in front of it. Since, from (3.11) and (3.14), corresponding points on the real and virtual images are located at equal distances from the plane of the hologram, the real image has the curious property that its depth is inverted. Such an image is called a pseudoscopic image, as opposed to a normal, or orthoscopic, image [Leith & Upatnieks, 1964; Rosen, 1967].

This depth inversion results in conflicting visual clues that make viewing of the real image psychologically unsatisfactory. Thus, if $O_1$ and $O_2$ are two elements in the object field, and if $O_1$ blocks the light scattered by $O_2$ at a certain angle, the hologram records information only on the element $O_1$ at this angle and records no information on this part of $O_2$. An observer viewing the real image from the corresponding direction cannot see this part of $O_2$, which, contrary to normal experience, is obscured by $O_1$, even though $O_2$ appears to be in front of $O_1$.

### 3.3.1 Production of an orthoscopic real image

A hologram that reconstructs an orthoscopic real image of an object can be produced in two steps [Rotz & Friesem, 1966].

In the first step, as shown in fig. 3.3, a hologram (H1) is recorded of the object with a collimated reference beam. When H1 is illuminated once again with the same collimated reference beam, it reconstructs two images of the object at unit magnification, one of them being an orthoscopic virtual image, the other a pseudoscopic real image. A second hologram (H2) is then recorded of this real image with a collimated reference beam.

When H2 is illuminated with a collimated beam, it reconstructs a pseudoscopic virtual image located in the same position as the real image formed by H1. However, the real image formed by H2 is orthoscopic. Since collimated reference beams are used throughout, the final real image is the same size as the object and free from aberrations.

A simpler method is to record a hologram of an orthoscopic real image of the object formed by a lens, or by a concave mirror. When this hologram is illuminated with the original reference beam, it reconstructs the original object wave, producing an orthoscopic real image.

### 3.4 Image aberrations

If the processed hologram is replaced in its original position in the arrangement in which it was recorded and illuminated once again with the reference beam used to record it, $z_P = z_R$, and $\mu = 1$, and the primary image coincides with the object. In any other case, the image may exhibit aberrations.

Although the locations of the image points were calculated in section (3.1) on the assumption that the reconstructed wavefronts were spherical, this as-

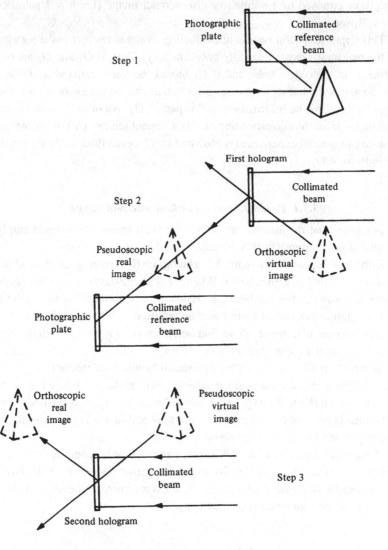

Fig. 3.3. Production of an orthoscopic real image by recording two holograms in succession [Rotz & Friesem, 1966].

sumption is not strictly true. The aberrations of these wavefronts can then be defined as the phase differences between the reference spheres centred on these points and the actual wavefronts in the hologram plane.

Hologram aberrations can be classified in the same manner as lens aberrations

[Hopkins, 1950]. It is convenient to use polar coordinates $(\rho, \theta)$ in the hologram plane, instead of cartesian coordinates, and write the third-order aberration as

$$
\begin{aligned}
\Delta\phi_3 = (2\pi/\lambda_2)[&-(1/8)\rho^4 S + (1/2)\rho^3(C_x \cos\theta + C_y \sin\theta) \\
&-(1/2)\rho^2(A_x \cos^2\theta + A_y \sin^2\theta + 2A_x A_y \cos\theta \sin\theta) \\
&-(1/4)\rho^2 F + (1/2)\rho(D_x \cos\theta + D_y \sin\theta],
\end{aligned} \tag{3.25}
$$

where $S$ is the coefficient of spherical aberration, $C_x$ and $C_y$ are the coma coefficients, $A_x$ and $A_y$ are the coefficients of astigmatism, $F$ is the coefficient for the field curvature, and $D_x$ and $D_y$ are the distortion coefficients.

These coefficients can be calculated if we retain the third-degree terms in the expansion for the phase of a spherical wavefront [Meier, 1965; Leith, Upatnieks & Haines, 1965]. In the discussion that follows, only the conjugate (real) image will be considered, and it will be assumed for simplicity that the object lies on the $x$ axis $(y_O = 0)$. The expressions for the aberration coefficients of the primary (virtual) image can be obtained by changing the signs of $z_O$ and $z_R$ in the corresponding expressions for the conjugate image.

Enlarging or reducing the hologram can make it easier to eliminate some of the aberrations. However, we will not consider this possibility, since the very close spacing of the carrier fringes in an off-axis hologram makes it impracticable.

### 3.4.1 Spherical aberration

It can be shown that the coefficient of spherical aberration is

$$
S = (1/z_P^3) - (\mu/z_O^3) + (\mu/z_R^3) - (1/z_4^3). \tag{3.26}
$$

When $z_R = z_O$, $z_3 = z_P$, and the spherical aberration is zero for both the reconstructed wavefronts. The reason for this is easily seen: because the two interfering wavefronts have the same curvature, the phase difference $(\phi_O - \phi_R)$ between them changes linearly across the field. The resulting straight hologram fringes cannot introduce spherical aberration.

### 3.4.2 Coma

The coefficient of coma is

$$
C_x = (x_P/z_P^3) - (\mu x_O/z_O^3) + (\mu x_R/z_R^3) - (x_4/z_4^3). \tag{3.27}
$$

Coma can be eliminated only if $z_R = z_O$ and $z_P = \pm z_O$; it then disappears for both images.

### 3.4.3 Astigmatism

The coefficient of astigmatism is

$$A_x = (x_P^2/z_P^3) - (\mu x_O^2/z_O^3) + (\mu x_R^2/z_R^3) - (x_4^2/z_4^3). \tag{3.28}$$

For astigmatism to be eliminated, it is necessary to make $z_R = z_O$, $(x_P/z_P) = -(\mu x_R/z_R)$, and $z_P = \mu z_O$. If $\mu \neq 1$, the last of these conditions is incompatible with the condition for coma to vanish, so that coma and astigmatism can be eliminated simultaneously only when $\mu = 1$.

### 3.4.4 Curvature of field

$$F = [(x_P^2 + y_P^2)/z_P^3] - [\mu(x_O^2 + y_O^2)/z_O^3]$$
$$+ [\mu(x_R^2 + y_R^2)/z_R^3] - [(x_4^2 + y_4^2)/z_4^3]. \tag{3.29}$$

This coefficient also disappears when the astigmatism is reduced to zero.

### 3.4.5 Distortion

$$D_x = [(x_P^3 + x_P y_P^2)/z_P^3] - [\mu(x_O^3 + x_O y_O^2)/z_O^3]$$
$$+ [\mu(x_R^3 + x_R y_R^2)/z_R^3] - [(x_4^3 + x_4 y_4^2)/z_4^3]. \tag{3.30}$$

Normally, distortion cannot be eliminated when $\mu \neq 1$.

### 3.4.6 Longitudinal distortion

It is apparent from (3.23) and (3.24) that, unless $M_{lat} = 1$ and $\mu = 1$, the longitudinal magnification is not, in general, the same as the lateral magnification, resulting in a distortion in depth. If the recording and reconstruction wavelengths are not the same ($\mu \neq 1$), longitudinal distortion can be minimized by a proper choice of the recording and reconstruction geometry [Hariharan, 1976a].

### 3.4.7 Nonparaxial imaging

The geometrical relations between the original object and the reconstructed image have been discussed in more detail by Neumann [1966], who has described a graphical method for locating the conjugate image. He has shown that, depending on the conditions of illumination, the hologram may produce two virtual images, two real images, or one virtual image and one real image. He has also shown that, under some conditions, only one image may exist, the energy of the second image then forming an evanescent field along the hologram.

The case of nonparaxial imaging was first studied by Champagne [1967], who obtained expressions that can be used where paraxial theory is unsatisfactory. Computer-based ray tracing methods based on these expressions have been investigated by Latta [1971a,b,c].

More precise imaging formulas that do not involve approximations for the inclination of the reference beam in recording or reconstruction have been derived by Miles [1972]. His results show that where the reference wave used in reconstruction differs appreciably from that used to record the hologram, the magnification in different azimuths is not, in general, the same, resulting in an anamorphic image. These formulas permit precise calculations of the wavefront aberration by a method based on exact ray tracing [Miles, 1973]. A simple method of ray tracing based on an elementary plane-wave model has also been described by Olson [1989], which is not limited to the paraxial case and can be applied to almost all thin transmission geometries.

### 3.4.8 Intensity distribution in the image

Since the hologram is illuminated with coherent light, a spot diagram cannot be used to calculate the intensity distribution in the image. Numerical calculations of the influence of hologram aberrations on the intensity distribution in the image of a point object have been carried out by Nowak and Zajac [1983] for typical hologram recording configurations. The results are consistent with third-order aberration theory.

### 3.5 Misalignment, source size, and spectral bandwidth

As described in the previous section, an image with no aberrations and the highest possible resolution is obtained when the hologram is replaced in its original position and the same reference beam used to record the hologram is used to illuminate it. However, there are many situations in which it is not possible to do so.

The effects of inaccurate repositioning of the hologram or inaccurate adjustment of the reference beam become quite serious at high numerical apertures. Banyasz, Kiss, and Varga [1988] have studied the imaging of a point source by a hologram with a numerical aperture of 0.85 and have shown that mismatch angles of 1 mrad can generate noticeable geometric aberrations, whereas smaller errors result in a broadening of the intensity distribution.

Similarly, the use of a source of finite size and spectral bandwidth to illuminate the hologram affects the resolution in the reconstructed image. For simplicity, we shall consider the effects in the $xz$ plane only.

### 3.5.1 Source size

Consider a hologram illuminated by a monochromatic point source of the same wavelength as that used to record it. From (3.9), the $x$ coordinate of the virtual image of an object point located at $(x_O, y_O, z_O)$ is

$$x_3 = \frac{x_P z_O z_R + x_O z_P z_R - x_R z_P z_O}{z_O z_R + z_P z_R - z_P z_O}. \tag{3.31}$$

The displacement of the reconstructed image for a small shift in the position of the source along the $x$ axis is then given by the relation

$$(dx_3/dx_P) = \frac{z_O z_R}{z_O z_R + z_P z_R - z_P z_O}. \tag{3.32}$$

Accordingly, if the source used to illuminate the hologram occupies very nearly the same position as the reference source used to record it, so that $z_P \approx z_R$, the image blur for a source size $\Delta x_P$ can be written as

$$\Delta x_3 = (z_O/z_P)\Delta x_P. \tag{3.33}$$

The acceptable value of the image blur for a display is determined by the resolution of the eye, which is about 0.5 mrad, or 0.5 mm at a viewing distance of 1 m. Accordingly, if the image is formed at a distance of 100 mm from the hologram, it can be illuminated with an extended monochromatic source (a mercury vapour lamp) with a diameter of 5 mm, located at a minimum distance of 1 m from the hologram.

### 3.5.2 Spectral bandwidth of the source

To calculate the effect of the spectral bandwidth of the source on the reconstructed image, it is convenient to assume that the hologram is recorded with a collimated reference beam of wavelength $\lambda_1$ but is illuminated at the same angle with a collimated reference beam of wavelength $\lambda_2$ when the reconstructed image is viewed. Under these conditions, $z_P = z_R = \infty$, but the quantities $(x_P/z_P) = (x_R/z_R)$ are still finite, and (3.9) and (3.11), the expressions for the coordinates of an image point, can be written as

$$x_3(\lambda_1, \lambda_2) = x_O + (x_P/z_P)(z_O/\mu) - (x_R/z_R)z_O, \tag{3.34}$$

and

$$z_3(\lambda_1, \lambda_2) = z_O/\mu, \tag{3.35}$$

where, as before, $\mu = \lambda_2/\lambda_1$.

The displacements of the image for a small change in the wavelength of the source used to illuminate the hologram are then given by the relations

$$(dx_3/d\lambda_2) = -\frac{x_P z_O}{z_P \mu \lambda_2}, \tag{3.36}$$

and

$$(dz_3/d\lambda_2) = -\frac{z_O}{\mu \lambda_2}. \tag{3.37}$$

If the source used to illuminate the hologram has a mean wavelength $\lambda_2$ approximately equal to $\lambda_1$, so that $\mu \approx 1$, and a spectral bandwidth $\Delta\lambda_2$, the transverse image blur due to the finite spectral bandwidth of the source is

$$|\Delta x_3| = (x_P/z_P)z_O(\Delta\lambda_2/\lambda_2), \tag{3.38}$$

and the longitudinal image blur is

$$|\Delta z_3| = z_O(\Delta\lambda_2/\lambda_2). \tag{3.39}$$

Although the longitudinal image blur is greater in magnitude, it is the transverse image blur that is usually more noticeable and limits the spectral bandwidth of the source that can be used to illuminate a hologram. The permissible spectral bandwidth decreases if either the depth of the image or the interbeam angle is increased.

Typically, if green light from a high-pressure mercury vapour lamp, which has a mean wavelength $\lambda_2$ of 546 nm and a spectral bandwidth of about 5 nm, is used to illuminate a hologram made with an argon-ion laser ($\lambda_1 = 514$ nm) and an interbeam angle of $30°$, in which case $(x_P/z_P) = \tan 30° = 1/\sqrt{3}$, (3.38) shows that the transverse image blur $|\Delta x_3|$ will be equal to 0.5 mm for an object point at a distance $z_O = 95$ mm from the hologram.

### 3.5.3 The image hologram

If the central plane of the image lies in the hologram plane, as in an image hologram (see section 2.5), the restrictions on the size and the spectral bandwidth of the source used to illuminate the hologram are minimized. In fact, if the interbeam angle and the depth of the image are both small, it is even possible to use an extended white-light source to illuminate the hologram. Image points in the hologram plane ($z_O = 0$) are then quite sharp and free from colour; other points in the image exhibit increased colour dispersion and blur as their distance from this plane increases.

## 3.6 Image luminance

The luminance of the image reconstructed by a hologram depends, in the first instance, on its diffraction efficiency, which can be defined as the ratio of the energy diffracted into the desired image by an element of the hologram to that incident on it from the source used to illuminate it. As we will see later, the diffraction efficiency of a hologram is determined by the recording medium used and the visibility of the carrier fringes. However, the luminance of the image also depends on the recording and reconstruction geometry [Hariharan, 1978].

### 3.6.1 The off-axis hologram

Consider a conventional off-axis hologram which, as shown in fig. 3.4, reconstructs a virtual image at a distance $d$ behind it. This image is viewed by an observer located in front of the hologram.

With a diffusely reflecting object, the flux that is incident on the photographic plate, while recording the hologram, can be taken as very nearly uniform over its whole area. As a result, the visibility of the carrier fringes and, therefore, the diffraction efficiency $\varepsilon$ can be assumed to be constant over the hologram. Accordingly, if the hologram is illuminated by a monochromatic beam of intensity $I$, the total energy diffracted into the image is $\varepsilon I A_H$, where $A_H$ is the area of the hologram.

As can be seen from fig. 3.4, the flux from any element of the reconstructed image is spread over a solid angle $\Omega_H = A_H/d^2$. The luminance of the image is, therefore,

$$L_v = \varepsilon I A_H (K_\lambda / \Omega_H A_I),$$
$$= \varepsilon I K_\lambda d^2 / A_I, \tag{3.40}$$

where $K_\lambda$ is the spectral luminous efficacy of the radiation, and $A_I$ is the area of the image.

Equation (3.40) shows that the luminance of the image increases with its distance from the hologram; however, this increase is at the expense of the solid angle over which the image can be viewed.

### 3.6.2 The image hologram

When a hologram is recorded of a real image projected either by an optical system or by another hologram (see section 3.3.1), the hologram reconstructs an image not only of the object but also of the optical system, including any aperture (or pupil) that limits the angular spread of the object beam. If the

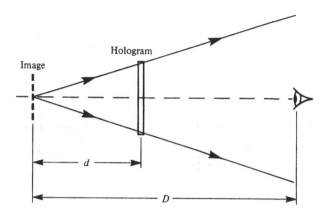

Fig. 3.4. Image reconstruction by an off-axis hologram: the hologram itself is the pupil [Hariharan, 1978].

source used to illuminate the hologram has the same wavelength and occupies the same position with respect to the hologram as the reference source used to record it, the location and size of this reconstructed pupil correspond to those of the pupil of the imaging system, or the boundaries of the primary hologram. As shown in fig. 3.5, the flux from any element of the reconstructed image is then confined within a solid angle $\Omega_P = A_P/D_P^2$, where $A_P$ is the area of the reconstructed pupil and $D_P$ is its distance from the image.

If the image is located at an appreciable distance from the hologram, and the numerical aperture of the imaging system used when recording the hologram is large enough, the flux in the object beam will be nearly uniform, and the diffraction efficiency of the hologram can be taken to be constant over its entire area. The luminance of the image is then

$$L_v = \varepsilon I A_H (K_\lambda / \Omega_P A_I).\qquad(3.41)$$

For a given hologram size and viewing distance, the luminance of the image is now independent of its distance from the hologram and depends only on the dimensions of the reconstructed pupil. A comparison of (3.41) with (3.40) shows that the luminance of the image has increased by a factor $(\Omega_H/\Omega_P)$, which is the reciprocal of the ratio of the solid angles of viewing available in the two cases. A substantial improvement in image luminance is therefore possible if the aperture of the imaging system, or the primary hologram, is masked so as to form an external pupil whose shape and size match the range of angles over which the hologram is actually to be viewed.

As described in section 3.5.3, a major advantage of image holograms is that

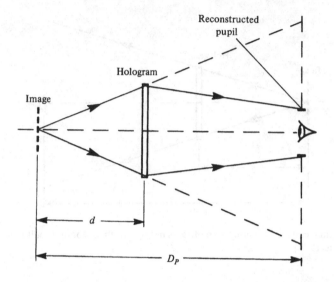

Fig. 3.5. Image formation with an external reconstructed pupil [Hariharan, 1978].

the image can be made to occupy a position straddling the hologram plane: this
minimizes the image blur when a source of finite size and spectral bandwidth
is used to illuminate the hologram. In this case, only the area of the hologram
corresponding to the image diffracts light and (3.41) reduces to

$$L_v = \varepsilon I K_\lambda / \Omega_P,  \qquad (3.42)$$

so that the luminance of the image is independent of its area. However, it is
apparent from (3.41) and (3.42) that the luminance of the image must be less
than the luminance of an image of the same object formed at a distance from
the hologram. Accordingly, to maximize image luminance, the object should
be at a sufficient distance to ensure that, while recording the hologram, the flux
from the object is spread out over the photographic plate. This distance is, of
course, limited by the size of the source to be used to illuminate the final image
hologram and the acceptable image blur.

### 3.7 Image speckle

When a diffusely reflecting object is illuminated, each of the microscopic elements making up the surface gives rise to a diffracted wave. With coherent
illumination, these diffracted waves can interfere with each other. Accordingly,
if the point-spread function of an imaging system is broader than these micro-

scopic elements, the diffracted amplitudes from many elements will contribute to the resultant amplitude at each point in the reconstructed image. Since the scale of the surface irregularities is usually such that the optical paths to neighbouring elements exhibit random differences amounting to several wavelengths, the interference of these diffracted components gives rise to local fluctuations in the image intensity (see Appendix 4). As a result, the image exhibits a granular, or speckled, appearance.

Speckle is a serious problem in holographic imaging. A number of methods have been described to reduce speckle in the reconstructed image (see, for example, the review by McKechnie [1975*c*]), and some of them will be discussed here.

If the subject is a two-dimensional transparency, illuminated with a uniform plane or spherical wave, the image should, in principle, be free from speckle. However, in practice, any imperfections, such as dust particles, on the surface of the transparency give rise to annoying diffraction patterns in the reconstructed image. Although these diffraction patterns can be minimized by recording a Fourier transform hologram, most of the power in the object beam is then concentrated in a small spot, resulting in poor diffraction efficiency and nonlinearity (see section 6.5), unless a relatively intense reference beam is used.

These problems can be avoided by using a diffuser to illuminate the transparency when recording the hologram. A further advantage then is that each point on the hologram receives light from every point of the transparency, so that the information stored on the hologram is no longer localized. As a result, local damage to the hologram results in only a slight loss of image luminance and an increase in scattered light, whereas with a photograph it can result in complete loss of information at that point. A spectacular consequence is that if the hologram is broken, any fragment reconstructs the complete image, though with reduced resolution. The penalty for this is that the image appears modulated on the random speckle pattern due to the diffuser.

If, however, the point-spread function of the imaging system is narrow enough to resolve individual elements on the diffuser, the complex amplitude at the diffuser is reproduced perfectly in the image plane. The image then exhibits a pure phase modulation and is free from speckle. The same result is obtained if the angular spread of light from the diffuser is small enough for all the light from it to pass through the aperture of the imaging system.

It is possible, therefore, to minimize speckle, while retaining most of the advantages of diffuse illumination, by using, instead of a random diffuser, a quasi-random phase plate [Upatnieks, 1967] or a phase grating [Gerritsen, Hannan & Ramberg, 1968; Gabor, 1970; Kato & Okino, 1973; Matsumura, 1975] in contact with the transparency.

Unfortunately, these techniques can be used only with nondiffusing objects, and then only for a single plane in the object. Accordingly, a number of methods of reducing speckle with diffusing objects have been proposed. All of them have some disadvantages, and the choice is usually determined by the sacrifices that can be made.

### 3.7.1 Low-pass filtering

The simplest method of speckle reduction when photographing the image is to use a hologram (or lens) aperture large enough that the average size of the speckles is less than the resolution limit of the photographic material used to record the image. The exposure at any point then corresponds to the intensity in the image averaged over a number of speckles, so that the intensity fluctuations are reduced significantly.

### 3.7.2 Partially coherent illumination

Because speckle is a consequence of the coherence of the illumination, it can be reduced by using partially coherent illumination to reconstruct the image. With an image hologram, it is possible to use a white-light source [Golbach, 1973]. Alternatively, the spatial coherence of the illumination can be reduced by means of moving diffusers [Lowenthal & Joyeux, 1971; Ih & Baxter, 1978].

### 3.7.3 Redundancy

In one technique [Martienssen & Spiller, 1967], the object is illuminated through a diffuser, and a number of holograms are made with the diffuser in different positions. If the images reconstructed by $N$ such holograms are superimposed on the same photographic film by successive exposures, each exposure records the same information about the object modulated on a different speckle pattern. As a result, the contrast of the speckle pattern is reduced by a factor of $N^{1/2}$. An alternative technique which gives the same result is the use of wavelength diversity [George & Jain, 1973].

### 3.7.4 Time-averaging

Instead of superimposing the images from a number of holograms, as described in the previous section, it is possible to sample a single large hologram by means of a small moving pupil [Dainty & Welford, 1971]. As the pupil uncovers different parts of the hologram, the speckle pattern in the image changes, so that the exposure records the sum of the irradiance distributions in a number of uncorrelated speckle patterns. This method was extended by Yu and Wang [1973], who proposed the use of a moving mask with randomly distributed openings covering the entire hologram, and by Som and Budhiraja [1975], who claimed that a further reduction in speckle could be obtained by the use of a moving mask, at some distance from the hologram, both in recording and reconstruction. However, it can be shown that while there is a reduction of speckle, it is always accompanied by a reduction in either the resolution or the contrast of the image [Hariharan & Hegedus, 1974*a,b*; McKechnie, 1975*a,b*; Östlund & Biedermann, 1977].

# 4

# Types of holograms

A hologram recorded on a photographic plate and processed normally is equivalent to a grating with a spatially varying transmittance. However, with suitable processing, it is possible to produce a spatially varying phase shift. In addition, if the thickness of the recording medium is large compared to the fringe spacing, volume effects are important. In an extreme case, it is even possible to produce holograms in which the fringes are planes running almost parallel to the surface of the recording material; such holograms can reconstruct an image in reflected light.

Based on these characteristics, holograms recorded in a thin recording medium can be divided into amplitude holograms and phase holograms. Holograms recorded in relatively thick recording media can be classified either as transmission amplitude holograms, transmission phase holograms, reflection amplitude holograms, or reflection phase holograms.

In the next few sections we shall examine some of the principal characteristics of these six types of holograms.

## 4.1 Thin holograms

Any hologram in which the thickness of the recording material is small compared to the average spacing of the interference fringes can be classified as a thin hologram. Such a hologram can be characterized by a spatially varying complex amplitude transmittance

$$\mathbf{t}(x, y) = |\mathbf{t}(x, y)| \exp[-i\phi(x, y)]. \tag{4.1}$$

### 4.1.1 Thin amplitude holograms

In an amplitude hologram, $\phi(x, y)$ is essentially constant while $|\mathbf{t}(x, y)|$ varies over the hologram. To calculate the complex amplitude of the diffracted waves

45

from such a hologram and, hence, its diffraction efficiency, consider a grating formed in a suitable thin recording medium by a plane object wave and a plane reference wave.

If we assume that the resulting amplitude transmittance is linearly related to the intensity in the interference pattern, the amplitude transmittance of the grating can be written as

$$|\mathbf{t}(x, y)| = \mathbf{t}_0 + \mathbf{t}_1 \cos Kx, \tag{4.2}$$

where $\mathbf{t}_0$ is the average amplitude transmittance of the grating, $\mathbf{t}_1$ is the amplitude of the spatial variation of $|\mathbf{t}(x)|$, and

$$K = 2\pi/\Lambda, \tag{4.3}$$

where $\Lambda$ is the spacing of the fringes.

Since the values of $|\mathbf{t}(x)|$ are limited to the range $0 \leq |\mathbf{t}(x)| \leq 1$, and the amplitudes of the diffracted waves are linearly proportional to the amplitude of the spatial variation of $|\mathbf{t}(x)|$, the diffracted amplitude is a maximum when

$$|\mathbf{t}(x)| = (1/2) + (1/2) \cos Kx,$$
$$= (1/2) + (1/4) \exp(iKx) + (1/4) \exp(-iKx). \tag{4.4}$$

The maximum amplitude in each of the diffracted orders is one fourth of that in the wave used to illuminate the hologram, so that the peak diffraction efficiency is

$$\varepsilon_{\max} = 1/16, \tag{4.5}$$

or 0.0625.

In practice, no recording medium has a linear response over the full range of transmittance values from 0 to 1; hence, this value of $\varepsilon$ cannot be achieved without running into nonlinear effects.

### 4.1.2 Thin phase holograms

For a lossless phase grating, $|\mathbf{t}(x)| = 1$, so that the complex amplitude transmittance is

$$\mathbf{t}(x) = \exp[-i\phi(x)]. \tag{4.6}$$

If the phase shift produced by the recording medium is linearly proportional to the intensity in the interference pattern,

$$\phi(x) = \phi_0 + \phi_1 \cos(Kx), \tag{4.7}$$

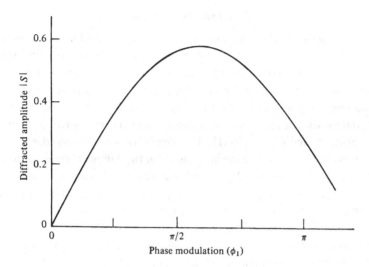

Fig. 4.1. Diffraction efficiency of a thin phase grating as a function of the phase modulation [Kogelnik, 1967].

and the complex amplitude transmittance of the grating is

$$\mathbf{t}(x) = \exp(-i\phi_0) \exp[-i\phi_1 \cos(Kx)]. \tag{4.8}$$

If we neglect the constant phase factor $\exp(-i\phi_0)$, the right-hand side of (4.8) can be expanded as a Fourier series to give the result

$$\mathbf{t}(x) = \sum_{n=-\infty}^{\infty} i^n J_n(\phi_1) \exp(inKx), \tag{4.9}$$

where $J_n$ is the Bessel function of the first kind of order $n$.

Such a thin phase grating diffracts a wave incident on it into a large number of orders, the diffracted amplitude in the $n$th order being proportional to the value of the Bessel function $J_n(\phi_1)$. Only the wave of order 1 contributes to the desired image. As shown in fig. 4.1, the amplitude diffracted into this order, which is proportional to $J_1(\phi_1)$, increases at first with the phase modulation and then decreases.

The diffraction efficiency of the grating is, accordingly,

$$\varepsilon = J_1^2(\phi_1), \tag{4.10}$$

and its maximum value is

$$\varepsilon_{\max} = 0.339. \tag{4.11}$$

## 4.2 Volume holograms

The medium in which a hologram is recorded can have a thickness of as much as a few millimetres, while the fringe spacing may be of the order of only 1 $\mu$m. The hologram is then a three-dimensional system of layers corresponding to a periodic variation of absorption or refractive index, and the diffracted amplitude is a maximum only when the Bragg condition is satisfied. The characteristics of generalized volume gratings have been discussed in detail by Russell [1981] and by Solymar and Cooke [1981]. For simplicity, we will consider only a grating produced by recording the interference of two infinite plane wavefronts in a thick recording medium. We will also assume that initially the recording medium is perfectly transparent but, after processing, develops a sinusoidal variation of the absorption or the refractive index in the direction perpendicular to the interference surfaces. In addition, although the interference surfaces can assume any orientation, we will consider only two limiting cases in which they are either perpendicular or parallel to the hologram plane.

The first case arises when the two interfering wavefronts make equal but opposite angles to the surface of the recording medium and are incident on it from the same side. Holograms recorded in this fashion produce a reconstructed image in transmitted light. The second case arises when the wavefronts are symmetrical with respect to the surface of the recording medium but are incident on it from opposite sides. Holograms of this type produce a reconstructed image by reflection.

The spacing between successive fringe planes is a minimum, and the volume effects are most pronounced, when the angle between the two interfering wavefronts is a maximum ($\approx 180°$). This has made it possible to produce reflection holograms with a wavelength selectivity high enough to reconstruct an image of acceptable quality when illuminated with white light.

## 4.3 The coupled wave theory

When analysing the diffraction of light by such thick gratings, it is necessary to take into account the fact that the amplitude of the diffracted wave increases progressively, whereas that of the incident wave decreases, as they propagate through the grating. One way of doing this is by means of a coupled-wave approach, such as that developed by Kogelnik [1967, 1969].

Consider a coordinate system in which, as shown in fig. 4.2, the $z$ axis is perpendicular to the surfaces of the recording medium and the $x$ axis is in the plane of incidence, while the fringe planes are oriented perpendicular to the plane of incidence. The grating vector $\mathbf{K}$ is perpendicular to the fringe planes.

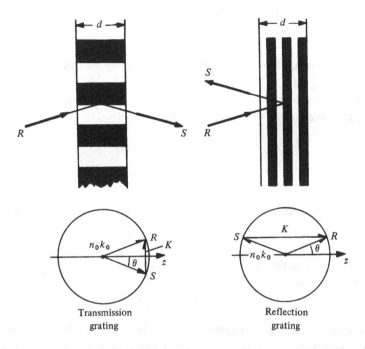

Fig. 4.2. Volume transmission and reflection gratings and their associated vector diagrams for Bragg incidence [Kogelnik, 1967].

It is of length $|\mathbf{K}| = 2\pi/\Lambda$, where $\Lambda$ is the grating period, and makes an angle $\psi$ ($\psi = 90°$ or $0°$ in the cases shown) with the $z$ axis. The refractive index $n$ and the absorption constant $\alpha$ are assumed to vary sinusoidally, their values at any point being given by the relations

$$n = n_0 + n_1 \cos \mathbf{K} \cdot \mathbf{x}, \tag{4.12}$$

$$\alpha = \alpha_0 + \alpha_1 \cos \mathbf{K} \cdot \mathbf{x}, \tag{4.13}$$

where the radius vector $\mathbf{x} = (x, y, z)$, and, for simplicity, the refractive index of the surrounding medium is also assumed to be $n_0$.

If monochromatic light is incident on the hologram grating at, or near, the Bragg angle, and if the thickness of the medium is large enough, only two waves in the grating need be taken into consideration; they are the incoming reference wave $R$ and the outgoing signal wave $S$. Since the other diffraction orders violate the Bragg condition strongly, they are severely attenuated and can be neglected. If we also assume that these two waves are polarized with their electric vectors perpendicular to the plane of incidence, their interaction in the grating can be described by the scalar wave equation

$$\nabla^2 E + k^2 E = 0, \tag{4.14}$$

where $E$ is the total electric field and $k$ is the (spatially varying) propagation constant in the grating.

We assume that the absorption per wavelength as well as the relative variations in refractive index of the medium are small, so that

$$n_0 k_0 \gg \alpha_0,$$
$$n_0 k_0 \gg \alpha_1,$$
$$n_0 \gg n_1, \tag{4.15}$$

where $k_0 = 2\pi/\lambda$. The propagation constant can then be written in the form

$$k^2 = B^2 - 2i\alpha_0 B + 4\kappa B \cos \mathbf{K} \cdot \mathbf{x}, \tag{4.16}$$

where $B = n_0 k_0$ is the average propagation constant, and $\kappa$ is the coupling constant defined as

$$\kappa = (\pi n_1/\lambda) - i\alpha_1/2. \tag{4.17}$$

This coupling constant describes the interaction between the reference wave $R$ and the signal wave $S$. If $\kappa = 0$, there is no modulation of the refractive index or the absorption and, hence, no diffraction.

The propagation of the two coupled waves through the grating can be described by their complex amplitudes $R(z)$ and $S(z)$, which vary along $z$ as a result of the energy interchange between them as well as energy losses due to absorption. The total electric field $E$ in the grating is then the sum of the fields due to these two waves, so that

$$E = R(z) \exp(-i\boldsymbol{\rho} \cdot \mathbf{x}) + S(z) \exp(-i\boldsymbol{\sigma} \cdot \mathbf{x}), \tag{4.18}$$

where $\boldsymbol{\rho}$ and $\boldsymbol{\sigma}$ are the propagation vectors for the two waves, defined by the propagation constants and the directions of propagation of $R$ and $S$. The quantity $\boldsymbol{\rho}$ is assumed to be equal to the propagation vector of the free reference wave in the absence of coupling, while $\boldsymbol{\sigma}$ is determined by the grating and is related to $\boldsymbol{\rho}$ and the grating vector $\mathbf{K}$ by the expression

$$\boldsymbol{\sigma} = \boldsymbol{\rho} - \mathbf{K}. \tag{4.19}$$

For the special case of incidence at the Bragg angle $\theta_0$, the lengths of both $\boldsymbol{\rho}$ and $\boldsymbol{\sigma}$ are equal to the free propagation constant $n_0 k_0$, and the Bragg condition, which can be written in the form

$$\cos(\psi - \theta_0) = K/2n_0 k_0, \tag{4.20}$$

is obeyed. If (4.20) is differentiated, we obtain the result

$$d\theta_0/d\lambda_0 = K/4\pi n_0 \sin(\psi - \theta_0). \tag{4.21}$$

It follows from (4.21) that small changes in the angle of incidence or the wavelength have similar effects.

A useful parameter for evaluating the effects of deviations from the Bragg condition is the dephasing measure $\zeta$, which can be defined from (4.19) as

$$\begin{aligned}
\zeta &= (|\rho|^2 - |\sigma|^2)/2|\rho|, \\
&= (B^2 - |\sigma|^2)/2B, \\
&= K\cos(\psi - \theta) - K^2\lambda/4\pi n_0. \tag{4.22}
\end{aligned}$$

For small deviations $\Delta\theta$ and $\Delta\lambda$ from the Bragg condition, (4.22) becomes

$$\zeta = \Delta\theta \cdot K \sin(\psi - \theta_0) - \Delta\lambda \cdot K^2/4\pi n_0. \tag{4.23}$$

To derive the coupled wave equations, (4.14) and (4.16) are combined, and (4.18) and (4.19) are inserted. If we compare the terms involving $\exp(-i\rho \cdot \mathbf{x})$ and $\exp(-i\sigma \cdot \mathbf{x})$, we get

$$R'' - 2iR'\rho_z - 2i\alpha BR + 2\kappa BS = 0, \tag{4.24}$$

and

$$S'' - 2iS'\sigma_z - 2i\alpha BS + (B^2 - |\sigma|^2)S + 2\kappa BR = 0, \tag{4.25}$$

where the primes denote differentiation with respect to $z$. If, in addition, we assume that the energy interchange between $S$ and $R$, as well as the energy absorption in the medium are slow, the second differentials $R''$ and $S''$ can be neglected. From (4.23) these equations can then be rewritten in the form

$$R'\cos\theta + \alpha R = i\kappa S, \tag{4.26}$$

and

$$[\cos\theta - (K/B)\cos\psi]S' + (\alpha + i\zeta)S = i\kappa R. \tag{4.27}$$

The coupled wave equations (4.26) and (4.27) show that the amplitude of a wave changes along $z$ because of coupling to the other wave ($\kappa R$, $\kappa S$) or absorption ($\alpha R$, $\alpha S$). Deviations from the Bragg condition result in $S$ being forced out of synchronism with $R$, due to the term involving $\zeta S$, and the interaction decreases.

The coupled wave equations, (4.26) and (4.27), can be solved for the appropriate boundary conditions. These are $R(0) = 1$, $S(0) = 1$, for transmission gratings, and $R(0) = 1$, $S(d) = 0$, for reflection gratings.

In the next few sections we will discuss the solutions for the most important cases, namely lossless phase gratings and pure absorption gratings, the grating planes being assumed to run either normal to the surface (for transmission gratings) or parallel to the surface (for reflection gratings). The method of solution of the coupled wave equations, as well as solutions for the cases of slanted gratings, lossy phase gratings, and mixed gratings are to be found in the original paper by Kogelnik [1969]. This paper also gives an extension of the theory to light polarized with the electric vector in the plane of incidence.

## 4.4 Volume transmission holograms

### 4.4.1 Phase gratings

In a lossless phase grating $\alpha_0 = \alpha_1 = 0$. Diffraction is caused by the spatial variation of the refractive index. The diffracted amplitude is then

$$S(d) = \frac{-i\exp(-i\chi)\sin(\Phi^2 + \chi^2)^{1/2}}{(1 + \chi^2/\Phi^2)^{1/2}}, \tag{4.28}$$

where

$$\Phi = \pi n_1 d/\lambda \cos\theta, \tag{4.29}$$

and

$$\chi = \zeta d/2 \cos\theta. \tag{4.30}$$

From (4.23), it follows that for incidence at the Bragg angle $\zeta = 0$, so that $\chi = 0$, and (4.28) becomes

$$S(d) = -i\sin\Phi. \tag{4.31}$$

Since the incident amplitude $R(0)$ is assumed to be unity, the diffraction efficiency is

$$\varepsilon = |S(d)|^2 = \sin^2\Phi. \tag{4.32}$$

As either $d$, the thickness, or $n_1$, the variation of the refractive index, increases, the diffraction efficiency increases until the modulation parameter $\Phi = \pi/2$. At this point $\varepsilon = 1.00$, and all the energy is in the diffracted beam. When $\Phi$ increases beyond this point, energy is coupled back into the incident wave, and $\varepsilon$ decreases.

When the angle of incidence or the wavelength of the incident beam deviates from the value required to satisfy the Bragg condition, the diffraction efficiency drops to

$$\varepsilon = \frac{\sin^2(\Phi^2 + \chi^2)^{1/2}}{(1 + \chi^2/\Phi^2)}. \tag{4.33}$$

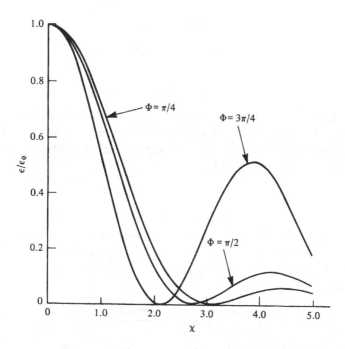

Fig. 4.3. Volume transmission holograms. Angular and wavelength selectivity of a phase grating showing the normalized diffraction efficiency ($\varepsilon/\varepsilon_0$) as a function of the parameter $\chi$, which is a measure of the deviation from the Bragg condition, for three different values of the modulation parameter $\Phi$ [Kogelnik, 1969].

The effect of an angular deviation $\Delta\theta$ or a wavelength deviation $\Delta\lambda$ from the Bragg condition can be studied readily, since they influence the diffraction efficiency mainly through the parameter $\chi$, which is a measure of the deviation from the Bragg condition. From (4.23) it can be rewritten in the form

$$\chi = \Delta\theta \, Kd/2, \tag{4.34}$$

or, alternatively,

$$\chi = \Delta\lambda \, K^2 d/8\pi n_0 \cos\theta_0, \tag{4.35}$$

while the modulation parameter $\Phi$ can be taken as constant.

Curves showing the normalized diffraction efficiency of a lossless transmission phase hologram as a function of the parameter $\chi$ are plotted in fig. 4.3 for three values of the modulation parameter $\Phi$. For a hologram with $\Phi = \pi/2$, which diffracts with maximum efficiency (1.00) at the Bragg angle, the diffraction efficiency drops to zero when $\chi = 2.7$.

### 4.4.2  Effect of losses

For a lossy phase grating (absorption constant $\alpha_0$), the diffracted amplitude at the Bragg angle is

$$S(d) = -\text{i}\exp(-\alpha_0 d/\cos\theta)\sin\Phi. \tag{4.36}$$

This expression has the same form as (4.31), except for an additional exponential term containing the absorption coefficient $\alpha_0 d$. This term mainly decreases the peak diffraction efficiency and has only a small effect on the angular and wavelength selectivity.

### 4.4.3  Amplitude gratings

In an amplitude grating, the refractive index does not vary, so that $n_1 = 0$. However, the absorption constant varies with an amplitude $\alpha_1$ about its mean value $\alpha_0$. In this case, the coupling constant $\kappa = -\text{i}\alpha_1/2$, and the diffracted amplitude is

$$S(d) = -\exp\left(-\frac{\alpha_0 d}{\cos\theta}\right)\exp(-\text{i}\chi)\frac{\sinh(\Phi_a^2 - \chi^2)^{1/2}}{(1 - \chi^2/\Phi_a^2)^{1/2}}, \tag{4.37}$$

where $\Phi_a = \alpha_1 d/2\cos\theta$, and $\chi$ is defined, as before, by (4.30).

For incidence at the Bragg angle, $\chi = 0$, and the diffraction efficiency can be written as

$$\varepsilon = \exp(-2\alpha_0 d/\cos\theta_0)\sinh^2(\alpha_1 d/2\cos\theta_0). \tag{4.38}$$

The diffracted amplitude increases with $\alpha_1$, but, since negative values of the absorption are excluded, $\alpha_1 \leq \alpha_0$. The highest diffraction efficiency is therefore obtained when $\alpha_1 = \alpha_0$, for a value of $\alpha_1 d\cos\theta_0 = \ln 3$; this maximum has a value

$$\varepsilon_{\max} = 1/27, \tag{4.39}$$

or 0.037.

Figure 4.4 shows the diffracted amplitude computed from (4.38) as a function of two parameters, $D_0 = \alpha_0 d/\cos\theta_0$ and $D_1 = \alpha_1 d/\cos\theta_0$. The quantity $D_1$ is a measure of the amplitude of the modulation, while $D_1/D_0 = \alpha_1/\alpha_0$ is a measure of the relative depth of modulation. The curves for constant $D_0$ show the behaviour for a constant background absorption and varying relative depth of modulation. A linear response and relatively good efficiency are obtained for a background absorption $D_0 = 1$.

Curves showing the angular and wavelength selectivity of amplitude gratings computed from (4.37) are presented in fig. 4.5. These curves are plotted as

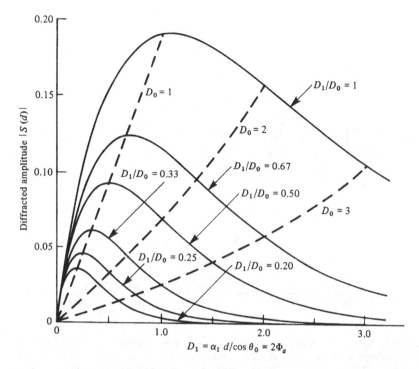

Fig. 4.4. Volume transmission holograms. Amplitude diffracted by an amplitude grating as a function of the modulation $D_1 = \alpha_1 d / \cos \theta_0 = 2\Phi_a$, for various bias levels $D_0 = \alpha_0 d / \cos \theta_0$ (broken lines) and modulation depths $D_1/D_0$ (solid curves) [Kogelnik, 1969].

functions of the parameter $\chi$, which is a measure of the deviation from the Bragg condition, for the special case when $\alpha_1 = \alpha_0$, and for values of the modulation parameter $\Phi_a = \alpha_1 d / 2 \cos \theta$ of 0.55 (corresponding to the peak diffraction efficiency of 0.037) and 1.0. The curves are similar, and, in both cases, the diffraction efficiency drops to zero when $\chi \approx 3.3$.

## 4.5 Volume reflection holograms

As shown in fig. 4.2, the fringe planes in a reflection hologram run more or less parallel to the surface of the recording medium, so that $\psi = 0$. When the hologram is illuminated by a wave incident from the left, the diffracted wave starts with an amplitude $S(d) = 0$ at the rear face of the hologram but gains in amplitude as it propagates towards the left.

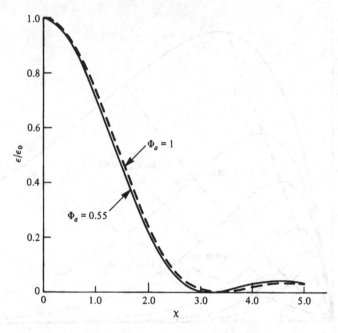

Fig. 4.5. Volume transmission holograms. Angular and wavelength selectivity of an amplitude grating for $\alpha_1 = \alpha_0 (D_1 = D_0)$ and values of $\Phi_a = D_1/2 = 0.55 (\varepsilon_0 = 0.037)$ and $\Phi_a = D_1/2 + 1 (\varepsilon_0 = 0.025)$. The curves show the normalized diffraction efficiency $\varepsilon/\varepsilon_0$ as a function of the parameter $\chi$, which is a measure of the deviation from the Bragg condition [Kogelnik, 1969].

### 4.5.1 Phase gratings

In a lossless phase grating $\alpha_0 = \alpha_1 = 0$, and the coupling constant $\kappa = \pi n_1/\lambda$. For small deviations from the Bragg angle, the diffracted amplitude is given by the expression

$$S(0) = i[(i\chi_r/\Phi_r) + (1 - \chi_r^2/\Phi_r^2)^{1/2} \coth(\Phi_r^2 - \chi_r^2)^{1/2}]^{-1}, \qquad (4.40)$$

where

$$\Phi_r = \pi n_1 d/\lambda \cos\theta, \qquad (4.41)$$

and

$$\chi_r = \zeta d/2 \cos\theta. \qquad (4.42)$$

The diffraction efficiency is then

$$\varepsilon = [1 + (1 - \chi_r^2/\Phi_r^2)/\sinh^2(\Phi_r^2 - \chi_r^2)^{1/2}]^{-1}. \qquad (4.43)$$

For a wave incident at the Bragg angle, $\chi_r = 0$, and (4.43) can be simplified

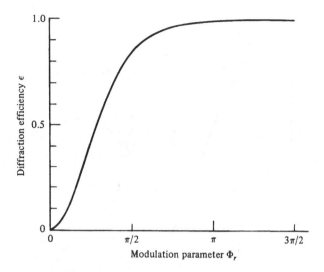

Fig. 4.6. Volume reflection holograms. Diffraction efficiency of a phase grating at the Bragg angle as a function of the modulation parameter $\Phi_r$.

and written as

$$\varepsilon = \tanh^2(\pi n_1 d/\lambda \cos\theta_0). \qquad (4.44)$$

As shown in fig. 4.6, the diffraction efficiency increases asymptotically to 1.00 as the value of $\Phi_r$ increases.

Curves showing the angular and wavelength selectivity of a lossless phase grating computed from (4.43) are presented in fig. 4.7 using the following relationships derived from (4.42) for the parameter $\chi_r$, which is a measure of the deviation from the Bragg condition.

$$\begin{aligned}\chi_r &= \Delta\theta(2\pi n_0 d/\lambda)\sin\theta_0,\\ &= (\Delta\lambda/\lambda)(2\pi n_0 d/\lambda)\cos\theta_0. \qquad (4.45)\end{aligned}$$

These curves are plotted for values of the modulation parameter $\Phi_r$ of $\pi/4$, $\pi/2$, and $3\pi/4$. While the diffraction efficiency drops to zero in all cases when $\chi_r \approx 3.5$, the selectivity curves become appreciably broader with increasing values of $\Phi_r$.

### 4.5.2 Amplitude gratings

In this case, as in section 4.4.3, $n_1 = 0$, while $\alpha_0$ and $\alpha_1$ are finite. The amplitude of the diffracted wave leaving the hologram is then

$$S(0) = -\{(\chi_{ra}/\Phi_{ra}) + [(\chi_{ra}/\Phi_{ra})^2 - 1]^{1/2}\coth(\chi_{ra}^2 - \Phi_{ra}^2)^{1/2}\}^{-1}, \qquad (4.46)$$

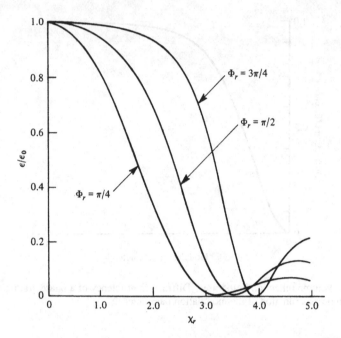

Fig. 4.7. Volume reflection holograms. Angular and wavelength selectivity of a phase grating, showing the normalized diffraction efficiency $(\varepsilon/\varepsilon_0)$ as a function of the parameter $\chi_r$, which is a measure of the deviation from the Bragg condition, for different values of the modulation parameter $\Phi_r$ [Kogelnik, 1969].

where

$$\Phi_{ra} = \alpha_1 d/2 \cos\theta, \qquad (4.47)$$

and

$$\chi_{ra} = (\alpha_0 d/\cos\theta_0) - (i\zeta d/2\cos\theta_0). \qquad (4.48)$$

If the incident wave is at the Bragg angle, $\zeta = 0$ and $\chi_{ra}/\Phi_{ra} = 2\alpha_0/\alpha_1$, so that (4.46) can be written as

$$\begin{aligned} S(0) = -\{(2\alpha_0/\alpha_1) + [(4\alpha_0^2/\alpha_1^2) - 1]^{1/2} \\ \times \coth[(d/\cos\theta_0)(\alpha_0^2 - \alpha_1^2/4)^{1/2}]\}^{-1}. \end{aligned} \qquad (4.49)$$

Curves of the diffracted amplitude as a function of the modulation parameter $D_1 = \alpha_1 d/\cos\theta_0 = 2\Phi_{ra}$ are presented in fig. 4.8 for different values of the modulation depth $D_1/D_0 = \alpha_1/\alpha_0$ and bias level $D_0 = \alpha_0 d/\cos\theta_0$. These curves show clearly that, for reflection holograms of this type, the best diffrac-

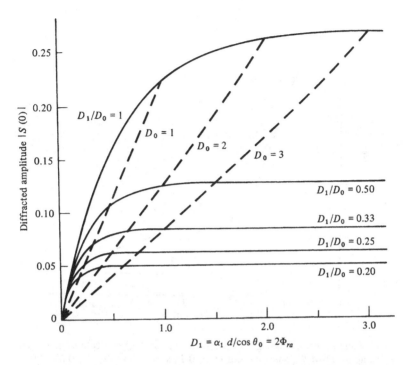

Fig. 4.8. Volume reflection holograms. Amplitude diffracted by an amplitude grating as a function of the modulation $D_1 = \alpha_1 d / \cos \theta_0 = 2\Phi_{ra}$, for various bias levels $D_0 = \alpha_0 d / \cos \theta_0$ (broken lines) and modulation depths $D_1/D_0$ (solid curves) [Kogelnik, 1969].

tion efficiency is obtained when $\alpha_0 d / \cos \theta_0 \geq 2$, corresponding to an optical density $\geq 1.7$. (See Appendix 5.)

For the special case when the modulation depth is a maximum ($\alpha_1 = \alpha_0$), (4.49) can be simplified further to

$$S(0) = -[2 + 3^{1/2} \coth(3^{1/2}\alpha_0 d/2 \cos \theta_0)]^{-1}, \qquad (4.50)$$

from which it is apparent that the maximum possible value of the diffraction efficiency is

$$\varepsilon_{max} = (2 + \sqrt{3})^{-2}, \qquad (4.51)$$

or 0.072.

Figure 4.9 shows curves for the angular and wavelength selectivity of an amplitude grating derived from (4.46), making use of the fact that (4.48) can

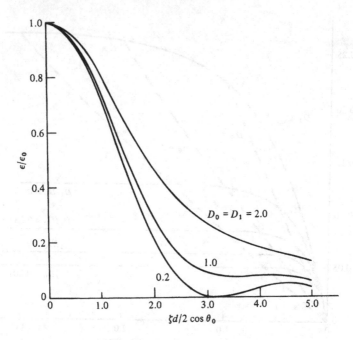

Fig. 4.9. Volume reflection holograms. Angular and wavelength selectivity of an amplitude grating for $D_1 = D_0$ and values of $D_1 = 0.2$ ($\varepsilon_0 = 0.007$), $D_1 = 1.0$ ($\varepsilon_0 = 0.05$), and $D_1 = 2$ ($\varepsilon_0 = 0.068$). The normalized diffraction efficiency ($\varepsilon/\varepsilon_0$) is plotted as a function of the term $\zeta d/2 \cos\theta_0$, which is a measure of the deviation from the Bragg condition [Kogelnik, 1969].

be rewritten in the form

$$\chi_{ra} = (\alpha_0 d / \cos\theta_0) - i\Delta\theta(2\pi n_0 d/\lambda_0) \sin\theta_0,$$
$$= (\alpha_0 d / \cos\theta_0) - i(\Delta\lambda/\lambda_0)(2\pi n_0 d/\lambda_0) \cos\theta_0. \quad (4.52)$$

In these curves, the diffracted amplitude is plotted as a function of the term $\zeta d / \cos\theta_0$, which is a measure of the deviation from the Bragg condition, for three values of the modulation parameter $D_1$ when the modulation depth is a maximum ($\alpha_1/\alpha_0 = 1$). The curves become broader as the loss parameter $D_0$ increases.

## 4.6 Discussion

The maximum diffraction efficiencies that can be obtained with the six types of gratings studied are summarized in Table 4.1.

While the maximum diffraction efficiency that can actually be obtained with a

Table 4.1. *Maximum theoretical diffraction efficiency for different types of gratings*

| Type of grating | Thin transmission | | Volume transmission | | Volume reflection | |
|---|---|---|---|---|---|---|
| Modulation | $\|t\|$ | $\phi$ | $\|t\|$ | $\phi$ | $\|t\|$ | $\phi$ |
| $\varepsilon_{max}$ | 0.0625 | 0.339 | 0.037 | 1.00 | 0.072 | 1.00 |

grating is limited by nonlinearities in the response of the recording medium and consequent distortions of the profile of the fringes, values approximating these theoretical limits have been obtained in a number of experiments, confirming the conclusions of the coupled wave theory.

On the other hand, the maximum diffraction efficiency that can be obtained with a hologram of a diffusely reflecting object is always much lower, because the spatial variations in the amplitude of the object wave (see Appendix 4) make it impossible to maintain optimum modulation over the entire area.

The average diffraction efficiencies of transmission phase holograms recorded with a diffuse object beam in both thin and thick recording media have been calculated by Upatnieks and Leonard [1970] assuming that the amplitude of the object wave has a Rayleigh probability distribution. These calculations show that the maximum diffraction efficiency is 0.22 with a thin recording medium, and 0.64 with a thick recording medium, against 0.339 and 1.00, respectively, for phase gratings recorded in the same media.

## 4.7 More accurate theories

Kogelnik's coupled wave theory assumes infinite plane wavefronts as well as a slow energy interchange and only a small absorption loss per wavelength. Two other major assumptions are that the modulation is sinusoidal and only the zero-order and Bragg-diffracted waves propagate within the grating. However, even if the first of these assumptions is satisfied, the existence of higher-order diffracted waves in phase gratings cannot be ignored, even for fairly small values of the refractive index modulation ($n_1 \geq 0.005$).

More accurate theories for the diffraction of light by volume holograms were therefore developed by a number of authors, who followed two other basic approaches. These were a rigorous modal approach [Burckhardt, 1966a; Chu & Tamir, 1970; Kaspar, 1973] and a thin-grating decomposition approach [Alferness, 1975a, 1976]. The equivalence of these two approaches and the coupled wave theory has been discussed by Alferness [1975b], Magnusson and Gaylord [1978b] and Gaylord and Moharam [1982]. More general treatments based on the coupled-wave theory have also been formulated by Su and Gaylord [1972], Kessler and Kowarschik [1975], Kowarschik and Kessler [1975], Magnusson and Gaylord [1977], Moharam and Gaylord [1981] and Gaylord and Moharam [1985]. These theories cover gratings with nonsinusoidal profiles and lead to equations that can be solved with the aid of a digital computer. As expected, they confirm that at larger values of the refractive index modulation a substantial amount of power can be diffracted into higher orders, and the diffraction efficiency at the second- and third-order Bragg angles can be quite high (> 0.2).

Effects due to the sharp boundaries between the sinusoidally modulated grating and the surrounding medium have also been ignored in Kogelnik's theory. These effects have been taken into consideration in a treatment using the modal theory [Langbein & Lederer, 1980; Lederer & Langbein, 1980], and reveal a slight overall enhancement of diffraction efficiency, as well as an oscillatory dependence on the thickness of the grating.

An assumption implicit in the preceding treatments is that the structure of the holographically recorded grating is perfectly uniform throughout the depth of the recording medium. However, with any recording material having an appreciable thickness and a finite absorption, the light waves used to produce the grating are attenuated inside the material, so that this assumption is no longer valid.

Diffraction by a grating in which the modulation term decreases exponentially with depth has been discussed by Kermisch [1969, 1971], Uchida [1973], Kowarschik [1976], Lederer and Langbein [1977] and Killat [1977a]. For incidence at the Bragg angle, the attenuation acts as a factor decreasing the effective thickness of the storage medium and reducing the maximum attainable diffraction efficiency. Away from the Bragg angle, there is an increase in the values of the secondary minima as well as a shift of the secondary maxima and minima.

Detailed experiments to study the effects of absorption in the recording medium using photographic emulsions as well as layers of dichromated gelatin have been described by Kubota [1978].

In phase holograms, it is also necessary to take into account the fact that there can be a change in the average dielectric constant. The effect of such a change has been analysed by Jordan and Solymar [1978], Solymar [1978], and Owen and Solymar [1980] on the assumption that the permittivity of the recording material changes during the recording process by an amount proportional to the square of the electric field. They have shown that, in general, the Bragg condition is no longer satisfied, resulting in a loss in diffraction efficiency. This loss is larger for higher beam-ratios, and much greater for reflection holograms than for transmission holograms.

An experimental study by Syms and Solymar [1983] with volume phase transmission gratings formed in bleached photographic materials has shown that if account is taken of all factors such as amplitude modulation (in addition to phase modulation), a second harmonic in the grating profile, and higher diffraction orders, the results are in good agreement with those expected from coupled-wave theory. This model was then used to deduce the major characteristics of the gratings, including saturation of the modulation with exposure. The results confirmed that while the material saturates at higher exposures, the diffraction efficiency is limited essentially by scatter and the consequent formation of noise gratings. A further study by Heaton and Solymar [1985a] with volume phase reflection gratings, in which the effects of dispersion are more important, has provided data on the variation of the grating properties with wavelength.

## 4.8 Criteria for thin holograms and volume holograms

As we have seen earlier, it is convenient to classify holograms into two broad categories, thin holograms and volume holograms, which correspond to the Raman–Nath and Bragg diffraction regimes, respectively.

The distinction between these two regimes is commonly made on the basis of a parameter $Q$ [Klein & Cook, 1967], which is defined by the relation

$$Q = 2\pi\lambda_0 d / n_0 \Lambda^2. \tag{4.53}$$

Small values of $Q(Q < 1)$ correspond to thin gratings, while large values of $Q(Q > 1)$ correspond to volume gratings. However, this criterion is not always adequate, since experimental observations [Alferness, 1976; Magnusson & Gaylord, 1977] have shown that several diffracted orders can be obtained from holographic gratings recorded in lithium niobate crystals and dichromated gelatin layers under some conditions, even with relatively large values of $Q$.

The criteria for distinguishing between thin holograms and volume holograms have been studied in detail by several authors [Magnusson & Gaylord, 1977, 1978a; Moharam & Young, 1978; Benlarbi, Cooke & Solymar, 1980]. They have shown that for small levels of modulation only a single wave is diffracted, even if $Q$ is small, so that either theory can be used, while for a large modulation higher-order waves cannot be neglected, even when $Q$ is quite large. Consequently, as the modulation amplitude increases, an intermediate regime appears and widens. While the corresponding simple formulas can still be used in the "thin" and "volume" regimes, the transition boundaries between the regimes depend on the modulation amplitude as well as the value of $Q$. It is possible to allow for this effect by the introduction of another parameter $P$

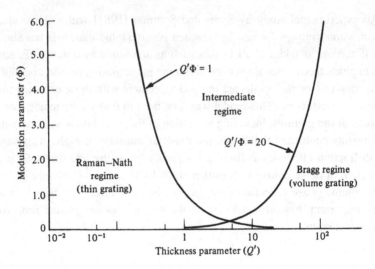

Fig. 4.10. Raman–Nath, intermediate and Bragg diffraction regimes for phase holograms [Moharam, Gaylord & Magnusson, 1980*b*].

[Nath, 1938], defined as

$$P = \lambda_0^2/\Lambda^2 n_0 n_1. \tag{4.54}$$

It can be shown that the relative power diffracted into higher-order modes is of the order of $P^{-2}$. Accordingly, the formulas applicable to the "volume" regime can be used with confidence only when $P \gg 1$.

On the basis of these criteria, Moharam, Gaylord, and Magnusson [1980*a*, *b*] have shown that the boundaries between the three diffraction regimes can be defined, as shown in fig. 4.10, by two curves. The boundary between the Raman–Nath and intermediate diffraction regimes corresponds closely to the curve

$$Q'\Phi = 1, \tag{4.55}$$

where $Q' = Q/\cos\theta$, $\theta$ being the angle of incidence within the grating, and $\Phi$ the modulation parameter given by (4.29).

On the other hand, the boundary between the intermediate and Bragg-diffraction regimes follows the curve

$$Q'/\Phi = 20, \tag{4.56}$$

which, from (4.54), reduces to

$$P = 10. \tag{4.57}$$

## 4.9 Anomalous effects with volume gratings

Some anomalous effects are also observed with volume holographic gratings which are not immediately apparent from the coupled-wave theory.

Thus, Leith *et al.* [1966] observed a peak in the intensity of the transmitted wave at the same angle of illumination for which the diffracted power was also a maximum. This effect, which is analogous to the Borrmann effect in x-ray diffraction, has also been observed in gratings recorded in KBr crystals [Aristov, Shekhtman & Timofeev, 1969] and in photochromic materials [Tomlinson & Aumiller, 1975]. Forshaw [1974] found that illumination of a thick holographic grating with a narrow laser beam resulted in the emergence of four diffracted beams.

The theory of these phenomena has been worked out in detail by Russell and Solymar [1980], who have shown that, for exact Bragg incidence, Kogelnik's formulas can be rewritten to show the existence of two standing wave patterns parallel to the grating vector, but with a phase difference of $\pi/2$. With an amplitude grating, one of these standing waves, whose maxima coincide with the absorption maxima, is attenuated rapidly along the depth of the grating. As a result, the amplitudes of the transmitted and diffracted waves leaving the other face, which, in the limit, are only due to the other standing wave, converge to the same value.

A two-dimensional generalization of the coupled-wave theory [Solymar, 1977a] permits a more exact analysis for bounded beams and has been used to explain the effects observed more fully.

Other diffraction phenomena observed in very thick holograms, such as scattering rings, have been discussed by Forshaw [1975] and by Ragnarsson [1978].

## 4.10 Multiply-exposed volume holographic gratings

Beacause of the high angular and wavelength selectivity of volume holograms, it is possible to record several holograms in the same thick recording medium and read them out separately, if their Bragg angles are sufficiently far apart. The minimum angular separation of the grating vectors for this corresponds to the condition that the maximum of the angular selectivity curve for one grating coincides with the first minimum for the other.

However, with amplitude transmission and reflection holograms, multiple recordings result in a considerable loss in diffraction efficiency. On the assumption that the available dynamic range is divided equally between $N$ gratings, it can be shown that the diffraction efficiency of each grating drops to $1/N^2$ of that for a single grating recorded in the same medium using the entire dynamic range [Collier, Burckhardt & Lin, 1971].

With volume phase gratings the situation is more involved. Diffraction at two or more phase gratings recorded in a thick material by multiple exposures has been analysed using a thin-grating decomposition method [Alferness & Case, 1975], as well as an approach based on the coupled-wave theory [Case 1975; Kowarschik, 1978*a,b*]. Their results show that with two gratings whose Bragg angles are far enough apart for coupling between the gratings to be neglected, and with linear recording, each of the gratings diffracts as if the other one did not exist. However, if $M$ gratings with very nearly the same spatial frequency are recorded sequentially, the diffraction efficiency of each grating is $1/M$ of that for a single grating [Couture & Lessard, 1984]. This loss in diffraction efficiency can be avoided by recording the gratings simultaneously. In this case, though, intermodulation effects as well as higher diffraction orders from individual gratings must be considered in computing the efficiency and signal-to-noise ratio (SNR) of the hologram [Solymar,1977*b*; Slinger, Syms & Solymar, 1987]. A numerical comparison of three different models for such multiplexed holograms has been made by Kostuk [1989], which shows that while the predicted total diffraction efficiency is very nearly the same for all three models, the effect of higher diffraction orders is a significantly lower SNR.

Coupling effects between the gratings are maximized if a common reference wave is used for all the exposures. It is then possible, by a suitable choice of modulation and inclination factors, to transfer the entire energy of the reconstructed wave into two or more diffracted beams and to vary their intensity ratio over a wide range. Similarly, energy can be cross coupled from one diffracted wave into the other.

## 4.11 Imaging properties of volume holograms

In the case of volume holograms, which diffract strongly at the Bragg angle, the amplitude of the reconstructed wavefront is affected by any changes of wavelength or geometry between recording and reconstruction. In addition, changes in the thickness of a photographic emulsion due to processing can result in a rotation of the fringe planes as well as a change in their spacing [Vilkomerson & Bostwick, 1967] (see fig. 4.11).

With gratings recorded with plane wavefronts it is possible to compensate for these changes by changing either the angle of illumination or the wavelength [Belvaux, 1975]. However, complete compensation is not possible with a hologram of a point at a finite distance or of an extended object, resulting in variations of amplitude across the reconstructed wavefront and reduced diffrac-

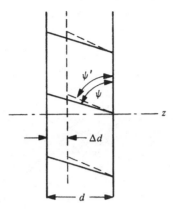

Fig. 4.11. Changes in the orientation and spacing of the fringe planes in a hologram due to a change in the thickness of the photographic emulsion.

tion efficiency. The effects of emulsion shrinkage in thick holograms have been considered by Latta [1971*c*] and by Forshaw [1973], who have shown that it is possible, in this case also, to define a pupil function; this pupil function differs from that for a thin hologram defined in section 3.4, since it involves spatial modulation of the amplitude as well as the phase.

# 5

# Optical systems and light sources

A typical optical system for recording transmission holograms of a diffusely reflecting object is shown in fig. 5.1; one for recording a reflection hologram is shown in fig. 5.2.

A simpler system for making reflection holograms is shown in fig. 5.3. This arrangement is essentially the same as that described originally by Denisyuk [1965] in which, instead of using separate object and reference beams, the portion of the reference beam transmitted by the photographic plate is used to illuminate the object. It gives good results with specular reflecting objects and with a recording medium, such as dichromated gelatin, which scatters very little light.

Making a hologram involves recording a two-beam interference pattern. The principal factors that must be taken into account in a practical setup to obtain good results are discussed in the next few sections.

## 5.1 Stability requirements

Any change in the phase difference between the two beams during the exposure will result in a movement of the fringes and reduce modulation in the hologram [Neumann, 1968].

In some situations, the effects of object movement can be minimized by means of an optical system in which the reference beam is reflected from a mirror mounted on the object [Mottier, 1969]. Alternatively, if the consequent loss in resolution can be tolerated, a portion of the laser beam can be focused to a spot on the object, producing a diffuse reference beam [Waters, 1972]. Stability requirements for reflection holograms can be minimized with a system similar to that shown in fig. 5.3, in which the hologram plate is rigidly attached to the object ("piggyback" holography) [Neumann & Penn, 1972]. The surface of the

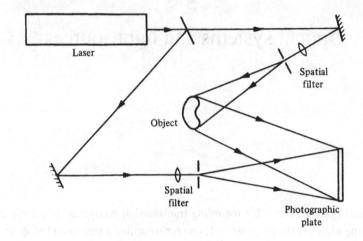

Fig. 5.1. Optical system for recording a transmission hologram.

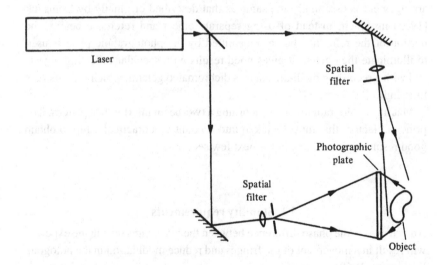

Fig. 5.2. Typical optical arrangement for recording a reflection hologram.

object can be treated with a retroreflective paint to increase the intensity of the object beam.

More commonly, mechanical disturbances are avoided by mounting all the optical components as well as the object and the recording medium on a stable surface. Acceptable results can be obtained with a concrete slab resting on inflated scooter inner tubes, but most laboratories now use a rigid optical table

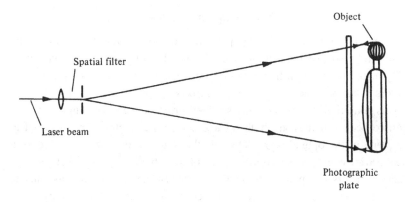

Fig. 5.3. Simple optical arrangement for making a reflection hologram.

supported by a pneumatic suspension system, so that it has a low natural frequency of vibration ($\approx$ 1 Hz). A steel table has the advantage that components can be mounted on magnetic bases or bolted down to its surface.

Air currents, acoustic waves, and temperature changes also cause major problems. Their effects are usually minimized by enclosing the working area.

Residual disturbances can be eliminated almost completely by a feedback system in which any motion of the interference fringes in the hologram plane is picked up by a photodetector [Neumann & Rose, 1967; MacQuigg, 1977]. Variations in its output are amplified and applied to a piezoelectric element which controls the position of a mirror in the reference-beam path to restore the optical path difference between the two beams to its original value.

## 5.2 Fringe visibility

As we have seen in Chapter 3, the amplitude of the wave diffracted by a grating increases with the modulation depth. Accordingly, to produce a hologram that reconstructs a bright image, the interference pattern formed at the recording medium by the object and reference waves should have as high a contrast as possible.

The contrast of the interference pattern at any point in the hologram plane is measured by the fringe visibility (see Appendix A1.1), which is given by the relation

$$\mathcal{V} = \frac{(I_{\max} - I_{\min})}{(I_{\max} - I_{\min})}, \tag{5.1}$$

where $I_{\max}$ and $I_{\min}$ are the local maximum and minimum values of the intensity.

### 5.3 Beam polarization

Most gas lasers used for holography have Brewster-angle windows on the plasma tube so that the output beam is linearly polarized. Maximum visibility of the fringes will then be obtained if the angle $\psi$ between the electric vectors in the two interfering beams is zero (see Appendix A1.1).

This condition is automatically satisfied, irrespective of the angle between the two beams, if they are both polarized with the electric vector normal to the plane of the optical table. On the other hand, if they are polarized with the electric vector parallel to the surface of the table, the angle $\psi$ between the electric vectors is equal to the angle $\theta$ between the two beams, and, in the extreme case where the two beams intersect at right angles, the visibility of the fringes drops to zero.

With a diffusely reflecting or metallic object, it is also necessary to take into account the changes in the polarization of the light scattered by it [Rogers, 1966; Ghandeharian & Boerner, 1978], which can result in a significant decrease in the visibility of the fringes. This decrease can be minimized either by rotating the polarization of the reference beam so that the cross-polarized light can also interfere with it, or by using a sheet polarizer in front of the hologram plate to eliminate the cross-polarized component. Another alternative is to use a circularly polarized reference beam [Vanin, 1979].

### 5.4 Beam splitters

If we assume that the interfering beams are polarized with the electric vector perpendicular to the plane of incidence, the fringe visibility is given by the relation

$$\mathcal{V} = 2|\gamma_{12}(\tau)|or/(o^2 + r^2), \tag{5.2}$$

where $r$ and $o$ are the amplitudes of the reference and object beams, and $\gamma_{12}(\tau)$ is the degree of coherence between them. If the beam ratio, $R = (r/o)^2$, is defined as the ratio of the irradiances of the reference and object beams, (5.2) can be rewritten as

$$\mathcal{V} = 2|\gamma_{12}(\tau)|R^{1/2}/(1 + R). \tag{5.3}$$

The fringe visibility is obviously a maximum when $R = 1$. However, the wave scattered by a diffusely reflecting object exhibits quite strong local variations in amplitude (see Appendix 4). Hence, in hologram recording, it is usually necessary to work with a value of $R \gg 1$ to avoid nonlinear effects. To optimize the visibility of the fringes at the hologram plane, it should be possible

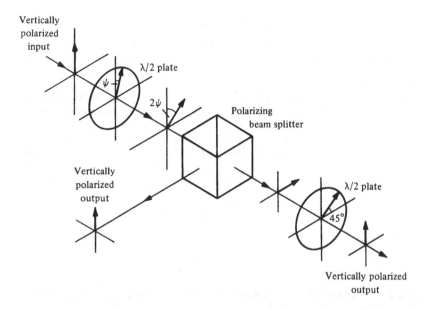

Fig. 5.4. Variable-ratio beam splitter using a polarizing prism.

to vary the ratio of the power in the beam illuminating the object to that in the reference beam.

A convenient way to do this is to use a beam splitter consisting of a disc coated with a thin aluminium film whose reflectivity is a linear function of the azimuth. Such a variable-ratio beam splitter must be used in the unexpanded laser beam because of the variation in the reflectivity across a larger beam. Since an aluminium film typically absorbs about 30 per cent of the energy incident on it, it can be used only at moderate laser powers ($< 500$ mW).

A much better variable-ratio beam splitter, with a uniform field and very low insertion loss, uses a polarizing prism to divide the incident beam into two orthogonally polarized components [Caulfield & Beyen, 1967]. A typical optical system is shown in fig. 5.4. The ratio of the transmitted and reflected powers is given by the relation

$$I_{\text{trans}}/I_{\text{refl}} = \cot^2 \psi, \tag{5.4}$$

where $\psi$ is the angle which the incident electric vector makes with the vertical. This ratio is controlled by using a half-wave plate to rotate the direction of polarization of the input beam. A fixed half-wave plate in the transmitted beam is used to bring the electric vector of this beam back to the vertical.

## 5.5 Beam expansion

The laser beam has to be expanded to illuminate the object and the plate on which the hologram is recorded. Usually, microscope objectives are used for this purpose.

If the laser is oscillating in the $TEM_{00}$ mode, the beam has a Gaussian profile, so that the amplitude at a point at a radial distance $r$ from the centre of the beam is

$$a(r) = a(0) \exp(-r^2/w^2), \qquad (5.5)$$

where $a(0)$ is the amplitude at the centre of the beam, and $w$ is the beam radius (the radius at which the amplitude drops to $1/e$ of its maximum value).

Accordingly, $P(r)$, the power within a circle of radius $r$, is given by the relation

$$P(r) = \int_0^r I(r) 2\pi r \, dr, \qquad (5.6)$$

where $I(r) = |a(r)|^2$ is the intensity at a radial distance $r$ from the centre of the beam; or

$$P(r) = P_{tot}[1 - \exp(-2r^2/w^2)], \qquad (5.7)$$

where $P_{tot}$ is the total power in the beam.

From (5.5) and (5.7) it follows that

$$P(r)/P_{tot} = 1 - [I(r)/I(0)]. \qquad (5.8)$$

This relation gives the loss in power that must be tolerated for a given degree of uniformity of illumination.

Due to the high coherence of laser light, the expanded beam invariably exhibits unwanted diffraction patterns (spatial noise) due to scattering from dust particles on the optical surfaces in the beam path. Spatial noise can be eliminated, and a clean beam obtained, by placing a pinhole at the focus of the microscope objective. If this pinhole has a radius $\rho$, spatial frequencies higher than $\xi = \rho/\lambda f$, where $\lambda$ is the wavelength of the light and $f$ is the focal length of the microscope objective, are blocked. Since these higher spatial frequencies represent noise, the transmitted beam has a smooth Gaussian profile.

A proper choice of the size of the pinhole can ensure that the power loss is minimal. To evaluate the loss, we make use of the fact that the amplitude in the focal plane of the microscope objective, obtained from the two-dimensional Fourier transform of (5.5), is

$$A(\rho) = A(0) \exp(-\pi^2 w^2 \rho^2 / \lambda^2 f^2), \qquad (5.9)$$

where $\rho$ is the radial distance from the centre of the beam. Typically, with an argon-ion (Ar$^+$) laser, $\lambda = 514$ nm and $w = 0.8$ mm, and, with a 16 mm microscope objective and a 10 $\mu$m diameter pinhole, the amplitude at the edge of the pinhole is only 0.08 of that at the centre. Hence, from (5.8), more than 99 per cent of the power in the beam is transmitted through the pinhole.

## 5.6 Exposure control

An accurate spot photometer is required to measure the irradiances due to the object and reference beams in the hologram plane, so as to set the beam ratio $R$ to a suitable value. Because of the limited dynamic range of photographic materials used for holography, the object illumination should be adjusted so that the irradiance in the hologram plane due to the object beam is reasonably uniform. In addition, precise control of the exposure is required to ensure good diffraction efficiency and avoid nonlinear effects. However, it is not enough to maintain a specified exposure time, because the laser output can fluctuate during the exposure. To overcome this problem it is convenient to use an electronic exposure-control unit which integrates the irradiance in the hologram recording plane and closes the shutter at a preset value of the radiant exposure [Lin & Beauchamp, 1970a]. A more sophisticated system which is suitable for multicolour holography has been described by Oreb and Hariharan [1981].

## 5.7 Coherence requirements

In order to obtain maximum fringe visibility, it is essential to use coherent illumination. Lasers provide an intense source of highly coherent light and, therefore, are used almost universally in optical holography.

Spatial coherence is automatically ensured if the laser oscillates in a single transverse mode, preferably the lowest order or TEM$_{00}$ mode, since this is inherently the most stable and gives most uniform illumination over the field. Most gas lasers are designed to operate in this mode [Bloom, 1968]. However, they are not usually designed for single-frequency operation, which would imply that they should also oscillate in only one longitudinal mode. As a result, the temporal coherence of the light from most lasers is limited by their longitudinal mode structure.

To obtain a satisfactory hologram, the maximum optical path difference between the object and reference beams in the recording system must be much less than the coherence length (see Appendix A1.3) of the light from the laser. With an extended object, the holodiagram [Abramson, 1969] (see section 15.9) can be used to optimize the layout.

76                        *Optical systems and light sources*

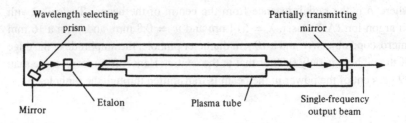

Fig. 5.5. Optical system of a typical argon-ion laser.

## 5.8 Temporal coherence of laser light

The simplest form of resonant cavity for a laser is made up of a pair of mirrors
separated by a distance $L$, though in some cases, where operation on more
than one line is possible, a wavelength selector prism may also be necessary,
as shown in fig. 5.5. The resonant frequencies of such a cavity are given by the
expression

$$\nu_n = (n + 1/2)(c/2L), \qquad (5.10)$$

where $n$ is an integer and $c$ is the speed of light. The laser can, therefore,
oscillate at all those frequencies within the gain curve of the active medium at
which, as shown in fig. 5.6, the gain is adequate to overcome the cavity losses.

The frequency spread of the individual modes depends on the losses as well
as the mechanical stability of the cavity structure and is typically about 3 MHz.
Accordingly, if a laser oscillates in a single longitudinal mode, the coherence
length of the output would be around 100 metres.

Since, in frequency space, the width of the individual modes is much less
than their separation (which may range from 600 MHz for a 25 cm cavity down
to 75 MHz for a 2 m cavity), the power spectrum of a laser oscillating in $N$
longitudinal modes can be represented by $N$ equally spaced delta functions. If
we assume that the total power is equally divided between the $N$ modes, the
power spectrum can be written as

$$S(\nu) = \sum_{n}^{n+N-1} \delta(\nu - \nu_n). \qquad (5.11)$$

The degree of temporal coherence of the output can then be evaluated as
described in Appendix A1.4, and, for an optical path difference $p$, is

$$\gamma_{12}(p) = \left| \frac{\sin(N\pi p/2L)}{N \sin(\pi p/2L)} \right|. \qquad (5.12)$$

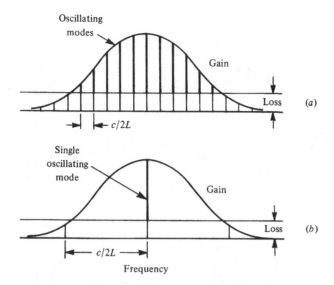

Fig. 5.6. Oscillation frequencies of a laser with (*a*) a long resonant cavity and (*b*) a short resonant cavity.

This is a periodic function whose first zero occurs when

$$p = 2L/N, \qquad (5.13)$$

corresponding to the effective coherence length. Interference fringes with acceptable visibility can usually be obtained for optical path differences that are less than half the coherence length.

It is apparent from (5.13) that the existence of more than one longitudinal mode reduces the coherence length severely. A short coherence length is troublesome, since it makes it essential to equalize the mean optical paths of the object and reference beams and restricts the maximum depth of the object field that can be recorded.

The simplest way to force a laser to operate in a single longitudinal mode is to use a very short cavity, so that the spacing of the longitudinal modes is greater than the width of the gain profile over which oscillation is possible. This occurs when

$$c/2L \geq \Delta \nu, \qquad (5.14)$$

where $\Delta \nu$ is the width of the gain profile (typically 1.7 GHz for a helium-neon (He-Ne) laser, 3.5 GHz for an $Ar^+$ laser). However, the power available from such a short cavity is extremely limited.

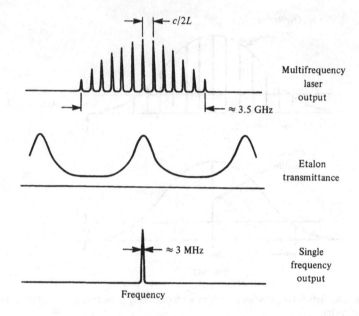

Fig. 5.7. Normal multifrequency output, etalon transmittance, and single-frequency output with an intracavity etalon for a typical argon-ion laser.

The most common method of ensuring single-frequency operation is to use an intracavity etalon as shown in fig. 5.5. In effect, the laser is made up of two resonant cavities, and only those modes which are common to both cavities, as shown in fig. 5.7, have low enough losses for oscillation to be possible. If the length of the etalon is made short enough that it satisfies (5.14), it can support only one mode, and single-frequency operation is obtained.

If the etalon is to act as a simple transmission filter, it must be tilted to decouple it from the laser cavity. It can then be tuned to maximize the output power by tilting it further so that its resonant frequency corresponds to the peak of the gain curve [Hercher, 1969]. An alternative method of tuning, which gives better frequency stability and efficiency, is to mount the etalon in an oven which can be maintained at any desired temperature. Decoupling is easier with an etalon having concentric spherical surfaces [Hariharan, 1982].

## 5.9 Laser safety

When working with any laser, adequate safety measures must be taken to avoid eye damage (see ANSI, 1986; BSI, 1983). Even with relatively low-power

Table 5.1. *Output wavelength and output power of gas lasers*

| Wavelength nm | Laser | Typical power mW | Colour |
|---|---|---|---|
| 442 | He-Cd | 25 | Violet |
| 458 | Ar$^+$ | 200 | Blue-violet |
| 476 | Kr$^+$ | 50 | Blue |
| 477 | Ar$^+$ | 400 | Blue |
| 488 | Ar$^+$ | 1000 | Green-blue |
| 514 | Ar$^+$ | 1400 | Green |
| 521 | Kr$^+$ | 70 | Green |
| 633 | He-Ne | 2-50 | Red |
| 647 | Kr$^+$ | 500 | Red |

lasers ($\approx 1$ mW) the beam should not enter the eye directly. This is because the laser beam is focused into a very small spot on the retina, resulting in a power density about $10^5$ times that at the cornea. With medium-power lasers ($< 100$ mW) care must also be taken to avoid stray reflections, though the risk of eye damage is less once the beam has been diffused or expanded. However, at power levels $> 1$ W, viewing even a light-coloured diffusing surface illuminated by the unexpanded beam is dangerous, since the surface intensity is comparable to that of the sun. With such lasers, as well as with pulsed lasers (see section 5.13), safety glasses must be worn.

## 5.10  Gas lasers

The light source most commonly used for holography is a gas laser. Gas lasers, in general, are cheaper and easier to operate and have better coherence characteristics than other types of lasers. The range of useful wavelengths and typical output powers available at these wavelengths with the most commonly used gas lasers are summarized in Table 5.1.

For a simple holographic system, the He-Ne laser is the most economical choice. It operates on a single line at 633 nm, does not require water cooling, and has a long life. However, depending on their power, commercial He-Ne lasers oscillate in two to five longitudinal modes, and the coherence length of the output is limited.

In contrast to the He-Ne laser, the Ar$^+$ laser has a multiline output but can be made to operate on a single line by replacing the reflecting end mirror by a prism and mirror assembly. It is also easy to obtain single-frequency operation with an etalon. Argon-ion lasers can give high-power output and an extended coherence length in the blue or green region of the spectrum, the two strongest laser lines being at 488 nm and 514 nm.

The krypton-ion ($Kr^+$) laser is very similar in its construction and character-istics to the $Ar^+$ laser and is a useful replacement for the He-Ne laser, where high output power and an extended coherence length are required at the red end of the spectrum (647 nm).

The helium-cadmium (He-Cd) laser provides a stable output at a relatively short wavelength (422 nm). It is very useful with recording materials such as photoresists (see section 7.3), whose increased sensitivity at this wavelength makes up for the lower power available.

## 5.11  Dye lasers

Even though dye lasers have not been used widely for holography, they have many advantages. Early dye lasers were built for pulsed operation, but contin-uous-wave operation can be obtained by using a thin flowing layer of a dye as the active medium and focusing the beam from a high-power $Ar^+$ laser on it.

The operating wavelength of a dye laser can be changed over a wide range by switching dyes. In addition, with a given dye, it is possible to tune the output over a range of wavelengths (about 50–80 nm) by incorporating a wave-length selector, such as a diffraction grating or a birefringent filter, in the laser cavity. Operation in a single longitudinal mode can be obtained by means of an intracavity etalon, to yield a single-frequency output of a few hundred milliwatts.

## 5.12  Diode lasers

Diode lasers can also be used for holography [Tatsuno & Arimoto, 1980]. They have the advantage that they are a diffraction-limited point source and do not require a spatial filter; in addition, they can be made to operate in a single longitudinal mode. Their main disadvantage, initially, was the low sensitivity of commercial photographic materials at the operating wavelength of 750 nm [Hart, Mendes, Bazargan & Xu, 1988]. However, diode lasers operating at shorter wavelengths are now available.

An attractive alternative is the use of diode lasers to pump a Nd:YAG laser with a frequency doubler. Such a system provides a compact cw source of green light ($\lambda = 532$ nm), with a coherence length of a few metres and an output power up to 100 mw.

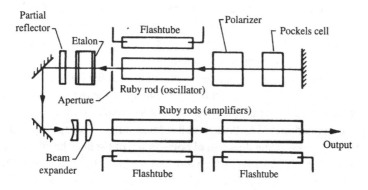

Fig. 5.8. Optical system of a $Q$-switched ruby laser for holography.

## 5.13 Pulsed lasers

With a pulsed laser, problems connected with the stability of the recording system are virtually eliminated [Jacobson & McClung, 1965; Siebert, 1967]. The ruby laser is still the most widely used type of pulsed laser for optical holography [Koechner, 1979a,b] mainly because of the large output energy available (up to 10 J per pulse) and the wavelength of the light emitted (694 nm), which is fairly well matched to the peak sensitivity of commercial photographic materials for holography.

While neodymium-doped yttrium aluminium garnet (Nd:YAG) lasers can operate more efficiently and at a higher pulse repetition rate, and have also been used for holography [Gates, Hall & Ross, 1970; Bates, 1973], they have the disadvantages that the energy per pulse is usually less, and a frequency-doubling crystal must be used to convert the infrared output to visible light (530 nm). However, an Nd:YAG laser system for pulsed holography of large scenes has been described by Andreev, Vorzobova, Kalintsev, and Staselko [1980]. A detailed comparison of the design and performance characteristics of ruby and frequency-doubled Nd:YAG lasers for holography has been made by Pitlak and Page [1985].

The active medium in a ruby laser is a rod, typically 5–10 mm in diameter and 75–100 mm in length, made of synthetic sapphire ($Al_2O_3$) doped with 0.05 per cent of ($Cr_2O_3$). When this rod is mounted in a suitable optical resonator, as shown in fig. 5.8, and pumped with a xenon flash lamp, it emits a series of pulses lasting approximately 250 $\mu$s, and polarized with the electric vector normal to the axis of the rod and the crystal $c$-axis [Lengyel, 1971].

For holography it is necessary to use a $Q$-switch in the cavity. This is a fast-acting optical shutter which is normally closed, so that the laser cannot oscillate when the flash lamp is fired and a large population is built up in the upper energy level. When the switch opens, the stored energy is released in a very short pulse with very high peak power.

The most useful type of $Q$-switch for holography is the Pockels cell, since it permits precise control of the timing and number of the output pulses. This device uses an electro-optic crystal which exhibits birefringence when an electric field is applied to it. The Pockels cell is located in the laser cavity with its principal axes at 45° to the direction of polarization of the laser beam, between the end reflector and a polarizer which is oriented to pass the laser beam. When the flash lamp is fired, a voltage is applied to the Pockels cell, producing a phase shift of $\pi/2$ between the two transmitted components. Any light from the laser rod transmitted by the Pockels cell and reflected back by the end mirror then returns with its plane of polarization rotated through 90° and is blocked by the polarizer. When the voltage on the Pockels cell is switched off, towards the end of the flash lamp pulse, the combination of the Pockels cell and the polarizer transmits freely, allowing laser oscillations to build up, so that a very short pulse, typically with a duration of 15 ns, is produced.

Because of the high gain of a ruby rod in an optical resonator, such a laser normally oscillates in a large number of transverse modes, and the spatial coherence of the output is very poor. To ensure operation in the $TEM_{00}$ mode, an aperture with a diameter of about 2 mm is inserted at a suitable point in the resonator.

Again, because of the relatively large width of the fluorescence line ($\approx 400$ GHz), a ruby laser normally oscillates in as many as 100–200 longitudinal modes and has a very short coherence length ($< 1$ mm). To obtain an increased coherence length, it is necessary to use a Fabry-Perot etalon in the cavity (a plate of fused silica, or sapphire, about 2–3 mm thick). True single-mode operation can be obtained with resonant reflectors consisting of two plates separated by a suitable distance, along with the use of a $Q$-switch with a relatively slow rise time, which prevents the modes with lower gain from building up to any appreciable extent before laser action occurs [McClung, Jacobson & Close, 1970; Young & Hicks, 1974].

Since the single-mode output from a ruby oscillator is limited ($\approx 0.04$ J), high-power ruby lasers use one or more additional ruby rods as amplifiers to boost the output while preserving spatial and temporal coherence. For maximum gain with stable operation, the amplifier rod is set so that its $c$-axis is parallel to that of the oscillator and is optically isolated from it by a spatial filter consisting of a sapphire pinhole placed at the common focus of two lenses. The focal lengths of these two lenses are chosen so that the beam from the oscillator is expanded to fill the amplifier rod, which may have a diameter of 20 mm and a length of 200 mm. An output of 1 J is possible with a single amplifier stage, and 10 J with two stages of amplification, with a coherence length of 5–10 metres.

## 5.14 Holography with pulsed lasers

The very short duration of the output pulse from a $Q$-switched ruby laser ($\approx$ 15 ns) makes it possible to record holograms of moving objects.

The drop in the visibility of the hologram fringes due to the movement of the object can be evaluated either from the change in the optical path or from the shift in the optical frequency due to the Doppler effect [Mallick, 1975]. The permissible object movement depends very much on the geometry of the recording system, but, typically, object velocities up to a few metres per second can be tolerated. The holodiagram [Abramson, 1969] (see section 15.9) can be used to minimize the sensitivity of the system to movement of the object in a particular direction.

Because of the high power density ($>$ $10^{12}$ W/m$^2$) in the undiverged laser beam, it is essential to use a beam splitter with low insertion loss and multilayer dielectric mirrors instead of metallized mirrors, which have a damage threshold of only about $10^{10}$ W/m$^2$. It is not possible to use microscope objectives with pinhole spatial filters to expand the beams; instead, simple negative lenses are used. All optical surfaces must be kept clean and free of dust particles, which can lower the damage threshold appreciably. Adequate safety precautions, including the use of protective eye wear, are essential [see ANSI, 1986; BSI, 1983].

A $Q$-switched ruby laser can be used to record a hologram of a living human subject [Siebert, 1968]. To avoid eye damage, it is essential to see that the energy density at the retina of the subject is well within the recommended safety limit. This is possible if, instead of illuminating the subject directly with the expanded laser beam, it is allowed to fall on a large diffuser which scatters the light, effectively constituting a diffuse source. A typical optical arrangement is shown in fig. 5.9 [Ansley, 1970]. The dimensions of the diffuser should be chosen so that the integrated radiance of the diffuser, measured at the cornea of the subject, is less than the maximum permissible exposure for an extended source of 700 J/m$^2$/sr; the undeviated energy density should also be less than 5 mJ/m$^2$ [see ANSI, 1986, BSI, 1983]. In addition, it is necessary to ensure that the specular reflection of the reference beam from the surface of the photographic plate is directed away from the subject. A detailed description of a professional studio setup for making holographic portraits has been given by Bjelkhagen [1992].

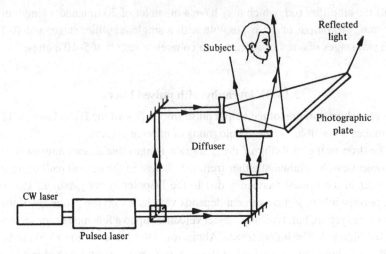

Fig. 5.9. Typical setup for recording a hologram of a human subject using a $Q$-switched ruby laser [Ansley, 1970].

Closer control of the illuminated area with less loss of light is possible if a holographic scatter plate is used instead of a diffuser [Webster, Tozer & Davis, 1979]. However, such scatter plates are not completely effective as diffusers; if they are used for portraiture, it is essential to ensure that the directly transmitted beam does not fall on the subject.

# 6

# The recording medium

Before we look at the different recording materials that have been used for optical holography, it is necessary to define the most important characteristics of a holographic recording medium.

## 6.1 Macroscopic characteristics

Any material used to record a hologram must respond to exposure to light (after additional processing, where necessary) with a change in its optical properties. The complex amplitude transmittance of such a material can be written as

$$\mathbf{t} = \exp(-\alpha d)\exp[-i(2\pi n d/\lambda)],$$
$$= |\mathbf{t}|\exp(-i\phi), \qquad (6.1)$$

where $\alpha$ is the absorption constant of the material, $d$ is its thickness, and $n$ is its refractive index. Accordingly, as mentioned earlier, holographic recording materials can be classified, for convenience, either as pure amplitude-modulating materials if only $\alpha$ changes with the exposure, or as pure phase-modulating materials if $\alpha \approx 0$, and either $n$ or $d$ changes with the exposure.

The response of the recording material, defined by (6.1), can be described in these two limiting cases by curves of amplitude transmittance versus exposure ($|\mathbf{t}| - E$ curves) as shown in fig. 6.1, or by curves of the effective phase shift against exposure ($\Delta\phi - E$ curves) as shown in fig. 6.2.

## 6.2 The modulation transfer function

Although curves such as those in figs. 6.1 and 6.2 describe the behaviour of a recording material on a macroscopic scale, they are not, by themselves, adequate to predict the response of the material on a microscopic scale.

85

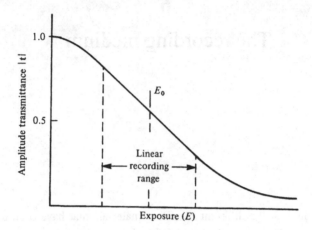

Fig. 6.1. Curve of amplitude transmittance ($|t|$) against exposure ($E$) for a recording material.

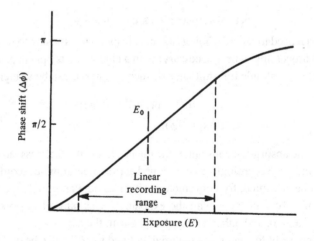

Fig. 6.2. Curve of phase shift ($\Delta\phi$) against exposure ($E$) for a recording material.

As discussed in section 5.2, the intensity at any point on the hologram recording medium is given by a relation of the form

$$I = \langle I \rangle [1 + \mathcal{V} \cos(\phi_r - \phi_o)], \tag{6.2}$$

where $\langle I \rangle$ is the average intensity, $\mathcal{V}$ is the visibility of the interference fringes formed by the reference and object beams, and $\phi_r$ and $\phi_o$ are their phases at this point.

The actual intensity distribution to which the material is exposed is different, because the light spreads laterally within the recording medium, to an extent determined by its scattering properties and its coefficient of absorption. As a result, the actual modulation of the intensity within the recording material is always less than that in the original interference pattern.

For any spatial frequency $s$, the ratio of $\mathcal{V}'(s)$, the actual modulation of the intensity distribution within the material, to $\mathcal{V}(s)$, the input modulation, is termed the modulation transfer function (or MTF) $M(s)$, so that

$$M(s) = \mathcal{V}'(s)/\mathcal{V}(s). \tag{6.3}$$

The parameter $M(s)$, which is normally less than unity, serves to characterize the relative response of the material at different spatial frequencies.

## 6.3 Diffraction efficiency

The diffraction efficiency $\varepsilon$ of a hologram has been defined (see section 3.6) as the ratio of the power diffracted into the desired image to that illuminating the hologram.

For an exposure time $T$, the effective exposure of the recording material is, from (6.2) and (6.3),

$$E = TI,$$
$$= T\langle I\rangle[1 + \mathcal{V}(s)M(s)\cos(\phi_r - \phi_o)],$$
$$= \langle E\rangle[1 + \mathcal{V}(s)M(s)\cos(\phi_r - \phi_o)]. \tag{6.4}$$

If we assume linear recording, as in (2.2), the amplitude transmittance of the recording medium can be written as

$$\mathbf{t} = \langle \mathbf{t}\rangle + \beta dE, \tag{6.5}$$

where $\langle \mathbf{t}\rangle$ is the amplitude transmittance for the average exposure $\langle E\rangle$, and $\beta$ is the value of $(d\mathbf{t}/dE)$ at $\langle E\rangle$. Accordingly, the amplitude transmittance of the hologram is

$$\mathbf{t} = \langle \mathbf{t}\rangle + \beta\langle E\rangle\mathcal{V}(s)M(s)\cos(\phi_r - \phi_o). \tag{6.6}$$

When the hologram is illuminated once again with the original reference wave $r$, the amplitude of the reconstructed image is

$$u_3 = (1/2)r\beta\langle E\rangle\mathcal{V}(s)M(s), \tag{6.7}$$

and its intensity is

$$I_3 = (1/4)|r|^2[\beta\langle E\rangle\mathcal{V}(s)M(s)]^2. \tag{6.8}$$

Accordingly, the diffraction efficiency of the hologram is

$$\varepsilon = (1/4)[\beta\langle E\rangle\mathcal{V}(s)M(s)]^2. \qquad (6.9)$$

Now, since $\beta E = (dt/dE)$,

$$\beta E = E(dt/dE)$$
$$= (dt/d \log E) \log_{10} e,$$
$$= 0.434\ \Gamma, \qquad (6.10)$$

where $\Gamma = (dt/d \log E)$.

The diffraction efficiency is proportional to the square of $\Gamma$, the gradient of the **t** *versus* $\log E$ curve, as well as to the squares of $\mathcal{V}(s)$, the input modulation, and $M(s)$, the MTF [Biedermann, 1969]. The maximum diffraction efficiency is obtained where the slope of the **t** *versus* $\log E$ curve is steepest. This is usually at a slightly higher exposure than that corresponding to the steepest part of the **t** *versus* $E$ curve.

A method of characterizing recording materials for holography based on (6.9) has also been described by Lin [1971]. For an ideal recording material, plots of $\sqrt{\varepsilon}$ *versus* $\langle E\rangle$ with $\mathcal{V}(s)$ as a parameter and plots of $\sqrt{\varepsilon}$ *versus* $\mathcal{V}(s)$ with $\langle E\rangle$ as a parameter should be straight lines. The departure of the characteristics of any real material from the ideal can be easily seen by comparing the actual measured curves with these straight lines. Apart from the maximum diffraction efficiency, or the exposure needed to obtain a given diffraction efficiency, these curves also make it possible to determine the range of fringe visibility $\mathcal{V}(s)$ or beam ratio $R$ for which the hologram recording is linear (indicated by the straight-line portion of the $\sqrt{\varepsilon}$ *versus* $\mathcal{V}(s)$ curve at constant $\langle E\rangle$, and the value of average exposure representing the best compromise between linearity and efficiency.

## 6.4 Image resolution

With an ideal recording material, the resolution of the image is determined only by the dimensions of the hologram. However, with any practical recording material, its MTF will, in general, affect the resolution as well as the intensity of the reconstructed image [Kozma & Zelenka, 1970].

This is because, in most holographic systems, the spatial frequency $s$ varies over the hologram. It then follows from (6.9) that, due to the corresponding variations in the MTF of the recording medium, the diffraction efficiency of the hologram will vary over its aperture. Wherever this variation is appreciable, it must be taken into account in evaluating the imaging properties of the hologram.

The only exceptions to this situation are when the image is formed at or near the hologram, or with a Fourier hologram (see section 2.3). In the latter case, the spatial frequency $s$ is constant over the entire hologram. As a result, the MTF of the recording medium affects only the intensity of the image; its resolution is determined by the aperture of the hologram.

At the other extreme, where the object and the reference source are at different distances from the hologram, and the hologram is quite large, it is possible for the spatial frequency $s$ to exceed the resolution limit of the recording material over part of the hologram. This limits the useful aperture of the hologram and, hence, the resolution of the image.

A simple method to visualize the effects of the MTF of the recording material, for a given point on the hologram, makes use of the concept of a fictitious mask located in the object beam [van Ligten, 1966]. The amplitude transmittance of this mask is proportional to the MTF, ranging from unity at a point on the object in line with the reference source (and, hence, corresponding to zero spatial frequency at the hologram) down to complete opacity at points at a large enough lateral separation from the reference source. More exactly, the effects of the MTF of the recording material can be taken into account, as was done for the effects of angular and wavelength selectivity in a thick hologram (see section 4.11), by defining a generalized space-variant pupil function, involving, in this case, the local spatial frequency and the MTF of the recording material [Kozma & Zelenka, 1970; Jannson, 1974].

## 6.5 Nonlinearity

Consider the interference of an object wave $o$ with a reference wave $r$ to produce an amplitude hologram. If we assume linear recording as in (2.2), the amplitude transmittance of the hologram is

$$\mathbf{t} = \mathbf{t}_0 + \beta T [rr^* + oo^* + r^*o + ro^*]. \tag{6.11}$$

However, in practice, the assumption of linear recording is not valid when the fluctuating terms on the right-hand side of (6.11) are comparable with the bias term.

This is usually the situation with phase holograms because of their intrinsic nonlinearity. Even if the phase shift produced by the recording medium is strictly proportional to the irradiance in the interference pattern, the complex amplitude transmittance at any point in a phase hologram is

$$\mathbf{t}(x, y) = \exp(-i\phi),$$
$$= 1 - i\phi - (1/2)\phi^2 + (1/6)i\phi^3 + \dots. \tag{6.12}$$

If the phase modulation is increased to obtain a high diffraction efficiency, the higher-order terms in (6.12) cannot be neglected.

The effects of nonlinear recording were first investigated by Kozma [1966] and Friesem and Zelenka [1967] for simple objects, and by Goodman and Knight [1968] for a diffusely reflecting object, using the characteristic function method of communication theory. A simpler approach [Bryngdahl & Lohmann, 1968c] is to assume that the amplitude transmittance of the recording material can be represented by a polynomial, which can be written as

$$\mathbf{t} = \mathbf{t}_0 + \beta_1 E + \beta_2 E^2 + \ldots,$$
$$= \mathbf{t}_0 + \beta_1 T [rr^* + oo^* + r^*o + ro^*]$$
$$+ \beta_2 T^2 [rr^* + oo^* + r^*o + ro^*]^2 + \ldots, \qquad (6.13)$$

if the higher-order terms are neglected. When the hologram is illuminated again by the reference wave $r$, which, for simplicity, can be assumed to be a plane wave of unit amplitude, the complex amplitude of the wave transmitted by the hologram is

$$u = \text{linear terms}$$
$$+ \beta_2 T^2 [(oo^*)^2 + o^2 + o^{*2} + 2o^2 o^* + 2oo^{*2}]. \qquad (6.14)$$

The immediate result of nonlinear recording is, therefore, the production of additional spurious terms.

A comparison of (6.14) and (6.11) shows that the term involving $(oo^*)^2$ in (6.14), like the term involving $oo^*$ in (6.11), corresponds to a halo surrounding the directly transmitted beam. However, the spatial frequency spectrum of this term, which is obtained by convolving the spectrum of $oo^*$ with itself, has twice the width of the spectrum of $oo^*$. This corresponds to a doubling of the angular width of the halo surrounding the directly transmitted beam.

The second term $o^2$ is the square of the complex amplitude of the object wave at the hologram. The mean direction of the corresponding diffracted wave makes an angle with the axis that is approximately twice that made by the object wave with it; this diffracted wave also has a curvature twice that of the object wave. Accordingly, it gives rise to a higher-order primary image. Similarly, the third term $o^{*2}$ can be interpreted as a higher-order conjugate image.

The fourth and fifth terms, which result in diffracted wavefronts with complex amplitudes proportional to $2o^2 o^*$ and $2oo^{*2}$, can be shown to be intermodulation terms giving rise to false images. With an extended diffusely reflecting object, the effect of these false images is to create a "noise halo" around the true image. This can be understood quite readily if we realize that the spatial-frequency spectrum of the wave corresponding to the fourth term is obtained by convolving the spatial-frequency spectrum of the object with the spatial-frequency spectrum of the speckle pattern produced by the object wave in the hologram plane.

Further studies using a power series technique have been made by Kozma [1968*b*] and by Kozma, Jull, and Hill [1970]. Their results can be used to calculate the ratio of the intensity in the reconstructed image to that due to the nonlinear noise background, as well as the shape of the noise distribution for a diffusely illuminated object. Unfortunately, the expressions obtained are quite complicated. This difficulty is partly overcome in an alternative treatment by Tischer [1970] based on the use of Chebyshev polynomials. With simple objects, such as lines and points, the terms of the series represent corresponding pairs of ghost images, so that, in principle, the optimum operational conditions can be evaluated readily.

However, the polynomial method has limitations, as pointed out by Velzel [1973], who was able to develop a complete theory of nonlinear holographic image formation using a modified transform method, and derive analytical expressions for the efficiency and the image contrast for recording media with an exponential characteristic (a linear phase hologram) and a binary characteristic. A generalization of this theory, which can be used where the transfer function of the recording medium is complex, has been outlined by Ghandeharian and Boerner [1977].

## 6.6 Effect of hologram thickness

Intermodulation effects would make it almost impossible to produce bright holographic images of good quality but for the fact that, in a volume hologram, intermodulation noise is reduced significantly by the angular selectivity of the hologram (see sections 4.4 and 4.5).

Consider a simple object consisting of only two points, which gives rise to two sets of interference fringes at the hologram plane having spatial frequencies $s_{01}$ and $s_{02}$. The intermodulation terms then have the general form $ps_{01} \pm qs_{02}$ where $p$ and $q$ are positive integers. Most of the resultant spatial frequencies are therefore significantly different from those corresponding to the object points. Qualitatively, it is apparent that when the hologram is illuminated once again with the same reference beam used to record it, the Bragg condition will not be satisfied for these intermodulation frequencies [Upatnieks & Leonard, 1970].

A detailed analysis of intermodulation noise in a volume hologram due to nonlinear recording has been made by Guther and Kusch [1974], using the coupled-wave theory (see section 4.3). Their results show that the noise is effectively limited by the thickness of the recording medium as well as by the aperture of the hologram. A simpler analysis by Hariharan [1979a] shows that if the angle between the two beams in the recording setup is large enough for the diffracted beams corresponding to different orders not to overlap, the signal-to-noise ratio should improve by a factor approximately equal to $(\psi/\Delta\theta)$ where $2\Delta\theta$ is the width of the passband of the angular selectivity function, and $\psi$ is the angular extent of the object.

## 6.7 Noise

In optical holography, noise is a convenient term for non-image-forming light that is diffracted or scattered in the same general direction as the reconstructed image. With recording materials such as photographic emulsions, there are basically two types of noise. One type of noise is due to noise gratings recorded by interference between the reference beam and light from the reference beam scattered by the individual silver halide crystals, which are distributed in the emulsion layer in a random manner [Syms & Solymar, 1982]; the other is replay scatter. Noise gratings are extremely sensitive to the reconstruction angle. Accordingly, if we exclude noise generated by nonlinear effects, which has been discussed in section 6.5, the main source of noise with photographic emulsions is light scattered from the individual grains during reconstruction [Goodman, 1967; Kozma, 1968a].

The noise spectrum of a photographic material can be studied with the optical system shown in fig. 6.3. If a uniformly exposed photographic plate with an amplitude transmittance $t(x, y)$ is placed in front of the lens and illuminated with a collimated beam of monochromatic light, the diffracted amplitude at any point $(x_f, y_f)$ in the back focal plane of the lens is (see Appendix 3)

$$a(x_f, y_f) = (ia/\lambda f) \exp[(-i\pi/\lambda f)(x_f^2 + y_f^2)]\mathbf{T}(\xi, \eta), \qquad (6.15)$$

where $a$ is the amplitude of the plane wave illuminating the plate, $f$ is the focal length of the lens and $t(x, y) \leftrightarrow \mathbf{T}(\xi, \eta)$; $\xi$ and $\eta$ are spatial frequencies defined by the relations $\xi = x_f/\lambda f$, $\eta = y_f/\lambda f$.

The intensity at this point is then

$$\begin{aligned} I(x_f, y_f) &= |a(x_f, y_f)|^2, \\ &= (a^2/\lambda^2 f^2)|\mathbf{T}(\xi, \eta)|^2. \end{aligned} \qquad (6.16)$$

If this intensity is averaged over a large enough area to eliminate local fluctuations due to speckle (see Appendix 4), it can be written as

$$I(\xi, \eta) = (a^2/\lambda^2 f^2)S(\xi, \eta), \qquad (6.17)$$

where $S(\xi, \eta)$ is the power spectrum of the transmittance of the plate.

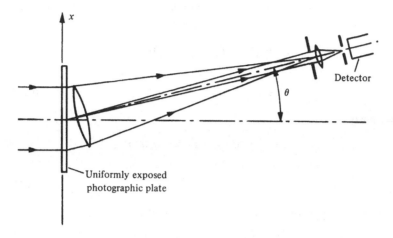

Fig. 6.3. Arrangement for measuring the noise spectrum of photographic materials used for holography.

To eliminate effects due to diffraction at the edges of the sample, a second lens is used to image the sample on to another aperture placed in front of the detector. The effective area of the sample is then the projection of this aperture on the sample. Since the latter is proportional to $1/\cos\theta$, this procedure also compensates for the decrease in the effective passband of the collecting aperture [Biedermann & Johansson, 1975].

## 6.8 Signal-to-noise ratio with coherent illumination

When computing the signal-to-noise ratio in the image reconstructed by a holo-gram, it is necessary to take into account the fact that it is the amplitudes of the signal and the noise that have to be added, since they are both encoded on the same coherent carrier [Goodman, 1967].

Consider the reconstructed image of an object consisting of a bright patch on a dark background. Let the intensity due to the nominally uniform signal be $I_i$, while that due to the randomly varying background is $I_N$. The noise $N$ in the bright area can then be defined as the variance of the resulting fluctuations in the intensity, and is given by the relation

$$N = [\langle I^2 \rangle - \langle I \rangle^2]^{1/2}, \tag{6.18}$$

where $I$ is the intensity at any point and $\langle I \rangle$ is the average intensity.

Since $a_N$ the complex amplitude of the background has circular statistics (see Appendix 4), the mean value of terms involving $a_N$ or $a_N^*$ as well as powers of $a_N$ or $a_N^*$, other than those involving only $|a_N|^2$, is zero. Hence, if $a_i$ is the complex amplitude of the signal, the average intensity is

$$
\begin{aligned}
\langle I \rangle &= \langle (a_i + a_N)(a_i^* + a_N^*) \rangle, \\
&= \langle I_i + I_N + a_i a_N^* + a_i^* a_N \rangle, \\
&= I_i + \langle I_N \rangle.
\end{aligned}
\tag{6.19}
$$

Similarly,

$$
\begin{aligned}
\langle I^2 \rangle &= \langle (I_i + I_N + a_i a_N^* + a_i^* a_N)^2 \rangle, \\
&= \langle I_i^2 + I_N^2 + 4 I_i I_N \rangle, \\
&= I_i^2 + \langle I_N^2 \rangle + 4 I_i \langle I_N \rangle.
\end{aligned}
\tag{6.20}
$$

However (see Appendix 4), $\langle I_N^2 \rangle = 2 \langle I_N \rangle^2$, so that

$$
\langle I^2 \rangle = I_i^2 + 2 \langle I_N \rangle^2 + 4 I_i \langle I_N \rangle.
\tag{6.21}
$$

Accordingly, from (6.18), (6.19) and (6.21) we have

$$
\begin{aligned}
N &= [\langle I_N \rangle^2 + 2 I_i \langle I_N \rangle]^{1/2}, \\
&= \langle I_N \rangle [1 + (2 I_i / \langle I_N \rangle)]^{1/2},
\end{aligned}
\tag{6.22}
$$

so that the signal-to-noise ratio is

$$
I_i / N = I_i / \langle I_N \rangle [1 + (2 I_i / \langle I_N \rangle)]^{1/2}.
\tag{6.23}
$$

In the limiting case when $I_i \gg \langle I_N \rangle$ (which is usually the situation for a hologram recording of good quality), the signal-to-noise ratio becomes

$$
I_i / N = (I_i / 2 \langle I_N \rangle)^{1/2}.
\tag{6.24}
$$

With coherent illumination, the signal-to-noise ratio is proportional to the square root of the ratio of the intensities of the signal and the scattered background. Even a weak scattered background leads to relatively large fluctuations in intensity in the bright areas of the image.

# 7

# Practical recording materials

The ideal recording material for holography should have a spectral sensitivity well matched to available laser wavelengths, a linear transfer characteristic, high resolution, and low noise. In addition, it should either be indefinitely recyclable or relatively inexpensive.

While several materials have been studied [Smith, 1977; Hariharan, 1980*b*], none has been found so far that meets all these requirements. However, a few have significant advantages for particular applications. This chapter reviews the properties of some of these materials in the light of the general considerations discussed in Chapter 6 (see Table 7.1 for a summary of their principal characteristics).

## 7.1 Silver halide photographic emulsions

Silver halide photographic emulsions are widely used for holography because of their high sensitivity and their commercial availability. In addition, they can be dye sensitized so that their spectral sensitivity matches the most commonly used laser wavelengths.

An apparent drawback of photographic materials is that they need wet processing and drying; however, development is actually a process, with a gain of the order of $10^6$, which amplifies the latent image formed during the exposure to yield high sensitivity as well as a stable hologram. Another advantage of the formation of a latent image is that the optical properties of the recording medium do not change during the exposure, unlike materials in which the image is formed in real time. This feature makes it possible to record several holograms in the same photographic emulsion without any interaction between them.

While photographic materials are commonly characterized by curves of optical density as a function of the logarithm of the exposure (see Appendix 5),

Table 7.1. *Recording materials for holography*

| Material | Reusable | Processing | Type of hologram | Exposure required $J/m^2$ | Spectral sensitivity nm | Resolution limit $mm^{-1}$ | Max. diffraction efficiency (sine grating) |
|---|---|---|---|---|---|---|---|
| Photographic emulsions | No | Wet chemical | Amplitude (normal) Phase (bleached) | $5 \times 10^{-3}$–$5 \times 10^{-1}$ | 400–700 | 1000–10000 | 0.05 0.60 |
| Dichromated gelatin | No | Wet chemical | Phase | $10^2$ | 350–580 | >10000 | 0.90 |
| Photoresists | No | Wet chemical | Phase | $10^2$ | uv–500 | 3000 | 0.30 |
| Photopolymers | No | Post exposure | Phase | $10$–$10^4$ | uv–650 | 200–1500 | 0.90 |
| Photochromics | Yes | None | Amplitude | $10^2$–$10^3$ | 300–700 | >5000 | 0.02 |
| Photothermoplastics | Yes | Charge and heat | Phase | $10^{-1}$ | 400–650 | 500–1200 (bandpass) | 0.30 |
| Photorefractive | | | | | | | |
| LiNbO$_3$ | Yes | None | Phase | $10^4$ | 350–500 | >1500 | 0.20 |
| Bi$_{12}$SiO$_{20}$ | Yes | None | Phase | $10$ | 350–550 | >10000 | 0.25 |

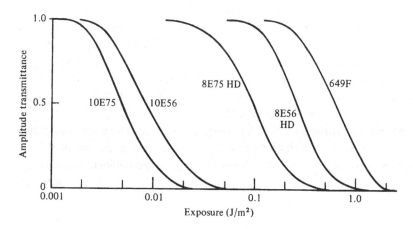

Fig. 7.1. Curves of amplitude transmittance |t| against exposure (E) for typical photographic emulsions used for holography.

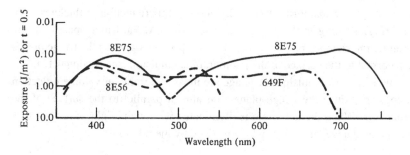

Fig. 7.2. Spectral sensitivity curves for typical photographic emulsions used for holography.

it is more convenient for holographic applications to use curves showing **t**, the amplitude transmittance of the material, as a function of the exposure. Typical |t| *versus* $E$ and spectral sensitivity curves for some of the silver halide photographic emulsions available for holography are presented in figs. 7.1 and 7.2. Most of these emulsions are available coated on glass plates or plastic film in a range of sizes and normally have an antihalation backing. Plates without any antihalation backing are available for making reflection holograms, though it is also possible to remove the antihalation backing from glass plates with alcohol.

Holograms recorded on these emulsions are processed using techniques similar to those used for normal photographic materials. Processing should be carried out immediately after exposure, since such fine-grain emulsions exhibit significant fading of the latent image. Another factor that should not be overlooked is reciprocity failure. While it is well known that reciprocity failure occurs at very short exposure times with pulsed lasers, it also occurs with continuous-wave lasers for exposure times greater than 1 second and is more noticeable with faster-acting developers [Johnson, Hesselink & Goodman, 1984; Binfield, Galloway & Watson, 1993].

Maximum diffraction efficiency is usually obtained at an average amplitude transmittance of about 0.45 (corresponding to an optical density of 0.7; see Appendix 5) for transmission holograms. For reflection holograms, maximum diffraction efficiency is obtained at much higher optical densities, around 2.0.

### 7.1.1 Emulsion shrinkage

The thickness of the processed photographic emulsion layer is usually about 15 per cent less than its original thickness due to the removal of the unexposed silver halide grains during fixing. Because of this reduction in thickness, $\Lambda'$, the fringe spacing in the hologram, is less than $\Lambda$, the fringe spacing in the original interference pattern, except for the special case when the fringe planes are normal to the surface of the photographic emulsion (see section 4.11).

The effects of emulsion shrinkage are most noticeable in volume reflection holograms, since the fringe planes run almost parallel to the surface of the emulsion. When the hologram is illuminated with white light, it reconstructs an image at a wavelength $\lambda'$ which is given by the relation

$$\lambda' = (\Lambda'/\Lambda)\lambda, \tag{7.1}$$

where $\lambda$ is the wavelength used to record the hologram. Typically, a reflection hologram recorded with a He-Ne laser ($\lambda = 633$ nm) reconstructs at a wavelength $\lambda = 530$ nm, giving a green image.

Because of emulsion shrinkage, amplitude holograms recorded in photographic emulsions also exhibit phase modulation that can modify the MTF. This phase modulation is mainly due to a surface relief image arising from local tanning (hardening) of the gelatin by the oxidation products of the developer in the immediate vicinity of the reduced silver [Smith, 1968].

As shown in fig. 7.3, after development and fixing, the wet emulsion layer is swollen to more than five times its normal thickness and is very soft. The tanned gelatin in the vicinity of a clump of developed silver grains absorbs less water than the nontanned areas and therefore dries more quickly. Shrinkage during drying pulls some of the gelatin from adjacent nontanned areas into the

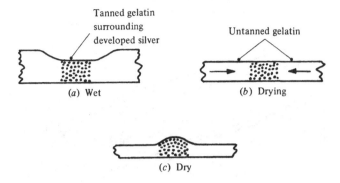

Fig. 7.3. Formation of a relief image due to local tanning of the gelatin [Smith, 1968].

image areas, and, when the emulsion has dried, the higher-density areas stand out in relief because both the image silver and the extra gelatin contribute to their bulk.

Since the relief image formed by local tanning is usually confined to low spatial frequencies ($< 200$ mm$^{-1}$), it normally contributes very little to the reconstructed image. However, it can give rise to intermodulation noise. The formation of such a relief image can be avoided by the use of a developer with a high sulphite content. Alternatively, the emulsion can be treated in a prehardening bath before processing.

### 7.1.2 Modulation transfer function

The MTF curves for the fine-grain photographic materials used for holography usually extend to quite high spatial frequencies. While the MTF can be measured with specially designed instruments [Biedermann & Johansson, 1972, 1975], it is derived most commonly in an indirect fashion from measurements of the diffraction efficiency of sinusoidal gratings with different spatial frequencies recorded on these materials [Buschmann, 1972]. In the latter case, the values obtained must be corrected for losses due to Fresnel reflection at the surfaces of the emulsion layer.

Typical curves of diffraction efficiency as a function of spatial frequency obtained for 8E75, 8E56, 10E75, and 10E56 emulsions for different wavelengths are presented in fig. 7.4. As can be seen, after an initial rapid drop, all of them exhibit a gently sloping region covering the range of spatial frequencies used in holography. However, beyond a certain limit, the diffraction efficiency again decreases rapidly with increasing spatial frequency. These effects are more

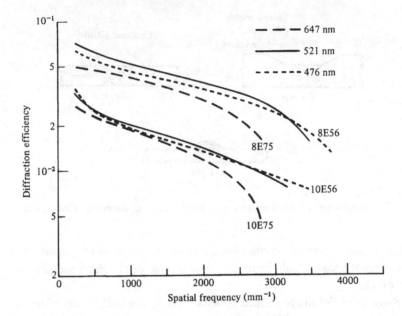

Fig. 7.4. MTF curves for typical photographic emulsions used for holography [Busch-mann & Metz, 1971].

pronounced for the more sensitive emulsions (10E75 and 10E56), which have larger grains, and can be attributed to scattering in the virgin emulsion during the exposure [Buschmann, 1972; Joly & Vanhorebeek, 1980].

### 7.1.3 Scattered flux spectra

Figure 7.5 shows normalized scattered flux spectra for three typical photo-graphic emulsions used for holography [Biedermann, 1970] exposed uniformly and processed to an amplitude transmittance $t \approx 0.45$ (optical density, $D \approx 0.7$).

These curves show that the scattered flux is higher for the most sensitive emulsion (10E70, an earlier version of 10E75) that has larger grains; in addition, exposure to laser light results in higher scattering than exposure to an incoherent source of the same wavelength. This is due to the speckle pattern produced in the emulsion layer during the exposure by coherent light scattered from individual grains. As expected, the section of the scattered flux spectrum extending to higher spatial frequencies, which is due to the grains, is very nearly a straight

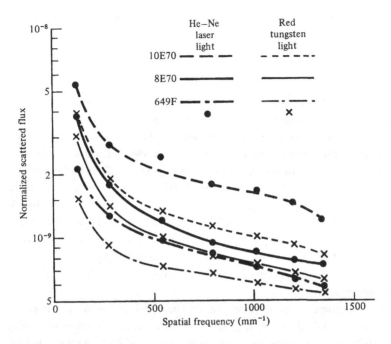

Fig. 7.5. Normalized scattered flux spectra for typical photographic emulsions used in holography [Biedermann, 1970].

line; however, there is a steep rise at lower spatial frequencies due to the relief image formed in the gelatin.

### 7.1.4 Monobath processing

Processing of holograms recorded on photographic emulsions for some purposes such as real-time hologram interferometry (see section 15.1) can be simplified and speeded up by the use of a monobath in which development and fixing take place simultaneously [Ragnarsson, 1970; Hariharan, Ramanathan & Kaushik, 1973].

A monobath is basically an active developer formula to which a silver halide complexing agent has been added. The image is formed initially by chemical development of the exposed silver halide grains. However, some of the unexposed silver halide dissolved by the complexing agent then migrates towards adjacent developing grains and is reduced there to metallic silver (physical development). A problem with monobaths is their tendency to form a sludge of metallic silver after use; this can be avoided either by discarding the monobath immediately after use or by using a sequestering agent [Dietrich, Raine & O'Brien, 1976].

### 7.1.5 Bleach techniques

Because of the low diffraction efficiency of amplitude holograms, holograms recorded in photographic emulsions are often processed to yield phase holograms, which have much higher diffraction efficiencies.

Two basic processes can be envisaged. In one, the variations in the optical density of the hologram are converted into thickness variations of the emulsion layer [Cathey, 1965; Altman, 1966]. In the other, the variations in the optical density are converted into a local modulation of the refractive index [Burckhardt, 1967]. Because the first method is effective only at relatively low spatial frequencies in the case of photographic materials and also has a lower maximum diffraction efficiency (0.339 for a surface relief grating against 1.00 for a volume phase grating), photographic phase holograms are almost exclusively of the latter type.

Early experiments to produce volume phase holograms were carried out with conventional rehalogenating bleach baths which, as shown in fig. 7.6(*a*), converted the image silver in the developed and fixed photographic emulsion into a transparent silver salt with a high refractive index [Upatnieks & Leonard, 1969]. Alternatively, the hologram was bleached after drying, using bromine vapour [Graube, 1974].

Unfortunately, the increase in diffraction efficiency obtained with these techniques was always accompanied by a high level of scattering. This was partly due to nonlinear effects arising from the formation of a relief image at low spatial frequencies [Upatnieks & Leonard, 1970] and partly due to scattering from the individual grains making up the recording. Since the scattering is proportional to the sixth power of the radius of the grains [Benton, 1971; Hariharan, Kaushik & Ramanathan, 1972], the size of the grains needs to be held to a minimum. Unfortunately, with most developers there is considerable grain growth due to solution physical development [Hariharan & Chidley, 1987*a*]; in addition, the oxidation products of the developer create a shell of hardened gelatin around each grain, so that its effective size is increased [van Renesse, 1980; Hariharan & Chidley, 1987*a*].

This drawback led to a considerable volume of work on other types of bleaches. One group consists of what may be called reversal or solvent bleaches [Chang & George, 1970; Hariharan, 1971; Lamberts & Kurtz, 1971; Buschmann, 1971] in which, as shown in fig. 7.6(*b*), the fixing bath is omitted and the developed silver is dissolved away leaving a phase hologram made up

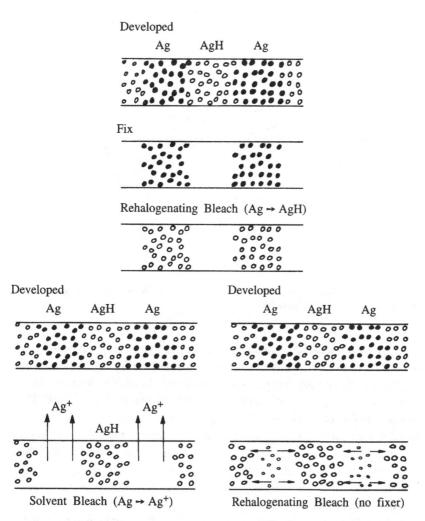

Fig. 7.6. Production of photographic volume phase holograms by (*a*) conventional, (*b*) solvent, or reversal, and (*c*) rehalogenating bleach techniques.

of the undeveloped silver halide grains. These crystals, being smaller, give rise to much less scattering. This technique also has the advantage that the relief image arising from local tanning by the developer depresses the MTF at low spatial frequencies resulting in a reduction in intermodulation noise [Lamberts & Kurtz, 1971]. Figure 7.7 shows the ratio of signal and noise powers for holograms of a diffusing object processed with conventional and reversal bleaches [Buschmann, 1971].

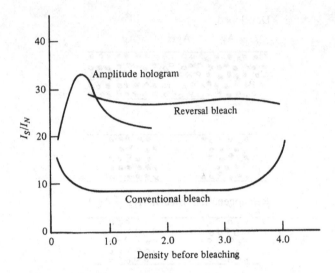

Fig. 7.7. Curves showing the ratio of the signal and noise powers for conventional and reversal bleaches [Buschmann, 1971].

Better bleach techniques have been developed by studying the actual processes taking place in the photographic emulsion [Hariharan, 1990a]. With any bleach, the primary process is the conversion of metallic silver into ionic silver, but the results are modified considerably by two other processes. The first is diffusion of the silver ions. The second is formation of silver ions and bromide ions due to the finite though small solubility of the unexposed silver bromide grains in the bleach bath. With a rehalogenating bleach used after fixing, both these processes result in the modulation of the refractive index and, therefore, the diffraction efficiency, dropping off as the fringe spacing decreases [Hariharan & Chidley, 1988a].

At small fringe spacings, material transfer due to the migration of silver ions dominates the bleach process. It then becomes possible to use a rehalogenating bleach immediately after development, without fixing, to produce phase holograms [Phillips, Ward, Cullen & Porter, 1980]. In this case, little or no silver or silver halide is removed from the emulsion layer and, as shown in fig. 7.6(c), a phase hologram is produced by material transfer from the exposed areas to adjacent unexposed areas [Hariharan, 1990b]. Material transfer by diffusion takes place because the solubility $S_r$ of small particles of radius $r$ is related to the solubility $S_\infty$ of large crystals by the Ostwald formula

$$\ln(S_r/S_\infty) = 2M\gamma/RT\rho r, \tag{7.2}$$

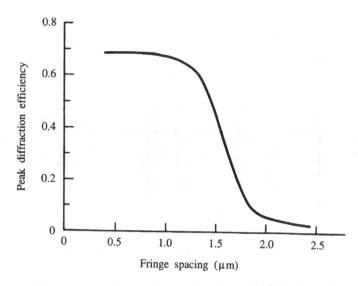

Fig. 7.8. Peak diffraction efficiency plotted as a function of the fringe spacing for transmission gratings processed with a rehalogenating bleach without fixing [Hariharan & Chidley, 1988*b*].

where $M$ is the molecular weight of the crystal, $\rho$ is its density, and $\gamma$ is the surface energy of the solid in the liquid. The solubility of small crystallites of silver bromide is, therefore, much higher than that of the bulk material. As a result, the freshly formed silver bromide nuclei in the unexposed areas go into solution, and the existing crystals in the unexposed areas grow at their expense. However, the excess of potassium bromide has to be controlled carefully to avoid the formation of clumps of grains by coalescence and a rapid increase in scattering [Joly, 1983; Cooke & Ward, 1984; Hariharan & Chidley, 1987*b*]. A factorial design approach that can be used to optimize the composition of a rehalogenating bleach for a specific purpose has been described by Kostuk [1991*a*].

When a rehalogenating bleach is used without fixing, its spatial frequency characteristics are completely different from those of the same rehalogenating bleach used after fixing. This is because material transfer through diffusion is effective over only a limited distance. As can be seen from fig. 7.8, the diffraction efficiency drops off rapidly for fringe spacings greater than a critical value corresponding to the diffusion length of the silver ion in the bleach bath [Hariharan & Chidley, 1988*b*]. With reflection holograms the fringes are very closely spaced ($d < 0.3\ \mu$m), and very good results can be obtained with a rehalogenating bleach without fixing. With such a bleach, the refractive-index

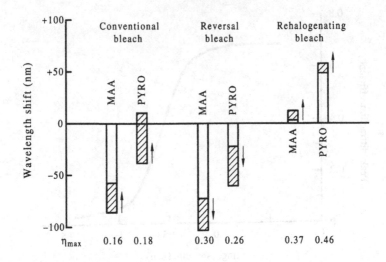

Fig. 7.9. Wavelength shifts (shaded areas) and variation with increasing exposure (arrows) obtained with reflection gratings using a nontanning Metol-ascorbic acid (MAA) developer and a strongly tanning pyrogallol (PYRO) developer with three different types of bleaches [Hariharan & Chidley, 1989].

modulation and, hence, the diffraction efficiency is determined essentially by the ratio of the intensity of the reference beam to that of the signal beam and is very nearly independent of the exposure over the linear portion of the H & D curve [Ward & Solymar, 1989].

As mentioned earlier, processing usually results in a change in the thickness of the photographic emulsion layer. With bleached reflection holograms, two competing processes are involved [Hariharan & Chidley, 1989]. The first is a loss of silver halide during processing, which leads to a decrease in the thickness of the emulsion layer. The other is overall tanning of the gelatin by the oxidation products of the developer. The gelatin layer swells to several times its original thickness when it is immersed in the developer, but tanning of the swollen gelatin inhibits its subsequent shrinkage on drying, resulting in an increase in thickness. As shown in fig. 7.9, the decrease in thickness is a maximum with a nontanning developer and a solvent bleach, whereas the use of a tanning developer and a rehalogenating bleach can actually result in an increase in thickness.

A major problem with bleached holograms is their tendency to darken when exposed to ambient light because of the formation of printout silver. Treatment with a desensitizing dye [Buschmann, 1971] helps but has the drawback that it stains the gelatin. However, the resistance of the hologram to printout also

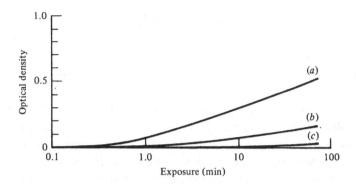

Fig. 7.10. Increase in optical density of bleached holograms when exposed to noon sunlight: (*a*) bleach containing KBr, (*b*) bleach followed by treatment in 0.2 per cent KI, (*c*) bleach containing 0.2 per cent KI [Hariharan & Ramanathan, 1971].

depends on the silver salt of which it is formed, AgCl being the poorest in this respect, while AgBr is considerably better and AgI is extremely stable [McMahon & Maloney, 1970]. It is possible to convert a hologram recording consisting of AgCl or AgBr to AgI by soaking it in a dilute solution of KI [Hariharan & Ramanathan, 1971]. A simpler method is to use a reversal bleach containing a small amount of KI [Hariharan, Ramanathan & Kaushik, 1971]. Comparative measurements of stability against exposure to light of untreated and treated holograms are presented in fig. 7.10.

## 7.2 Dichromated gelatin

Dichromated gelatin is, in some respects, an almost ideal recording material for volume phase holograms; it has large refractive-index-modulation capability, high resolution, and low absorption and scattering.

Hologram recording in dichromated gelatin makes use of the fact that a gelatin layer containing a small amount of a dichromate such as $(NH_4)_2Cr_2O_7$ becomes progressively harder on exposure to light. This hardening is due to the photochemically produced $Cr^{3+}$ ion forming localized cross-links between the carboxylate groups of neighbouring gelatin chains.

This effect has been used in the past to produce a relief image by washing off the hardened gelatin with warm water. However, much better phase holograms can be obtained if the gelatin film is processed to obtain a modulation of the refractive index [Shankoff, 1968; Lin, 1969].

### 7.2.1 Film preparation and processing

Since plates and films coated with dichromated gelatin are not commercially available, it is necessary to prepare them in the laboratory. It is possible to coat glass plates with a uniform gelatin film [Brandes, Francois & Shankoff, 1969; McCauley, Simpson & Murbach, 1973]. In this case, it is necessary to harden the gelatin film initially to a degree at which it does not dissolve in water during subsequent processing but remains soft enough for the photochemical reaction to produce a significant difference in the local hardness. Alternatively, a method that has been used widely is to dissolve and remove the silver halide in the emulsion layer in a photographic plate. Kodak 649F plates have been used commonly because they have a fairly thick ($\approx 15\ \mu$m) coating of gelatin that is not excessively hardened.

To sensitize the plates, they are soaked for about 5 min at 20°C in an aqueous solution of $(NH_4)_2Cr_2O_7$ to which a small amount of a wetting agent has been added. They are then allowed to drain on edge and dried at 25–30°C in darkness or under a red safelight. Since the $Cr^{6+}$ ion is reduced slowly to $Cr^{3+}$ even in the dark, the plates should be used within a well-defined period after they are sensitized (usually 8–24 hours) to ensure reproducible results.

After exposure in the holographic system using blue light from an $Ar^+$ laser ($\lambda = 488$ nm, $E \approx 10^2$ J/m$^2$), the gelatin film is soaked in a 0.5 per cent solution of $(NH_4)_2Cr_2O_7$ as well as in an additional hardening bath to control the overall hardness of the film. The film is then washed in water at 20–30°C for 10 min to remove the dichromate. Because the melting point of the gelatin film is higher than the temperature of the water, none of the gelatin goes into solution. However, the film absorbs a large amount of water and swells. Since the rigidity of the gel is higher in the hardened areas, they swell to a lesser extent than the unhardened areas.

The crucial stage in processing is the next step, in which the swollen film is immersed for 3 min in two successive baths of isopropanol at 20–30°C and then dried thoroughly. To avoid condensation of moisture, which can result in reduced diffraction efficiency, the use of a blast of warm air as the film is pulled out of the second bath has been recommended [Meyerhofer, 1972]. However, too rapid drying can result in a milky-white film which scatters strongly.

Two successive baths of isopropanol are necessary to ensure that dehydration is complete. Raising the temperature of the water and isopropanol baths increases the sensitivity of the film but can also result in an increase in scattering. Any residual water is eliminated by drying the plates in a vacuum for 1–2 hours. Since moisture, or high ambient humidity, rapidly degrades holograms recorded on dichromated gelatin, they should be protected by a cover glass cemented on with an epoxy cement.

Simplified procedures for processing dichromated gelatin holograms have been developed by Georgekutty and Liu [1987] and by Jeong, Song, and Lee [1991], while procedures for obtaining dichromated gelatin holograms from Agfa 8E75 HD plates have been described by Oliva, Boj, and Pardo [1984]. Uniformly bright holograms without blotches are ensured by spinning the plates during dehydration and drying [Bahuguna, Beaulieu & Arteaga, 1992].

If the sensitizing and processing sequence is carried out properly, it should result in a hologram with high diffraction efficiency and low scattering. However, if the results are not satisfactory, it is possible, within limits, to reprocess the hologram and obtain a better result [Chang, 1976]. With reflection holograms, the wavelength at which the hologram reconstructs an image can be controlled by regulating the degree of swelling of the gelatin [Coleman & Magariños, 1981]. A more effective method of controlling the reconstruction wavelength is to use a mixture of two different kinds of gelatin, one water soluble, and the other with high bloom strength, so that the former is washed out during processing [Kubota, 1989].

A critical factor in obtaining good results with dichromated gelatin is the degree of preliminary hardening of the gelatin layer. If it is too low, the resulting holograms are noisy and, if it is too high, the available refractive index modulation and the sensitivity decrease. The drop in sensitivity when the gelatin layer is hardened, prior to exposure, with a chromium salt occurs because many of the carboxyl groups have already reacted and are no longer available. This can be avoided if a noble metal salt or formaldehyde, both of which form cross-links with the amino groups and not with the carboxyl groups, is used to preharden the gelatin [Mazakova, Pancheva, Kandilarov & Sharlandjiev, 1982b].

### 7.2.2 Resolution and spectral sensitivity

The characteristics of dichromated gelatin as a holographic recording medium have been discussed in some detail by Chang [1979]. Dichromated gelatin has excellent resolution and an MTF which is almost flat out to a spatial frequency of 5000 mm$^{-1}$. It is therefore very suitable for making reflection holograms as well as transmission holograms.

Very high levels of refractive-index modulation can be obtained with dichromated gelatin layers: values as large as 0.08 with very low scattering are possible with careful processing [Chang & Leonard, 1979; Salminen & Keinonen, 1982]. Typical curves showing the diffraction efficiencies of transmission and reflection gratings recorded in dichromated gelatin are presented in fig. 7.11. These curves show that diffraction efficiencies close to the theoretical maximum for volume phase holograms can be obtained.

The spectral sensitivity of dichromated gelatin at 514 nm is normally only

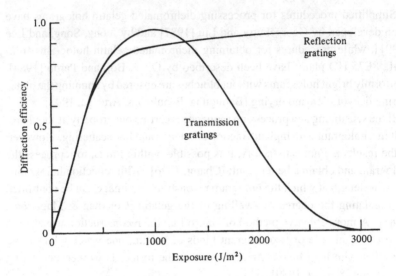

Fig. 7.11. Diffraction efficiency of gratings recorded in dichromated gelatin as a function of the exposure [Chang, 1979].

about a fifth of that at 488 nm and it drops to zero at around 580 nm. The possibility of extending it to longer wavelengths for holographic recording was first explored by Graube [1973], who used dichromated gelatin sensitized with triphenylmethane dyes to record holograms with a He-Ne laser. Good diffraction efficiency can be obtained with exposures of 500–1000 J/m² at 633 nm if methylene blue in an aqueous dichromate solution is used as a sensitizer [Kubota & Ose, 1979a]. More recently, Blyth [1991] has shown that tetramethylguanidine (an electron donor), used in conjunction with methylene blue sensitized dichromated gelatin, permits a further reduction in exposure.

### 7.2.3 *Mechanisms of image formation*

The chemical reactions and physical mechanisms involved in recording a volume hologram in a film of dichromated gelatin are quite complex.

The earliest explanation of the large local changes in refractive index observed was put forward by Curran and Shankoff [1970], who suggested that cracks were formed at the fringe boundaries and that the dimensions of these cracks increased with increasing exposure. While the presence of voids within the gelatin was confirmed by the drop in diffraction efficiency when the hologram was immersed in an index-matching fluid, this explanation was not entirely satisfactory, since such cracks would scatter light quite strongly. An alternative theory proposed by Meyerhofer [1972], in which the index modulation was caused by the bonding of isopropanol molecules to the chromium atoms at the sites of the cross-links, is contradicted by the fact that the hologram image is destroyed by very small amounts of water.

Case and Alferness [1976] suggested that instead of cracks, a very large number of small vacuoles, with dimensions much smaller than the wavelength of light, were formed in unhardened areas. This would allow a smooth variation in the refractive index across a fringe and account for the low scattering. Calculations of the modulation of the refractive index and the diffraction efficiency based on their model are in good agreement with experimental observations. In addition, an overall decrease in the refractive index, predicted by their model, has been found. It appears [Chang & Leonard, 1979; Graube, personal communication] that hologram formation in dichromated gelatin may take place in two stages, involving rearrangement of the gelatin chains as well as the formation of voids. Sjölinder [1981] has suggested that the observed refractive index modulation is at least partly due to the formation in the hardened areas of complexes with a high refractive index involving a $Cr^{3+}$ compound, gelatin and isopropanol.

Kubota [1989] has shown that the asymmetry observed in the wavelength selectivity curve of a volume reflection hologram recorded in dichromated gelatin is due to the formation of a chirped grating in which the modulation of the refractive index decreases with depth from the free surface, while the period increases. Work by Rallison and Schicker [1992] on the polarization properties of dichromated gelatin gratings also shows that the film expands during processing, so that the bulk index of the gelatin drops to a value between 1.4 and 1.2. This expansion is mostly in a region near the free surface, so that a volume reflection grating with a chirped microstructure is formed. In this region, the net expansion can be up to 1.5 times, with a correspondingly low value of the bulk index and a very high value of the index modulation. This conclusion is supported by work by Boj, Crespo, and Quintana [1992], who have shown how the reflected bandwidth can be controlled by varying the processing conditions.

### 7.2.4 Silver-halide sensitized gelatin

A technique that combines the relatively high sensitivity of photographic materials with the low scattering and high light-stability of dichromated gelatin involves exposing a silver halide photographic emulsion and then processing it so as to obtain a volume phase hologram consisting solely of hardened gelatin.

This technique, which was originally developed by Pennington, Harper, and

Laming [1971], was studied in more detail by Graver, Gladden, and Estes [1980] and by Chang and Winick [1980].

In this technique, the exposed photographic emulsion was developed in a metol-hydroquinone developer and then bleached in a bath containing $(NH_4)_2Cr_2O_7$. During the bleaching process, the developed silver was oxidized to $Ag^+$, while the $Cr^{6+}$ ions in the bleach were reduced to $Cr^{3+}$ ions. These $Cr^{3+}$ ions formed cross-links between the gelatin chains in the vicinity of the oxidized silver grains, causing local hardening. The emulsion was then fixed to remove the unexposed silver halide, washed and dehydrated with isopropanol, and dried exactly as for a dichromated gelatin hologram.

However, experiments by Hariharan [1986a,b] have shown that this is not the only mechanism involved, and that the oxidation products of the developer, which have a local tanning action, also contribute to the formation of the latent image. Optimized processing techniques have been described for Kodak 649F plates by Angell [1987] and for Agfa-Gevaert 8E75 HD and 8E56 HD plates by Fimia, Beléndez, and Pascual [1991]. Transmission gratings with diffraction efficiencies up to 80 per cent can be produced with this technique.

Fimia, Pascual, and Beléndez [1992] have also shown that volume reflection gratings with diffraction efficiencies up to 55 per cent can be produced with this technique by using a nonsolvent developer and holding the bleached plates for 24 hours at 90 per cent relative humidity before dehydrating them in isopropanol.

The resolution possible with this technique is limited by the grain size of the photographic emulsion. Higher resolution can be attained with specially prepared emulsion layers [Mazakova, Pancheva, Kandilarov & Sharlandjiev, 1982a]. Glass plates coated with a 20 $\mu$m thick layer of gelatin are soaked in a solution of silver nitrate and bathed in an alkali halide solution, so that very fine grains of silver halide are formed in the gelatin layer. The excess alkali halide is then washed away and the layer is sensitized by treating it with a solution of dichromate and an optical sensitizer.

### 7.3 Photoresists

Photoresists are light-sensitive organic films that yield a relief image after exposure and development. Several photoresists have been used to record holograms [Bartolini, 1977a], but they are all relatively slow, typically requiring an exposure of $10^2$ J/m$^2$ to blue light ($\lambda = 442$ nm), and because a thin phase hologram is formed, nonlinear effects are noticeable at diffraction efficiencies greater than about 0.05. However, they have the advantage that replication using thermoplastics is easy (see section 11.4).

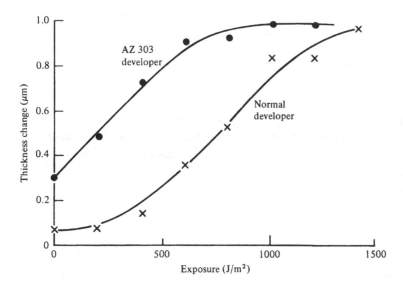

Fig. 7.12. Depth *versus* exposure characteristics obtained with AZ-1350 photoresist [Bartolini, 1974].

Two types of photoresists are available. In negative photoresists, the exposed areas become insoluble and the unexposed areas are dissolved away during development. Relatively long exposures are necessary with such photoresists, usually through the back of the plate, to ensure that the exposed photoresist adheres to the substrate during development. For this reason, positive photoresists in which the exposed areas become soluble and are washed away during development are preferable.

The most widely studied positive photoresist is Shipley AZ-1350. The sensitivity of this photoresist is a maximum in the ultraviolet and drops rapidly for longer wavelengths towards the blue. Hence, holograms are best recorded either with an $Ar^+$ laser at 458 nm, or with a He-Cd laser at 442 nm. The lower power available at these wavelengths is more than compensated for by the increase in sensitivity [Beesley, Castledine & Cooper, 1969].

The photoresist is usually coated on a glass substrate by spinning to give a layer 1–2 $\mu$m thick. This layer is then baked at 75°C for 15 min to ensure complete removal of the solvent. As shown in fig. 7.12, an approximately linear depth *versus* exposure characteristic is obtained if AZ-303 A developer, an alkaline solution, is used instead of the normal developer for this material. In addition, AZ-303 A developer gives an increase in sensitivity by a factor of 2 to 3 over normal processing. To control the etch rate, the AZ-303 A developer is diluted with four parts of distilled water. The exposure can also be reduced significantly, and improved linearity obtained, by uniformly pre- or post-exposing the photoresist to incoherent light from a fluorescent lamp [Bartolini, 1972, 1974; Norman & Singh, 1975; Livanos, Katzir, Shellan & Yariv, 1977].

The MTF curve for AZ-1350 photoresist is almost flat up to spatial frequencies of 1500 mm$^{-1}$, and does not appear to be affected significantly by small changes in the coating thickness. In addition, because the material has no grain structure, scattered light is negligible. These characteristics, as well as the convenience of replication, have led to the widespread use of photoresists in the production of blazed holographic diffraction gratings (see section 13.1).

## 7.4 Photopolymers

A number of organic materials are known that can be activated through a photosensitizer to exhibit thickness and refractive index changes due to photopolymerization or cross-linking. Thick layers can be made to yield volume phase holograms with high diffraction efficiency and high angular selectivity, which can be viewed immediately after exposure. After the exposure, a continuing dark reaction due to diffusion of the monomer into the zones of polymerization increases the refractive-index modulation. The hologram is then stabilized by a uniform post-exposure to complete the reaction. The characteristics of some of these materials have been reviewed by Booth [1975, 1977] and by Bartolini [1977b].

Two photopolymers are now available for holographic recording, Polaroid DMP 128 and Du Pont OmniDex. Polaroid DMP 128 uses dye-sensitized photopolymerization of a vinyl monomer incorporated in a polymer matrix, which is coated on a glass or plastic substrate. Coated plates or films can be stored for several months and exposed with blue, green, or red light. Processing involves a uniform exposure to white light followed by a developer/fixer bath. Diffraction efficiencies of 95 per cent are obtained with exposures of 40 to 100 J/m$^2$ for transmission holograms and 300 J/m$^2$ for reflection holograms [Ingwall & Fielding, 1985]. In the case of reflection holograms, processing in a single bath results in a wide spectral bandwidth (100–200 nm), while a narrow spectral bandwidth (60 nm) is obtained when the development and stabilization steps are separated [Ingwall, Troll & Vetterling, 1987]. Electron micrographs show that the high refractive-index modulation obtained is due to the formation of alternating solid and porous regions in the processed film [Ingwall & Troll, 1989].

Du Pont OmniDex holographic recording film consists of a polyester base film coated with a photopolymer layer. A removable polyester cover sheet is laminated on to the tacky photopolymer layer. Holograms are usually recorded by contact copying a master hologram (see section 11.4), since exposure energy requirements are typically around 300 J/m$^2$ at 514 nm. Close contact is ensured by removing the cover sheet and laminating the tacky film to the master hologram. The film is then exposed to UV light to cure the image, after which it is no longer tacky and can be separated from the master hologram. Finally, the film is baked in a forced-air convection oven at 100–120°C for 1–2 hours to obtain an increase in index modulation [Smothers, Monroe, Weber & Keys, 1990; Weber, Smothers, Trout & Mickish, 1990].

Processing as described above yields volume reflection holograms that reconstruct an image at nearly the same wavelength as that used to record them. The reconstruction wavelength can be changed by an additional processing step that involves laminating the holographic recording film, after exposure to UV light, to a colour tuning film and baking the sandwich. During baking, diffusion of the monomer from the colour tuning film to the holographic recording film swells the photopolymer layer, shifting the reconstruction permanently to a longer wavelength. Precise control of the final reconstruction wavelength, over the range from 514 to 650 nm, is possible by an appropriate choice of the composition and thickness of the colour tuning film [Zager & Weber, 1991].

## 7.5 Photochromics

Photochromic materials undergo reversible changes in colour when exposed to light. Several organic photochromics have been studied [Bartolini, 1977*b*], but they are prone to fatigue and have a limited life. Inorganic photochromics are usually crystals doped with selected impurities: photochromism is due to reversible charge transfer between two species of electron traps. A detailed survey of such materials has been made by Duncan and Staebler [1977].

Photodichroic crystals have also received some attention. In these crystals, the absorption centres are anisotropic, and selective alignment of the anisotropic absorption centres is induced by exposure to linearly polarized light. The result is a preferred orientation for absorption, which can be switched between two orthogonal directions. Photodichroic crystals have a higher sensitivity (typical exposure 10$^2$ J/m$^2$) than photochromic crystals (typical exposure 10$^3$ J/m$^2$). In addition, a single laser can be used for storage, readout, and erasure; only the direction of linear polarization is changed. The most commonly used photodichroic crystals are alkali halides; their characteristics have been reviewed by Casasent and Caimi [1977].

Inorganic photochromics are grain free and have high resolution. Because of their large thickness, a number of holograms can be stored in them. In addition, they require no processing and can be erased and re-used almost indefinitely. However, despite these advantages, their applications have been limited by their low diffraction efficiency (< 0.02) and their low sensitivity.

## 7.6 Photothermoplastics

A surface relief hologram can also be recorded in a thin layer of thermoplastic which is combined with a photoconductor and charged to a high voltage. On exposure, a spatially varying electrostatic field is created. The thermoplastic is then heated so that it becomes soft enough to be deformed by the field and, finally, cooled to fix the pattern of deformations [Urbach, 1977].

Such materials have a reasonably high sensitivity over the whole visible spectrum and yield a thin phase hologram with fairly high diffraction efficiency. In addition, they have the advantage that they do not require wet processing. If a glass substrate is used, the hologram can be erased and the material re-used a number of times. Hologram recording cameras using photothermoplastics are extremely useful for applications such as hologram interferometry [Ineichen, Liegeois & Meyrueis, 1982] (see Chapters 15 and 16).

The most widely used type of photothermoplastic is a multilayer structure consisting of a substrate (glass or Mylar) coated with a thin, transparent, conducting layer (usually indium oxide), a photoconductor, and a thermoplastic [Urbach & Meier, 1966; Lin & Beauchamp, 1970b].

As shown in fig. 7.13, the film is initially sensitized in darkness by applying a uniform electric charge to the top surface using a corona device that moves over the surface at a constant distance from it and sprays positive ions on to it. As a result, a uniform negative charge is induced on the conductive coating on the substrate.

In the second step, when the film is exposed, charge carriers are produced in the photoconductor wherever light is incident on it. These charge carriers migrate to the two oppositely charged surfaces and neutralize part of the charge deposited there during the sensitizing step. This reduces the surface potential but does not change the surface charge density and the electric field, so that the image is still not developable.

Accordingly, in the next step, the surface is charged once again to a constant potential, using the same procedure as in the first step. As a result, additional charges are deposited wherever the exposure had resulted in a migration of charge. The electric field now increases in these regions, producing a spatially varying field pattern and, hence, a developable latent image.

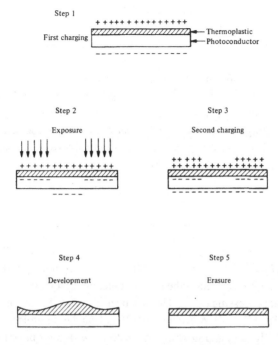

Fig. 7.13. Record-erase cycle for a photothermoplastic recording material [Lin & Beauchamp, 1970*b*].

In the fourth step, this latent image is developed by heating the thermoplastic uniformly to a temperature near its softening point. This is done most conveniently by passing a current briefly through the conductive coating on the substrate. The thermoplastic layer then undergoes local deformation as a result of the varying electric field across it, becoming thinner wherever the field is higher (the illuminated areas) and thicker in the unexposed areas. Once the thermoplastic layer has cooled to room temperature, this thickness variation is frozen in, so that the hologram is quite stable.

Because the latent image is relatively insensitive to exposure to light after the second charging, it is possible to monitor the diffraction efficiency of the hologram during development and to terminate the application of heat at the proper time.

Finally, when the plate is to be re-used, it is flooded with light, and the thermoplastic layer is heated to a temperature somewhat higher than that used for development. At this temperature, the thermoplastic is soft enough for surface tension to smooth out the thickness variations and erase the previously recorded hologram. A blast of compressed air or dry nitrogen is then used to cool the photothermoplastic rapidly to room temperature, in preparation for the next exposure.

An alternative method of development studied by Saito, Imamura, Honda, and Tsujiuchi [1980] is the use of solvent vapour to soften the thermoplastic. This has the advantage that it eliminates the need to heat the substrate. In addition it gives higher sensitivity and lower noise. Enhanced sensitivity can also be obtained by the use of double-layer and triple-layer photoconductor systems [Saito, Imamura, Honda & Tsujiuchi, 1981].

### 7.6.1 Materials

The most widely used photoconductor is poly-$n$-vinyl carbazole (PVK) sensitized with 2,4,7 trinitro-9-fluorenone (TNF). This material has good sensitivity, but suffers from ghost images – the reappearance of a previously erased image in a subsequent processing cycle. This has been shown to be due mainly to low electron mobility [Colburn & Dubow, 1973] and can be minimized by increasing the TNF/PVK ratio and flooding the material with light prior to or during erasure.

The life of the thermoplastic layer is limited by degradation caused by ozone formed during the charging process. When protected from degradation by ozone, materials such as a styrene-methacrylate copolymer appear capable of extended life with good spatial-frequency response and diffraction efficiency. Commercial photothermoplastics have a life of more than 300 cycles.

### 7.6.2 Characteristics

The MTF of all photothermoplastics is limited to a band of spatial frequencies. The spatial frequency corresponding to the peak response depends mainly on the thickness of the thermoplastic layer, but the usable bandwidth is determined by a number of parameters, including the properties of the material used, the field strength, and the manner of development. Commercial photothermoplastics have a peak response at spatial frequencies around $800\,\text{mm}^{-1}$. A typical curve showing the diffraction efficiency as a function of spatial frequency, for gratings recorded with two plane wavefronts of equal amplitude, is shown in fig. 7.14. A diffraction efficiency close to the theoretical maximum of 0.339 for a thin phase hologram can be obtained with an exposure of less than $0.1\,\text{J/m}^2$.

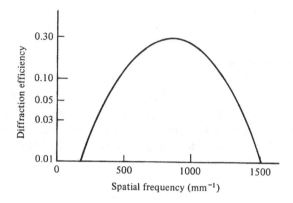

Fig. 7.14. Diffraction efficiency as a function of spatial frequency for gratings recorded in a typical photothermoplastic.

### 7.6.3 Noise

A problem encountered with photothermoplastics is "frost" – a random surface modulation whose power spectrum is very similar to the MTF of the material. The exact causes of frost are not known, but it can be minimized by using thinner thermoplastic films and lower charging voltages. It has also been shown [Reich, Rav-Noy & Friesem, 1977] that frost is reduced with a thermoplastic having a narrow distribution of molecular weights. In addition, the use of a thin, electrically insulating layer between the photoconductor and thermoplastic layers results in a substantial reduction in frost [Lee, Lin & Tufte, 1977].

A rather surprising feature of the image reconstructed by a hologram recorded on a photothermoplastic is the lack of intermodulation noise even at quite high diffraction efficiencies [Credelle & Spong, 1972]. This is because intermodulation noise arises from terms in (6.14) whose spectra involve multiple correlations and convolutions of the object spectrum and are therefore mainly confined to low spatial frequencies. Since the response of the photothermoplastic at such low spatial frequencies is virtually nil, these terms are severely attenuated.

## 7.7 Photorefractive crystals

In some electro-optic crystals, exposure to light frees trapped electrons, which then migrate through the crystal lattice and are again trapped in adjacent unexposed regions. This migration usually occurs through diffusion or an internal photovoltaic effect. The spatially varying electric field produced by the resulting space-charge pattern modulates the refractive index through the electro-optic effect, resulting in the formation of a volume phase hologram. This hologram can be erased by uniformly illuminating the crystal, which can be recycled almost indefinitely.

Heaton *et al.* [1984] have made a theoretical study of the diffraction efficiency and angular selectivity of volume phase gratings recorded in such crystals. Heaton and Solymar [1985*b*] have discussed the time dependence of the energy transfer between the fringes formed by the recording beams and the refractive-index grating that is produced and have shown that, in the limit, energy transfer is a maximum when the phase difference between the fields is $\pi/2$.

### 7.7.1 LiNbO₃

A number of photorefractive crystals have been studied for hologram recording [Staebler, 1977]. One of these is Fe-doped $LiNbO_3$. This material can give good diffraction efficiency and, because of the high angular sensitivity of such a thick recording medium, it is possible to store a large number of holograms in a single crystal [Burke, Staebler, Phillips & Alphonse, 1978; Glass, 1978]. However, the high angular sensitivity of the hologram also creates problems. To satisfy the Bragg condition, it is normally necessary to use the same wavelength for readout as for recording, and this results in degradation of the stored image. This problem can be avoided by using light of a longer wavelength, to which the crystal is insensitive, for readout, and satisfying the Bragg condition by making use of the crystal birefringence [Petrov, Stepanov & Kamshilin, 1979*a,b*].

A major drawback is the low sensitivity of $LiNbO_3$ (typical exposure $10^4 J/m^2$). Higher sensitivity can be obtained with other crystals such as $Sr_{0.75}Ba_{0.25}Nb_2O_6$ [Thaxter & Kestigian, 1974].

### 7.7.2 $Bi_{12}SiO_{20}$ and $Bi_{12}GeO_{20}$

Much higher sensitivity has been obtained with photoconductive electro-optic crystals such as $Bi_{12}SiO_{20}$ (BSO) and $Bi_{12}GeO_{20}$ (BGO) by the application of an external electric field (see Stepanov and Petrov [1984]). Photocarrier drift is minimized and maximum resolution obtained by using a transverse field, the hologram fringes then being perpendicular to the direction of the applied field as shown in fig. 7.15 [Huignard & Micheron, 1976]. The photosensitivity with this configuration is low (3 $J/m^2$ for BSO, for a diffraction efficiency of 0.01), but still comparable to that for slow photographic materials. A maximum diffraction efficiency of 0.25 is possible with a field of 900 V/mm, and the storage time constant in darkness is about 30 h. Erasure is achieved by uniform illumination of the crystal, which can be recycled indefinitely without fatigue

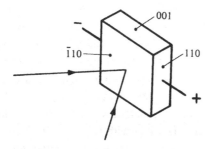

Fig. 7.15. Hologram recording configuration for BSO [Huignard, 1981].

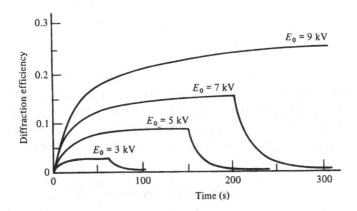

Fig. 7.16. Record-erase cycles in BSO at different applied fields. Sample size 10 × 10 × 10 mm, $\lambda = 514$ nm [Huignard & Micheron, 1976].

effects. Typical recording-erasure curves for BSO at different applied fields at an incident power density of 2.45 W/m$^2$($\lambda = 514$ nm) are presented in fig. 7.16. Experiments with a frequency-doubled Nd:Yag laser show that the response time depends on the incident power density and can be reduced to < 5 ns, provided the incident power density is high enough (> 10$^7$ W/m$^2$) [Hermann, Herriau & Huignard, 1981]. Higher diffraction efficiency and an increase in the holographic sensitivity can be obtained by preliminary infrared irradiation ($\lambda = 650$–800 nm) [Kamshilin & Miteva, 1981].

A feature of BSO, when used in this configuration, is that the form of the MTF depends on the applied field [Huignard, Herriau, Rivet & Günter, 1980]. With no applied field, a high-pass characteristic is obtained, while a field of 600 V/mm gives a low-pass characteristic. An applied field of 200 V/mm results in a flat MTF.

Interesting optical effects are observed with these crystals since they are also optically active [Miridonov, Petrov & Stepanov, 1978]. As a result, the transmitted beam, the diffracted beam, and the scattered light exhibit differences in polarization that can be exploited to minimize scatter noise, even in a nearly on-axis recording configuration, by the introduction of a sheet polarizer in the image plane [Herriau, Huignard & Aubourg, 1978].

The main problem with these materials is that optical readout is destructive. Nondestructive readout is possible with $LiNbO_3$, $Bi_{12}SiO_{20}$, and $BaTiO_3$ by thermal fixing, in which a compensating ionic charge grating (which cannot be erased optically) is formed at an elevated temperature. A more convenient procedure with $Bi_{12}SiO_{20}$, $BaTiO_3$, and $Sr_{0.75}Ba_{0.25}Nb_2O_6$ is electrical fixing by applying an external field at room temperature [Micheron & Bismuth, 1972, 1973; Herriau & Huignard, 1986; Qiao, Orlov, Psaltis & Neurgaonkar, 1993], which makes it possible to read out the stored image over a much longer time.

The unique characteristics of these materials have opened up several interesting possibilities, including the continuous generation of a phase-conjugate wavefront (see section 12.5), real-time interferometry (see section 16.1), and speckle reduction by real-time integration of a number of images with independent speckle patterns [Huignard, Herriau, Pichon & Marrakchi, 1980].

## 7.8 Spectral hole-burning

The width of the absorption bands of organic dye molecules in polymer films is quite large (typically, several terahertz), due to inhomogeneous broadening, and is only weakly dependent on the temperature. On the other hand, the homogeneous absorption of individual molecules at cryogenic temperatures may have a width of only 10 MHz. In such a situation, incident laser light excites only a small fraction of the molecules which absorb at the laser wavelength. Since the products of this photo-transformation, which can be quite long lived, absorb in a different spectral region, a dip, commonly called a spectral hole, is generated in the inhomogeneously broadened absorption band.

One application of spectral hole-burning has been in recording a number of holograms, with different applied electric fields and different optical frequencies, within the inhomogeneously broadened absorption band [Renn & Wild, 1987]. Interactions between the holograms can be eliminated by sweeping the laser frequency over a 1 GHz interval, and the phase over $2\pi$, while recording each hologram, permitting the storage of as many as 2000 holograms in a single sample [Kohler, Bernet, Renn & Wild, 1993].

Another interesting application of spectral hole-burning has been in holographic recording of the temporal and spatial characteristics of ultra-short light pulses [Rebane *et al.*, 1983; Saari, Kaarli & Rebane, 1986]. This technique can be extended to produce holographic interferograms which show, not only spatial changes, but also spectral and temporal changes occurring on an ultra-fast time scale [Rebane & Aaviksoo, 1988].

# 8

# Holograms for displays

An obvious application of holography is in displays. With the availability of suitable photographic emulsions coated on glass plates in sizes up to 1.5 m × 1.0 m, as well as film in rolls up to 1 m wide, several striking displays have been made using holograms. Some of the problems involved and techniques used to solve them have been described by Fournier, Tribillon and Vienot [1977] as well as by Bjelkhagen [1977a]. Even larger displays are possible with projection techniques using a reflecting lenticular screen (see, for example, Okoshi [1977]).

However, conventional holograms have several drawbacks [Benton, 1975], such as the limited angle over which the image can be viewed, low image luminance, and the need to use a monochromatic source to illuminate the hologram, as well as the necessity to illuminate the subject with laser light when recording the hologram. This chapter discusses some of the techniques developed for holographic displays that have overcome these limitations (see also Benton [1980]).

## 8.1 360° holograms

Holograms that give a 360° view of the object can be recorded using either four or more plates [Jeong, Rudolf & Luckett, 1966; Chau, 1970] or a cylinder of film to surround the object [Hioki & Suzuki, 1965; Jeong, 1967; Stirn, 1975; Upatnieks & Embach, 1980].

A very simple optical system for this purpose [Jeong, 1967] is shown in fig. 8.1. In this arrangement, the object is placed at the centre of a glass cylinder which has a strip of holographic film taped to its inner surface with the emulsion side facing inwards. The laser beam, which is expanded by a microscope objective and filtered by a pinhole, is incident on the object from above. The central portion of the beam illuminates the object, while the outer portions which fall directly on the film constitute the reference beam.

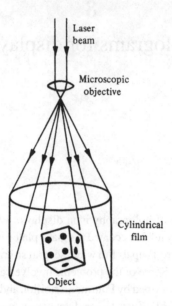

Fig. 8.1. Simple setup for making a 360° hologram [Jeong, 1967].

To view the reconstructed image, the processed film is replaced in its original position in the recording setup, after the object has been removed, and illuminated once again with the same laser.

A more efficient system is obtained if the object is mounted on a plane or convex mirror. In this case, the light reflected from the mirror constitutes the reference beam. However, the simple optical system described above has the advantage that it gives a very clean hologram, since it avoids noise due to defects in the mirror.

## 8.2 Double-sided holograms

Although cylindrical holograms reconstruct images that can be viewed over a complete circle, they have the disadvantage that they are quite bulky and need a special illumination system. A more compact alternative, which approximates to a full view of the image, is a flat hologram which, when viewed from one side, displays the front of the object and, when viewed from the other side, displays its back [George, 1970]. If the positions of the two reconstructed images are properly chosen, the original spatial perspective is preserved, and the hologram appears to contain the object.

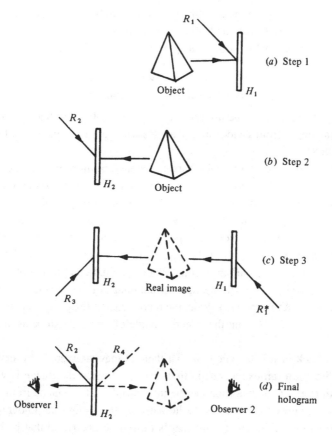

Fig. 8.2. Steps involved in the production of a double-sided reflection hologram [George, 1970].

A typical sequence of operations to produce a reflection hologram of this type is shown in fig. 8.2. In the first step, a transmission hologram is made of the object on plate 1 with the reference beam R1. Then, with the illuminated object still in position, plate 2 is exposed with the reference beam R2. After this, the object is removed and plate 1 is illuminated with the conjugate of R1, so that it forms a real image of the object in the same position. Another exposure is then made on plate 2 with another reference beam R3. Plate 2, after it is processed, is the final double-sided hologram. When it is illuminated, as shown in fig. 8.2(*d*), with two beams of white light, R2 and R4 (which is the conjugate of R3), it can be viewed from the two sides and reconstructs the front and back of the object in their correct spatial relationship to each other.

This technique has been combined with the rainbow hologram technique (see section 8.6) by Hariharan [1977b], to produce double-sided transmission holograms that reconstruct bright images with a white-light source.

## 8.3 Composite holograms

Another method that can be used to produce a flat hologram that reconstructs views of an object from a wide range of angles is to record a composite hologram [King, 1968].

This technique uses a normal off-axis reference beam setup, but the object is placed on a turntable that can be rotated about a vertical axis and a mask containing a vertical slit is placed just in front of the photographic plate so that only a narrow strip of the plate is exposed. A series of holograms is then recorded on the plate. Between successive exposures, the mask is translated along the horizontal axis by the width of the slit, while the object is rotated through a small angle in the opposite direction. When the final composite hologram is illuminated, the observer sees a single virtual image of the object. However, if he moves his head from side to side, the reconstructed image appears to rotate about a vertical axis so that the effective angle of viewing can be as much as 360°.

A drawback of this technique is that there is inevitably some hyperstereo-scopic distortion (an exaggerated effect of depth) when the image is viewed with both eyes. This distortion can be minimized if the distance from which the image is viewed is increased. Alternatively, the composite hologram can be recorded with a horizontal slit that is moved vertically, so that both eyes view the image through a single hologram strip. In this case, of course, vertical parallax is lost.

## 8.4 Holographic stereograms

It is also possible to synthesize a hologram that reconstructs an acceptable three-dimensional image from a series of two-dimensional views of an object recorded from different angles [McCrickerd & George, 1968; De Bitteto, 1969].

For this, a series of photographs of the subject is taken from equally spaced positions along a horizontal line. Contiguous, narrow, vertical strip holograms are then recorded of these photographs on a photographic plate, as shown in fig. 8.3. When this holographic stereogram is illuminated with a point source of monochromatic light, the viewer sees a three-dimensional image. The image lacks vertical parallax, but it exhibits horizontal parallax over the range of angles covered by the original photographs.

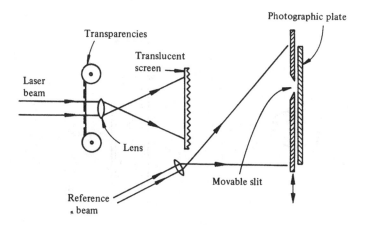

Fig. 8.3. Optical system used to record a holographic stereogram from a series of two-dimensional transparencies [De Bitteto, 1969].

The obvious advantage of this technique over direct recording is that a laser is required for the second step only. White light can be used to illuminate the subject in the first step, so that holographic stereograms can be made of quite large scenes and even of living subjects.

Variations of this technique have been described by a number of authors. In one [Redman, Wolton & Shuttleworth, 1968], successive frames are stored as image holograms covering the whole photographic plate. Angular multiplexing is achieved by rotating the plate, along with the reference beam, between exposures by an angle corresponding to the original angular movement of the camera. Haig [1973] used a modified setup in which the reference beam is incident at an angle in the vertical plane, permitting it to move with the plate and allowing the use of a white-light source to illuminate the final image hologram. However, with these techniques, the number of frames that can be multiplexed is limited by the drop in diffraction efficiency that occurs with multiple exposures (see section 4.10). A simple arrangement using focused images that requires only translation of an aperture stop has been described by Prikryl and Kvapil [1980]. A volume reflection hologram that reconstructs a synthesized three-dimensional image with white light can also be made from a holographic stereogram [Kubota & Ose, 1979*b*].

Holographic stereograms have applications in three-dimensional displays of x-ray and ultrasonic images. Techniques for this purpose have been studied by Kasahara, Kimura, Hioki, and Tanaka [1969], Groh and Kock [1970], Sopori and Chang [1971], and Kock and Tiemens [1973].

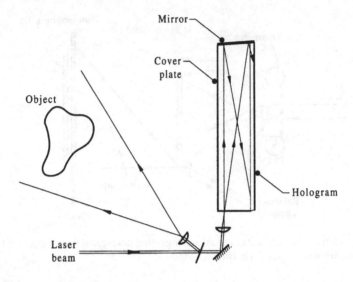

Fig. 8.4. Optical system used to record an edge-lit hologram [Upatnieks, 1992].

## 8.5 Edge-lit holograms

Because the illuminating beam has to be expanded to fill the hologram, most holographic displays require a considerable amount of space. One way to avoid this problem is to use edge illumination [Lin, 1970].

A typical arrangement that can be used to record an edge-lit hologram is shown in fig. 8.4 [Upatnieks, 1992]. In this setup, the reference beam is brought to a focus on the edge of the glass plate and expands entirely within it.

The image can be reconstructed by a laser diode placed on the edge of the cover plate at the point where the reference beam was brought to a focus. Alternatively, a compact optical system incorporating a diffractive element can be used to generate a collimated beam [Kubota, Fujioka & Kitagawa, 1992]. Edge illumination also makes it possible to recycle the illuminating beam by redirecting the light that is not diffracted to another portion of the hologram [Upatnieks, 1992].

The need for a monochromatic source to view the image can be overcome by a system in which a holographic optical element is also recorded on the same plate [Amitai *et al.*, 1993]. Light from a white-light source incident on the holographic optical element is conveyed along the plate by total reflection to the hologram, which couples it out to create the object wave. The chromatic dispersion of the holographic optical element compensates for the dispersion of the hologram.

An obvious application of edge-lit holograms is in compact portable displays. Another application is in sights, where they can be used to project an image of a reticle at infinity [Upatnieks, 1988].

## 8.6 Rainbow holograms

A major advance in the technology of holograms for displays was the development by Benton [1969, 1977] of a new type of transmission hologram capable of reconstructing a bright, sharp, monochromatic image when illuminated with white light. In this technique, part of the information content of the hologram is deliberately sacrificed to gain these advantages.

What is given up is parallax in the vertical plane; this is relatively unimportant, as pointed out earlier (see section 8.4), since depth perception depends essentially upon horizontal parallax. In return, a white-light source can be used for reconstruction, and there is a considerable gain in the brightness of the reconstructed image.

As shown schematically in fig. 8.5(*a*), the first step in making a rainbow hologram is to record a conventional transmission hologram of the object. When this hologram is illuminated, as shown in fig. 8.5(*b*), by the conjugate of the original reference wave, it generates a diffracted wave that is the conjugate of the original object wave and produces a real image of the object with unit magnification.

Vertical parallax is then eliminated from this image, as shown in fig. 8.5(*c*), by a horizontal slit placed over the primary hologram. In effect, this limits the range of angles in the vertical plane from which the real image can be viewed, without restricting the range of viewing angles in the horizontal plane.

A second hologram is now recorded of this real image, as shown in fig. 8.5(*d*). The reference beam for this hologram is a convergent beam inclined in the vertical plane, and the hologram plate is placed so that the real image formed by the first hologram is very close to it.

When the final hologram is illuminated with the conjugate of the reference beam used to make it, it forms an orthoscopic real image of the object near the hologram, as shown in fig. 8.6(*a*). In addition, it forms a real image of the slit placed across the primary hologram. All the light diffracted by the hologram passes through this slit pupil. Hence, if the observer's eyes move outside this slit pupil the image disappears. This corresponds to the almost complete elimination of vertical parallax. However, if the hologram is viewed from within the slit pupil, a very bright image of the object is seen.

With a white-light source, the slit image is dispersed in the vertical plane, as shown in fig. 8.6(*b*), to form a continuous spectrum. An observer whose eyes

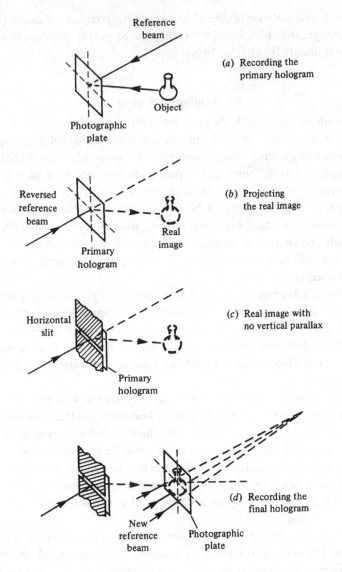

Fig. 8.5. Steps involved in the production of a rainbow hologram [Hariharan, 1983].

are positioned at any part of this spectrum then sees a sharp three-dimensional image of the object in the corresponding colour.

Figure 8.7 shows an optical system that permits both steps of the process to be carried out with a minimum of adjustments. A collimated reference beam is

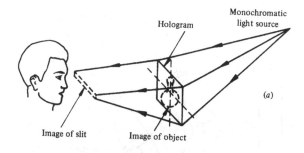

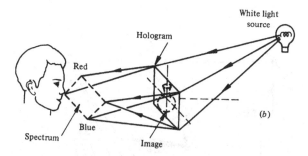

Fig. 8.6. Reconstruction of the image by a rainbow hologram (a) with monochromatic light, and (b) with white light [Hariharan, 1983].

used, as shown in fig. 8.7(a), to record the primary hologram. When the setup is modified, as shown in fig. 8.7(b), for the second stage of the process, it is necessary only to turn the primary hologram through 180° about an axis normal to the plane of the figure and replace it in the plate holder. An undistorted real image is then projected into the space in front of the primary hologram. Vertical parallax is eliminated by a limiting slit, a few millimetres wide, placed over the collimating lens with its long dimension normal to the plane of the figure (this corresponds to the horizontal in the final viewing geometry). A convergent reference beam is used to record the final hologram, which, after processing, is reversed for viewing. When it is illuminated with a divergent beam from a point source of white light, an orthoscopic image of the object is formed, and a dispersed real image of the limiting slit is projected into the viewing space.

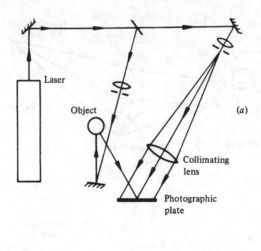

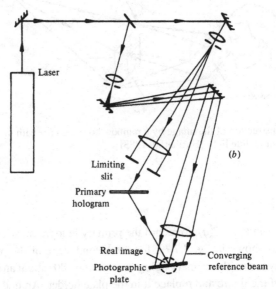

Fig. 8.7. Optical arrangement used to produce a rainbow hologram [Hariharan, 1977*b*].

### 8.6.1 Edge-lit rainbow holograms

Normally, rainbow holograms need to be illuminated with a distant point source to minimize image distortion. However, a modified recording process can be used to produce edge-lit rainbow holograms [Benton, Birner & Shirakura, 1990].

This is a three-step technique, in which the first step, as before, involves recording a conventional hologram (H1) of the object. In the second step the real image produced by H1 is projected beyond the original object plane and a second hologram (H2) is made. This hologram (H2) is then illuminated with the conjugate of the reference beam used to record it, so that the image wave retraces the path it has taken in the second step, but in the reverse direction. A hologram (H3) is now recorded in the object plane using a reference source located near the hologram.

The final rainbow hologram (H3) can be illuminated by a point source placed in the same position as the reference source used to record it, and, because each stage uses exact conjugate or direct illumination, the image produced is free from distortion.

A problem with this technique is the formation of spurious gratings resulting from Fresnel reflections at the various interfaces in the vicinity of the emulsion. This problem can be overcome by recording the final hologram in an immersion tank containing an index-matching liquid.

### 8.6.2 One-step rainbow holograms

A rainbow hologram can also be produced in a single step from real images of the object and the slit produced by an optical system [Benton, 1977; Chen & Yu, 1978]. An optical setup for this purpose is shown in fig. 8.8. In this case a lens is used to form an orthoscopic image of the object at unit magnification just in front of the hologram plate. Unit magnification must be used if a three-dimensional image is to be free from distortion in depth, since the longitudinal magnification is the square of the transverse magnification (see section 3.2.3). A narrow slit is placed between the object and the focal plane of the lens, so that it is imaged into the viewing space at a suitable distance from the hologram. A diverging reference beam is employed in this arrangement, corresponding to that used for reconstruction.

The only disadvantage of this technique is that the field of view is limited by the aperture of the lens. While various methods have been suggested to minimize this problem [Tamura, 1978*b*; Benton, Mingace & Walter, 1979], a simple solution is to use a large concave spherical mirror [Hariharan, Hegedus & Steel, 1979], as described in section 9.5.

### 8.6.3 Image blur in the rainbow hologram

The image formed by a rainbow hologram is free from speckle, because it is reconstructed with incoherent light, and usually appears quite sharp to the

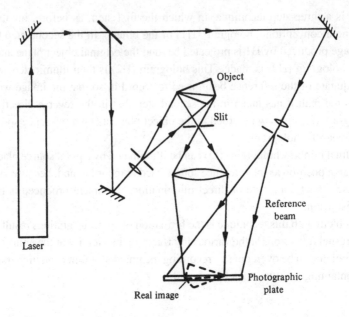

Fig. 8.8. Optical system used to record a rainbow hologram in a single step [Yu, Tai & Chen, 1980].

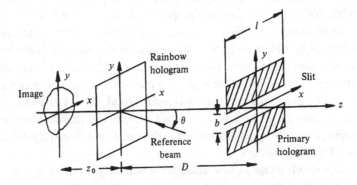

Fig. 8.9. Analysis of image blur in a rainbow hologram [Wyant, 1977].

naked eye. However, there is always some image blur. The extent of this blur depends on the recording geometry and the size of the source used to illuminate the hologram and has been analysed by Wyant [1977] and by Tamura [1978a].

The primary cause of image blur is the finite wavelength spread in the image. To calculate the image blur due to this cause, consider a rainbow hologram made with the setup shown in fig. 8.9. If the angles made with the axis by the rays from the primary hologram to the final rainbow hologram are small compared to the reference beam angle $\theta$, the wavelength spread observed when the rainbow hologram is illuminated with white light is

$$\Delta\lambda = (\lambda/\sin\theta)[(b/D) + (a/D)], \tag{8.1}$$

where $\lambda$ is the mean wavelength of the reconstructed image, $b$ is the width of the slit, $a$ is the diameter of the pupil of the eye, and $D$ is the distance from the primary hologram to the final rainbow hologram. The two terms, $(b/D)$ and $(a/D)$, in (8.1) correspond to the angular subtense of the slit and the eye pupil measured from the hologram during recording and reconstruction, respectively. Typically, if $\lambda = 500$ nm, $\theta = 45°$, $D = 300$ mm, and $a = b = 3$ mm, then $\Delta\lambda = 14$ nm.

The image blur $\Delta y_{\Delta\lambda}$ due to this wavelength spread is then given by the relation

$$\Delta y_{\Delta\lambda} = z_0(\Delta\lambda/\lambda)\sin\theta,$$
$$\approx z_0(a + b)/D, \tag{8.2}$$

where $z_0$ is the distance of the image from the hologram. If the image is formed at a distance of 5 cm from a hologram with the parameters listed earlier, the image blur is approximately 1 mm. This corresponds to an angular blur of about 3 mrad at the eye pupil, which is acceptable.

Again, when a light source of finite size is used to illuminate the final hologram, the image exhibits an angular blur equal to the angular spread $\psi_s$ of the source, as viewed from the hologram. The resultant image blur is

$$\Delta y_s = z_0\psi_s. \tag{8.3}$$

If this is not to exceed the blur due to the wavelength spread, the size of the source should satisfy the condition

$$\psi_s < (a + b)/D. \tag{8.4}$$

The final source of image blur is diffraction at the slit; this is noticeable only when the width of the slit is very small. If we assume that the slit is imaged on the eye pupil and its width $b \ll D$, the image blur due to diffraction is approximately

$$\Delta y_b = 2\lambda(z_0 + D)/b. \tag{8.5}$$

For the hologram parameters listed earlier, the image blur due to wavelength spread is greater than that due to diffraction, as long as the slit width $b$ is more than 1 mm.

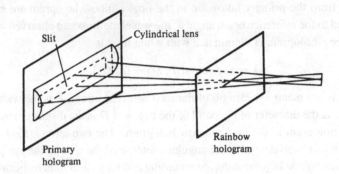

Fig. 8.10. Optical system used to produce a rainbow hologram with reduced image blur [Leith & Chen, 1978].

### 8.6.4 Rainbow holograms with reduced image blur

The image blur due to the finite spread of wavelengths in the image is mainly in the vertical plane and is a minimum when the image is formed in the hologram plane. A simple modification to the optical arrangement gives reduced image blur for images formed at an appreciable distance from the hologram [Leith & Chen, 1978].

In the modified arrangement, a cylindrical lens is placed over the slit aperture in front of the primary hologram, as shown in fig. 8.10, so that the image is brought to a focus in the vertical plane at the plate used to record the final hologram, without affecting the focus in the horizontal plane. While the reconstructed image has considerable astigmatism, it is sharp and free from colour blur. This technique does not, of course, increase the depth over which a sharp image can be obtained, but it permits a distant plane to be sharply imaged at the expense of a sacrifice in resolution for objects normally imaged in the plane of the hologram.

This technique has been extended to the one-step hologram process by Chen [1979] and by Yu, Ruterbusch, and Zhuang [1980]. Apart from permitting the use of a much wider slit, it also produces a higher object-beam intensity while recording the final rainbow hologram.

### 8.6.5 Image luminance in a rainbow hologram

It has been shown [Hariharan, 1978] that with a recording system of the type shown in fig. 8.9, the average luminance of the image reconstructed at a wavelength $\lambda$ by a rainbow hologram is given by the expression

$$L_v = [(D + z_0)/D][(A_H/A_I)(1/\psi)\Lambda\varepsilon E_\lambda K_\lambda], \tag{8.6}$$

where $A_H$ is the area of the hologram that diffracts light, $A_I$ is the area of the image, $\psi = l/(D + z_0)$ is the range of viewing angles in the horizontal plane, $\Lambda$ is the average spacing of the fringes in the hologram, $\varepsilon$ is the diffraction efficiency of the hologram, $E_\lambda$ is the spectral irradiance at the hologram due to the beam illuminating it, and $K_\lambda$ is the spectral luminous efficacy of the radiation. If the image is close to the hologram, $z_0$ is negligible and $A_H \approx A_I$, since only the area of the hologram corresponding to the image diffracts light. The image luminance then becomes

$$L_v = \Lambda\varepsilon E_\lambda K_\lambda/\psi. \tag{8.7}$$

To maximize its luminance, the image should be located at the maximum distance from the hologram consistent with the permissible image blur, and the interbeam angle should be made as small as possible, consistent with the required field of view in the vertical plane.

## 8.7 White-light holographic stereograms

A development of these techniques was the production by Cross of cylindrical holographic stereograms which, when illuminated with white light, reconstructed an almost monochromatic image (see Benton [1975]).

To produce such a holographic stereogram, the subject is placed on a slowly rotating turntable and a movie camera is used to make a record of a 120° or 360° rotation. Typically, three movie frames are exposed for each degree of rotation, so that the final movie sequence may contain up to 1080 frames.

The optical system used to produce a holographic stereogram from this movie film is shown schematically in fig. 8.11. In this system, each frame is imaged in the vertical plane on to the hologram film. However, in the horizontal plane, the cylindrical lens brings all the rays leaving the projector to a line focus. A contiguous sequence of vertical strip holograms is then recorded of successive movie frames, covering the full range of views of the original subject, with a reference beam incident from below.

When the processed film is formed into a cylinder and illuminated with white light, a monochromatic image is seen that changes colour, as with any rainbow hologram, when the observer moves his head up or down. Due to the very large number of frames recorded, a modest amount of subject movement can also be accommodated without destroying the stereoscopic effect.

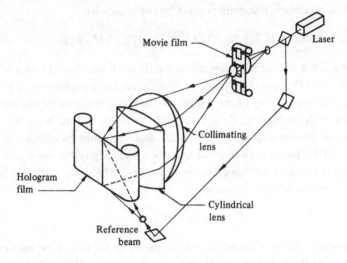

Fig. 8.11. Optical system used to produce a white-light holographic stereogram [Huff & Fusek, 1980].

### 8.7.1 *Large white-light holographic stereograms*

Large white-light stereograms usually have a flat format and, because of the difficulties involved in scaling up the system shown in fig. 8.11, are made by a two-step process [Newswanger & Outwater, 1985]. The first step is to record a range of perspective views of the scene, either by means of a camera moved sideways or by photographing a series of computer-generated images. These perspective views are then projected successively on a diffusing screen using a laser source, and a series of vertical strip holograms are recorded on a film (H1) at locations along the horizontal axis corresponding to the position of the camera for that particular view. In the second step, a narrow horizontal strip of H1 is illuminated with the conjugate of the reference beam used to record it, so that it projects a composite real image in the plane originally occupied by the diffusing screen. The final holographic stereogram (H2) is then recorded on a film placed in this plane.

### 8.7.2 *Image distortion in holographic stereograms*

The images reconstructed by holographic stereograms can exhibit quite serious distortion if a strict correspondence is not maintained between the recording geometry and the final viewing geometry. The anamorphic distortion observed with a plane holographic stereogram has been analysed by Glaser [1973] and

Glaser and Friesem [1977], who have shown that it vanishes when

$$b(a - d)/a(b - d) = 1, \qquad (8.8)$$

where $a$, $b$, and $d$ are the distances from the viewing plane to the input trans-parency, the apparent location of the image, and the hologram, respectively.

The case of a cylindrical holographic stereogram has been studied by Honda, Okada, and Tsujiuchi [1981]. In this case the distortion is a minimum when

$$d = 2r/m, \qquad (8.9)$$

where $d$ is the distance from the camera to the centre of rotation of the subject, $r$ is the radius of the cylindrical hologram, and $m$ is the image magnification ($m < 1$). Some optical techniques that can be used to compensate for distortion in the horizontal plane, yielding a wider field, have been outlined by Huff and Fusek [1981].

With a white-light stereogram, an additional cause of distortion is the fact that unless the hologram is viewed from a position corresponding to the focus of the field lens in the recording setup, different parts of the image are reconstructed in different colours. This distortion is particularly noticeable when the observer is too close to the hologram but is not so objectionable when he is farther away [Okada, Honda & Tsujiuchi, 1981].

Okada, Honda, and Tsujiuchi [1982] have also calculated the image blur in a white-light stereogram when a source of finite size is used to illuminate it. With a vertical line source, the image blur near the axis, as in a rainbow hologram, is influenced only by the width of the source; however, near the edges, the blur increases rapidly with the length of the source.

### 8.7.3  Holographic stereograms with reduced distortion

The basic reason for distortion with holographic stereograms is that vertical and horizontal details of the scene are displayed differently. Changes in the distance of the viewer from the hologram produce a natural change in perspective in the horizontal plane but have no effect on the perspective in the vertical plane.

The need to minimize distortion is a problem in the production of large holographic stereograms that may be viewed from a distance of several metres. In this case, the separation of the master hologram and the projection screen in the first step, and the distance between the master hologram and the final hologram in the second step, are dictated by the viewing distance, while the size of the master hologram defines the size of the viewing zone. A way to overcome these limitations is by predistortion of the individual images.

Optical predistortion is simple but limited in its applicability. A more powerful and flexible method is the use of computer-based image predistortion [Halle, Benton, Klug & Underkoffler, 1991]. In this case, the predistortion can be performed using, as the input, a sequence of views from a simple computer graphics camera, or a CAD database, and generating, as the output, another sequence of views with the desired distortion. It should be noted that while the input sequence has all the perspective views required, the data needed for each output image may have to be assembled from many input views.

Distortion-compensated stereograms have several advantages. Apart from allowing a more compact recording setup and an arbitrary choice of the viewing distance, the viewing zone can be made significantly wider. In addition, the final hologram can be illuminated with a beam having the same geometry as the reference beam used to record it, making possible the production of edge-lit holographic stereograms [Farmer, Benton & Klug, 1991].

## 8.8 Holographic movies

A moving three-dimensional image can be produced if a sequence of holograms of a scene is presented to an observer. Because the reconstructed image is not affected by movement of the hologram, provided an approximation to a Fourier transform setup is used (see sections 2.3 and 2.4), the film on which the holograms are recorded can be moved continuously.

In early attempts at holographic cinematography (see, for example, De Bitteto [1970]), the reconstructed image was viewed through the film by one person at a time. While this limitation is acceptable for technical studies such as microscopic studies of marine plankton [Heflinger, Stewart & Booth, 1978], it is a serious drawback in the entertainment field. Komar [1977] therefore used a lens with a diameter of 200 mm to record a series of image holograms of the scene on 70 mm film. After processing, the reconstructed image was projected with an identical lens on to a special holographic screen, equivalent to several superimposed concave mirrors. Each of these holographic concave mirrors then formed a real image of the projection lens in front of a spectator, so that when he looked through this pupil he saw a full-size three-dimensional image.

Further progress in the field of holographic movies has been reviewed by Smigielski, Fagot, and Albe [1985] and by Komar and Serov [1989]. More recently, two artistic movies using holographic stereograms (*Masks*, 4 minutes, and *The Dream*, 8 minutes) have been produced by Alexander [see Lucie-Smith, 1992].

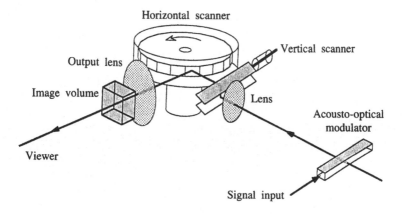

Fig. 8.12. Schematic of a holographic video display system [St.-Hilaire *et al.*, 1990].

## 8.9 Holographic video imaging

A real-time holographic display of three-dimensional information is now possible [St.-Hilaire *et al.*, 1990; St. Hilaire, Benton & Lucente, 1992].

The problem of the limited information bandwidth available, even with current technology, has been overcome by several steps to reduce the information content of the hologram to the minimum acceptable figure. Data on a synthetic three-dimensional object are transferred to a computer containing 16,000 microprocessors in a massively parallel architecture. The video signal transmitted from the frame buffer is used to drive the acousto-optic modulator (AOM) in the display system shown in fig. 8.12.

The expanded laser beam emerges from this AOM with a phase modulation across its width that is proportional to the input signal representing a section of one horizontal line of the hologram. A spinning polygonal mirror and a galvanometer scanner are used to multiplex these sections in the horizontal and vertical directions, respectively, to build up the holographic image, which appears to float a few centimetres in front of the output lens. Images up to 130 mm × 170 mm and 200 mm deep have been produced by this system.

Multicolour images have also been generated with such a system by using a three-channel AOM and illuminating the three channels with red (633 nm), green (532 nm), and blue (442 nm) lasers [St.-Hilaire, Benton, Lucente & Hubel, 1992]. A holographic grating after the AOM brings the three beams into alignment. The images exhibit a large gamut of colours and good colour registration.

# 9

# Colour holography

The fact that a multicolour image can be produced by a hologram recorded with three suitably chosen wavelengths was first pointed out by Leith and Upatnieks [1964].

The resulting recording can be considered as made up of three incoherently superposed holograms. When it is illuminated once again with the three wavelengths used to make it, each of these wavelengths is diffracted by the hologram recorded with it to give a reconstructed image in the corresponding colour. The superposition of these three images yields a multicolour reconstruction.

However, while multicolour holography was demonstrated at quite an early stage, its further development was held up initially by several practical problems. These problems, as well as later advances that have made multicolour holography practical, are described in this chapter (see also the review by Hariharan [1983]).

## 9.1 Light sources for colour holography

The most commonly used lasers for colour holography are the He-Ne laser ($\lambda$ = 633 nm) and the $Ar^+$ laser, which has two strong output lines ($\lambda$ = 514 nm and 488 nm; see Table 5.1). The range of colours that can be reconstructed with these three wavelengths as primaries can be determined by means of the C.I.E. chromaticity diagram [Optical Society of America, 1953]. In this diagram, as shown in fig. 9.1, points representing monochromatic light of different wavelengths constitute the horseshoe-shaped curve known as the spectrum locus; all other colours lie within this boundary. New colours obtained by mixing light of two wavelengths, such as 633 nm and 514 nm, lie on the straight line AB joining these primaries. If light with a wavelength of 488 nm is also used, any colour within the triangle ABC can be obtained.

A wider range of hues can be obtained if other laser lines are used, permitting

145

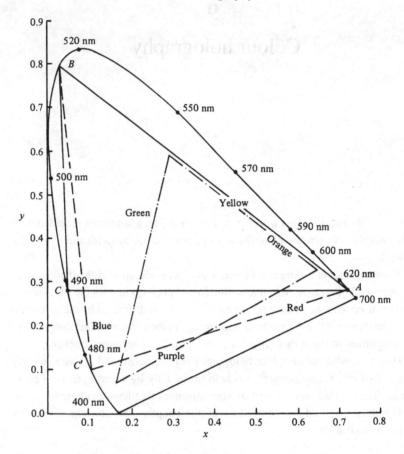

Fig. 9.1. C.I.E. chromaticity diagram. The triangle ABC shows the range of hues that can be produced by a hologram illuminated with primary wavelengths of 633 nm, 514 nm, and 488 nm, while the broken lines show the extended range possible with a blue primary at 477 nm. The chain lines enclose the range of hues that can be reproduced by a typical colour-television display [Hariharan, 1983].

a better choice of primaries. One of these is the He-Cd laser line ($\lambda = 422$ nm), which is very attractive as a blue primary but involves the use of one more laser. A simpler alternative is the blue line of the $Ar^+$ laser at a wavelength of 477 nm, which at some sacrifice of power can give a significant improvement in colour rendering, as shown by the broken lines in fig. 9.1.

The $Kr^+$ laser line at 647 nm is a useful alternative to the red line of the He-Ne laser for recording large holograms, because the power available is much higher, and single-mode output can be obtained with an etalon. The use of the

$Kr^+$ laser line at 521 nm, or the output from a frequency-doubled Nd:YAG laser ($\lambda = 532$nm), as the green primary has been found to give much better yellow images.

A detailed analysis of colour reproduction by volume reflection holograms has been made by Hubel and Solymar [1991]. This study involved recording holograms of a set of eight Munsell colour chips using eight combinations of recording wavelengths. Good agreement was found between the theoretically predicted and experimentally measured image colours. The optimum recording wavelengths were found to be 460, 530, and 615 nm.

## 9.2 The cross-talk problem

A problem in multicolour holography is that each hologram diffracts, in addition to light of the wavelength used to record it, the other two wavelengths as well. As a result, a total of nine primary images and nine conjugate images are produced. Three of these give rise to a full-colour reconstructed image at the position originally occupied by the object. The remaining images resulting from light of one wavelength diffracted by a component hologram recorded with another wavelength are formed in other positions and overlap with and degrade the multicoloured image.

Several methods have been tried to eliminate these cross-talk images, including spatial-frequency multiplexing [Mandel, 1965, Marom, 1967], spatial multiplexing or coded reference waves [Collier & Pennington, 1967], and division of the aperture field [Lessard, Som & Boivin, 1973]. However, all these methods suffer from drawbacks such as a restricted image field, a reduction in resolution, or a decrease in the signal-to-noise ratio. In addition, they need multiple laser wavelengths (or equivalent monochromatic light sources) to illuminate the hologram.

## 9.3 Volume holograms

The first methods to eliminate cross talk that did not involve such penalties were based on the use of volume holograms.

A hologram recorded with several wavelengths in a thick medium contains a set of regularly spaced fringe planes for each wavelength. When this hologram is illuminated once again with the original multiwavelength reference beam, each wavelength is diffracted with maximum efficiency by the set of fringe planes created originally by it, producing a multicoloured image. However, the cross-talk images are severely attenuated since they do not satisfy the Bragg condition.

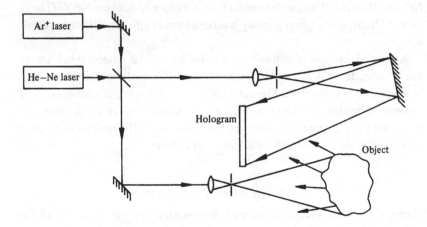

Fig. 9.2. Setup used to record a multicolour hologram of a diffusely reflecting object in a thick recording medium [Friesem & Federowicz, 1966].

This principle was first applied by Pennington and Lin [1965] to produce a two-colour hologram of a transparency, and subsequently extended by Friesem and Fedorowicz [1966, 1967] to three-colour imaging of diffusely reflecting objects. The optical setup used by them is shown schematically in fig. 9.2. In it, blue and green light ($\lambda = 488$ nm and $514$ nm) from an $Ar^+$ laser was mixed with red light ($\lambda = 633$ nm) from a He-Ne laser to produce two beams containing light of all three wavelengths. One beam was used to illuminate the object while the other was used as a reference beam, and the resulting hologram was recorded in a thick photographic emulsion. When this hologram was illuminated once again with a similar multicolour beam at the appropriate angle, a multicolour reconstructed image with negligible cross talk was obtained.

## 9.4 Volume reflection holograms

As we have seen in Chapter 4, the highest wavelength selectivity is obtained with volume reflection holograms. Such holograms can even reconstruct a monochromatic image when illuminated with white light. Their use for multi-colour imaging followed directly [Lin, Pennington, Stroke & Labeyrie, 1966; Upatnieks, Marks & Fedorowicz, 1966; Stroke & Zech, 1966].

To produce a multicolour image, a volume reflection hologram is recorded with three wavelengths, so that one set of fringe planes is produced for each wavelength. When such a hologram is illuminated with white light, each set of fringe planes, because of its high wavelength selectivity, diffracts a narrow band of wavelengths centred on the original laser wavelength used to record it. As a result, a multicolour reconstructed image, free from cross talk, is obtained.

However, early reflection holograms recorded on photographic emulsions and processed in the conventional manner had several drawbacks. The most serious of these was their low diffraction efficiency, which was aggravated by the fact that when more than one hologram was recorded in the same emulsion layer, the available dynamic range was shared between the recordings. As a result, the diffraction efficiency of each recording was reduced by a factor approximately equal to the square of the number of holograms [Collier, Burckhardt & Lin, 1971] (see section 4.10).

Another problem was the reduction in the thickness of a photographic emulsion layer that occurred during processing caused by the removal of the unexposed silver halide. This resulted in a reduction in the fringe spacing and a shift in the colour of the reconstructed image towards shorter wavelengths. This shift could be eliminated by soaking the emulsion in an aqueous solution of triethanolamine before drying [Lin & Lo Bianco, 1967], but this technique was not entirely satisfactory.

Several technical improvements have made it possible to obtain much better results. Thus, low-noise bleaches can be used to produce reflection holograms with improved diffraction efficiency [Hariharan, 1972; Phillips, Ward, Cullen & Porter, 1980; Cooke & Ward, 1984], and the use of relatively thin emulsion layers ($\approx 6\mu$m thick), which diffract a wider spectral bandwidth, gives as high an image luminance as a thicker layer, with the advantage of lower scattering [Hariharan, 1972, 1979b].

Triethanolamine cannot be used to correct the shrinkage in thickness of such bleached holograms, since they then darken rapidly when exposed to light, due to the formation of photolytic silver. A solution of D(-) sorbitol can be used instead [Hariharan, 1980a]; alternatively, a tanning developer can be used to eliminate this shrinkage [Joly & Vanhorebeek, 1980; Hariharan & Chidley, 1989].

Since bleached reflection holograms are completely transparent at wavelengths outside the relatively narrow band that is diffracted, it is also possible to record the three component holograms for different primary wavelengths on two separate plates and superimpose them to make up the final multicolour hologram. This technique permits the use of different emulsions to record the component holograms, one with optimum characteristics for the red, and the other with optimum characteristics for the green and blue. In addition, an improvement in image luminance by a factor of 2 or more is obtained if the three component holograms are divided between two plates in this manner, instead of being recorded in the same emulsion layer.

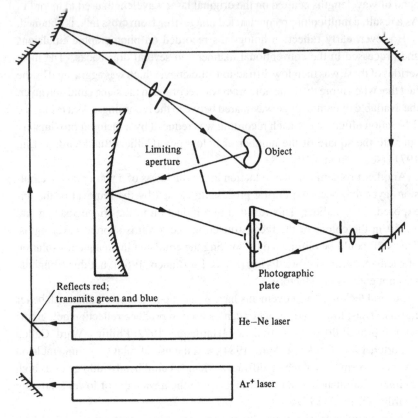

Fig. 9.3. Optical system used for recording multicolour reflection holograms with increased image luminance [Hariharan, 1980*a*].

A further improvement in image luminance can be obtained if the diffracted light from the hologram is concentrated into a smaller solid angle. For this, a hologram is recorded, not of the original object, but of a real image of the object projected either by another hologram, or by an optical system whose effective aperture is limited by a suitably shaped stop as shown in fig. 9.3. This is the same principle that has been exploited in the rainbow hologram. The gain in image luminance is proportional to the reciprocal of the available solid angle of viewing [Hariharan, 1978]. Normally, a gain in image luminance by a factor of 3 or 4 can be obtained, without any loss in convenience, if the range of viewing angles in the vertical plane is restricted to about 15°.

Typically, the red component hologram is recorded with a He-Ne laser ($\lambda =$ 633 nm) and 8E75 plates, while an $Ar^+$ laser is used with 8E56 plates for the

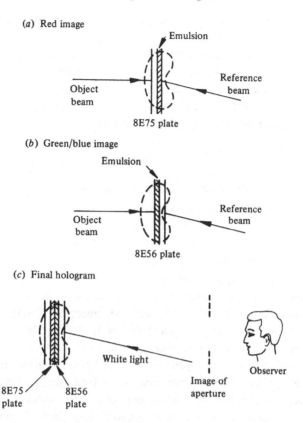

Fig. 9.4. Schematic showing how a multicolour reflection hologram is built up from exposures on two photographic plates [Hariharan, 1980a].

green ($\lambda = 514$ nm) and blue ($\lambda = 488$ nm) exposures. As shown in fig. 9.4, the 8E75 plate is exposed with the emulsion side towards the reference beam, while the 8E56 plate is exposed with the emulsion side facing the mirror.

After drying, the plates are assembled with the emulsion layers in contact and the reconstructed images are viewed, with the 8E56 plate in front and the 8E75 plate behind, to equalize the diffraction efficiencies of the holograms.

Even better results can be obtained by using other recording materials. One approach has been to record the green and blue components in a dichromated gelatin layer sensitized with methylene blue [Kubota & Ose, 1979c; Kubota, 1986]; another has been to record the three holograms in separate photopolymer layers [Hubel & Klug, 1992]. In the latter case, because of the low red sensitivity of the photopolymer (DuPont Holographic Recording Film), three primary holograms are recorded at appropriate laser wavelengths on silver halide emulsions. The emulsion used to record the red component is pre-swollen so that it reconstructs at a convenient, shorter wavelength. A contact copy made from the red master hologram (see section 11.4) using this shorter wavelength is then swollen during processing, by baking in contact with a sheet of DuPont Color Tuning Film, to reconstruct a red image. Finally, this contact copy is laminated to contact copies made from the other two master holograms using green and blue light.

### 9.5 Multicolour rainbow holograms

A completely different approach was opened up by the extension of the rainbow hologram technique to three-colour recording [Hariharan, Steel & Hegedus, 1977; Tamura, 1977; Suzuki, Saito & Matsuoka, 1978]. This made it possible to produce holograms that reconstruct very bright multicolour images when illuminated with a white-light source.

To produce a multicolour rainbow hologram, three primary holograms (colour separations) are made from the object with red, green, and blue laser light using an optical system similar to that shown in fig. 8.7. In the second stage, these primary holograms are used with the same laser sources to make a single hologram consisting of three superimposed recordings. When this multiplexed hologram is illuminated with a white-light source, it reconstructs three superimposed images of the object. In addition, three spectra are formed in the viewing space. These spectra are, as before, dispersed real images of the limiting slit. However, these spectra are displaced vertically with respect to each other, as shown in fig. 9.5, so that, in effect, each component hologram reconstructs an image of the limiting slit in its original position in the colour with which it was made. Accordingly, an observer viewing the hologram from the point where the spectra overlap sees three superimposed images of the object reconstructed in the colours with which the primary holograms were made.

Multicolour rainbow holograms can also be produced in a single step [Chen, Tai & Yu, 1978; Hariharan, Hegedus & Steel, 1979]. A typical optical system for this purpose, using a concave mirror, is shown in fig. 9.6. The object, turned sideways, was placed on one side of the axis of the mirror so that its image was formed on the other side, at the same distance from the mirror. A vertical slit was placed between the object and the mirror, at such a distance from the mirror that a magnified image of the slit was formed in the viewing space at a distance of about 1 m from the hologram.

Although in principle the three superimposed holograms making up the final multicolour rainbow hologram can be recorded on a single plate, there are

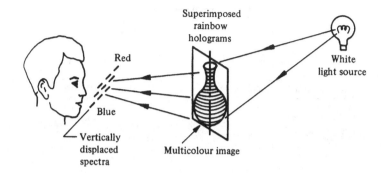

Fig. 9.5. Reconstruction of a multicolour image by superimposed rainbow holograms [Hariharan, 1983].

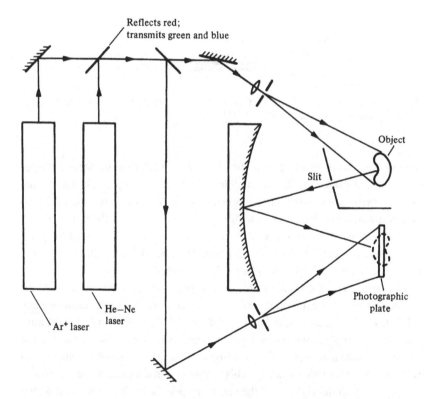

Fig. 9.6. Layout of the optical system used to produce multicolour rainbow holograms in a single step with a concave mirror [Hariharan, Hegedus & Steel, 1979].

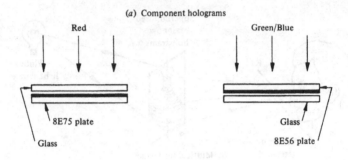

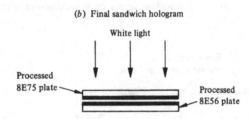

Fig. 9.7. Schematic of the sandwich technique used to make multicolour rainbow holograms [Hariharan, Steel & Hegedus, 1977].

several advantages in using a sandwich technique [Hariharan, Steel & Hegedus, 1977]. Besides making it possible to use different types of photographic plates, whose characteristics are optimized for the different wavelengths, it also makes it much easier to match the diffraction efficiencies of the three individual holograms. In addition, it also gives much brighter images, since, with bleached holograms, the loss in diffraction efficiency due to multiplexing three holograms on a single plate can be partially avoided.

The simplest method is to use two plates. In this case, as shown in fig. 9.7(*a*) on the left, the red component hologram is recorded on a red-sensitive plate (8E75), which is loaded into the plate holder with the emulsion facing forward and with a clear glass plate of the same thickness in front of it. Next, the blue and green component holograms are recorded on an orthochromatic plate (8E56) without any antihalation coating. This is loaded into the plate holder, as shown in fig. 9.7(*a*) on the right, with the emulsion side facing backward and with a clear glass plate of the same thickness behind it. Finally, as shown in fig. 9.7(*b*), the two processed plates are cemented together, with their emulsions in contact, to form the final multicolour hologram. Registration of the two plates need be done only to an accuracy determined by the residual image blur and is automatic if the plate holder is used as an assembly jig. Typically, this technique gives an improvement in diffraction efficiency by a factor of 2 or more over that obtained if all the holograms are recorded in the same emulsion.

Multicolour rainbow holograms give bright images even when illuminated with an ordinary tungsten lamp. In addition, the images exhibit high colour saturation and are free from cross talk. Problems with emulsion shrinkage are eliminated, since volume effects are not involved, and shrinkage primarily affects only the thickness of the emulsion. As with any rainbow hologram, the colours of the image change with the viewing angle in the vertical plane. This change can be utilized effectively in some types of displays, but, where necessary, it can be kept within acceptable limits by optimization of the length of the spectra projected into the viewing space and the use of baffles to define the available range of viewing angles in the vertical plane.

## 9.6 Pseudocolour images

The fact that the colour information in a hologram is recorded only in the form of different carrier fringe frequencies suggests the possibility of using a single laser wavelength and generating these carrier fringe frequencies by some other means.

One method is to record three superimposed rainbow holograms with different reference beam angles [Tamura, 1978a]. Alternatively, the position of the limiting slit can be changed between exposures [Vlasov, Ryabova & Semenov, 1977; Yan-Song, Yu-Tang & Bi-Zhen, 1978].

A problem with pseudocolour rainbow holograms is that the images reconstructed in a different colour from that used to record the hologram are displaced with respect to an image of the same colour. The displacement in the vertical plane is

$$\Delta y = z_0 (2\Delta\lambda/\lambda) \tan^3 \theta, \qquad (9.1)$$

while the longitudinal displacement is

$$\Delta z = z_0 (2\Delta\lambda/\lambda), \qquad (9.2)$$

where $z_0$ is the distance of the image from the hologram, $\lambda$ is the wavelength with which the hologram is recorded, and $(\lambda + \Delta\lambda)$ is the wavelength at which the image is reconstructed. These displacements can be brought within acceptable limits if the image is formed sufficiently close to the hologram.

With volume reflection holograms, changing the angle between the reference and object beams has little effect on the colour of the reconstructed image. However, its colour is affected by changes in the thickness of the recording medium, and these changes can be controlled and used to produce pseudocolour images [Hariharan, 1980c].

Typically, a He-Ne laser can be used to produce a pseudocolour reflection hologram. In this case, the red component hologram is recorded first on a plate exposed with the emulsion side towards the reference beam. This plate is bleached and processed to eliminate emulsion shrinkage. The green and blue component holograms are then recorded on another plate exposed with the emulsion side towards the object beam. The green component hologram is exposed first with the emulsion in its normal condition. After this, the emulsion is soaked in a 3 per cent solution of triethanolamine and dried in darkness. The blue component hologram is then exposed on the swollen emulsion. Normal bleach processing eliminates the triethanolamine and produces the usual shrinkage. Accordingly, the first exposure yields a green image, while the second produces one at an even shorter wavelength, that is to say, a blue image.

After drying, the plates are cemented together with the emulsion layers in contact. The images are viewed with the hologram reconstructing the green and blue images in front and the hologram reconstructing the red image behind.

Very high quality pseudocolour images can be produced from contact copies (see section 11.4) of three master holograms made on DuPont photopolymer Holographic Recording Film with a short wavelength (476 nm). Two of the copies are swollen during processing, by baking in contact with DuPont Color Tuning Film, so that they reconstruct green and red images, before the three films are sandwiched together [Hubel & Klug, 1992].

In this case also, the images reconstructed at wavelengths differing from that used to record the holograms undergo shifts that depend on $\mu$, the ratio of these wavelengths [Hariharan, 1976a, 1980c]. To avoid lateral misregistration, the lateral magnification $M_{lat}$ must be independent of $\mu$; this is possible if the hologram is illuminated with a collimated beam. While it is not possible to eliminate longitudinal misregistration and longitudinal distortion simultaneously, acceptable results can be obtained if the image is centred on the hologram plane.

## 9.7 Achromatic images

Holograms that reconstruct a black-and-white image when illuminated with white light have the advantage that they can produce very bright images, since they use the entire output of the source.

As pointed out in section 3.5.3, a very nearly achromatic image of an object of limited depth can be produced by an image hologram. However, to produce an achromatic image of an object with appreciable depth, some technique of dispersion compensation must be used.

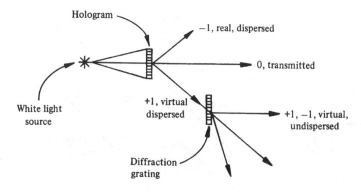

Fig. 9.8. Production of an achromatic image by dispersion compensation using a diffraction grating [De Bitteto, 1966].

One of the earliest methods [De Bitteto, 1966] used, as shown in fig. 9.8, a plane diffraction grating, with a line spacing equal to the average fringe spacing in the hologram, to provide equal but opposite angular dispersion. To avoid directly transmitted light spilling into the field of view, a light shield consisting of a set of baffles, like a venetian blind, was placed between the hologram and the grating [Burckhardt, 1966b].

Brighter reconstructed images can be obtained with techniques using rainbow holograms. The simplest method is merely to use a vertical line source to illuminate the hologram [Benton, 1969, 1977]. This results in a series of overlapping spectra projected into the viewing space, so that a near-white image is obtained.

However, there is still some residual colour blur because the images reconstructed by different wavelengths have different magnifications and are formed at different distances from the hologram. This is also true of the slit images constituting the spectrum projected into the viewing space. To produce a truly achromatic image, the red, green, and blue images must coincide; this means that the overlapping spectra must lie along the same line.

To obtain this result, Benton [1978] used a multiply-exposed holographic lens, which produced the effect of a series of point sources of light located at suitable angles and distances, in combination with a narrow strip of the first hologram, to make a second hologram. This hologram was then illuminated with monochromatic light, and a third hologram was made of the image reconstructed by it. When the final hologram was illuminated with white light, it reconstructed a set of overlapping spectra in the viewing space which coincided at appropriate wavelengths, so that an achromatic image was obtained over a wide range of viewing angles.

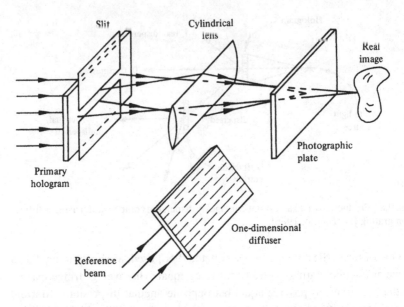

Fig. 9.9.  Setup using a one-dimensional diffuser to produce a hologram that reconstructs
an achromatic image [Leith, Chen & Roth, 1978].

Another technique described by Leith, Chen, and Roth [1978] used a one-
dimensional diffuser to generate multiple reference beams. As shown in fig. 9.9,
the primary hologram was masked by a horizontal slit and formed a real image
at some distance from the hologram plate. However, the introduction of a
cylindrical lens caused this image to be focused in the vertical plane at the
hologram itself. The phase plate in the reference beam diffused it in the vertical
direction and transmitted it, without scatter, in the horizontal direction. As a
result, light of any colour incident on the hologram was scattered through a
range of angles in the vertical plane determined by the extent of the diffuser.
This had little effect on the image resolution, since the image was focused in
the vertical plane at the hologram, but it washed out the colour effects normally
obtained with a rainbow hologram.

## 9.8  Achromatic and pseudocolour stereograms

Holographic stereograms have the advantage that three-dimensional displays
can be produced from a wide range of inputs, such as photographic, video, and
computer-generated images. Transmission holographic stereograms recorded
with a single wavelength can be made to produce achromatic, black-and-white

images by using a two-step process which compensates for spectral dispersion during viewing with white light [Benton, 1983].

With volume reflection holograms, their high Bragg selectivity makes it possible to reproduce several independent images in different colours, affording the capability of full-colour imaging. In addition, with materials that can be processed to yield a wide-band reconstruction, such as dichromated gelatin or photopolymers, achromatic images with extended depth can be produced [Benton, 1988].

The first step is to record a range of perspective views of the scene, either by means of a camera moved sideways or by photographing a series of computer-generated images. These perspective views are then projected successively, using a laser source, on a diffusing screen as shown in fig. 9.10(*a*), and a series of strip holograms are recorded on a plate (H1) at locations along the horizontal axis corresponding to the position of the camera for that particular view. The plate used to record H1 is set at the "achromatic angle" $\alpha$ with the axis defined by the relation

$$\alpha = \tan^{-1}(\sin\theta_{\text{ref}}) \tag{9.3}$$

where $\theta_{\text{ref}}$ is the reference beam angle for the final hologram.

In the second step, the master hologram (H1) is illuminated with the conjugate of the reference beam used to record it, as shown in fig. 9.10(*b*), so that it projects a composite real image in the plane originally occupied by the diffusing screen, and the final holographic stereogram (H2) is recorded using a reference beam incident from below.

When the final holographic stereogram (H2) is illuminated from above, as shown in fig. 9.10(*c*), with a beam of white light that retraces the path of the reference beam, the images of H1 formed by different wavelengths will be coplanar, and the images of each strip hologram in H1 will be collinear. An eye placed close to the image of H1 will therefore see only one of the perspective views projected into H2 reconstructed by all the wavelengths reaching the eye. As a result, an observer sees a succession of sharp achromatic three-dimensional images as he moves his head from side to side.

With a recording material that diffracts only a narrow spectral band, the same technique can be modified to achieve full colour reproduction. In this case, three sets of primary colour separation perspective views are used to produce three composite master holograms (H1s) on plates centred at different heights in the achromatic plane. The final hologram (H2) is then exposed in turn to the real image projected by each master, using the same laser, and the recording material is processed so that each exposure reconstructs an image of the appropriate colour. When H2 is illuminated with white light, the images

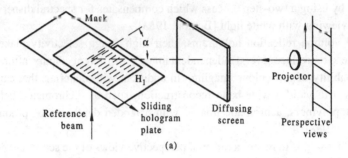

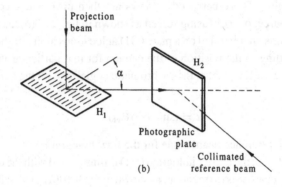

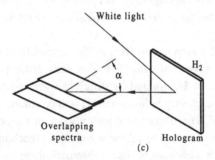

Fig. 9.10. Achromatic reflection stereograms: (*a*) and (*b*) production, and (*c*) image reconstruction [Benton, 1988].

of the three master holograms (H1s) overlap at the intended viewing point to produce a full colour reconstruction.

A one-step method for making full-colour holographic stereograms with a single laser, using computer-based image predistortion to correct for the variation in image magnification with wavelength, has also been described by Klug, Halle, and Hubel [1992].

Another way of making pseudocolour reflection holographic stereograms involves dividing each column of holographic pixels in an image-plane holographic stereogram into red, green, and blue subcolumns, which reconstruct images in the corresponding colour [Klug & Kihara, 1994]. These subcolumns are produced by spatially selective pre- or postexposure swelling of the subcolumns by the required amounts, using DuPont photopolymer Holographic Recording Film in conjunction with DuPont Color Tuning Film that has been subjected to precisely registered ultraviolet pre-exposures, through a mask, to control the swelling of each subcolumn.

# 10

# Computer-generated holograms

Holograms generated by means of a computer can be used to produce wave-fronts with any prescribed amplitude and phase distribution; they are therefore extremely useful in applications such as laser-beam scanning and optical spatial-filtering (see sections 13.3 and 14.2) as well as for testing optical surfaces.

The production of holograms using a digital computer has been discussed in detail by Lee [1978], Yaroslavskii and Merzlyakov [1980], and Dallas [1980], and involves two principal steps.

The first step is to calculate the complex amplitude of the object wave at the hologram plane; for convenience this is usually taken to be the Fourier transform of the complex amplitude in the object plane. It can be shown, by means of the sampling theorem, that if the object wave is sampled at a sufficiently large number of points (see Appendix 2), this can be done with no loss of information. Thus, if an image consisting of $N \times N$ resolvable elements is to be reconstructed, the object wave is sampled at $N \times N$ equally spaced points, and the $N \times N$ complex coefficients of its discrete Fourier transform are evaluated. This can be done quite easily with a computer program using the fast Fourier transform algorithm [Cochran *et al.*, 1967] for arrays containing as many as $1024 \times 1024$ points.

The second step involves using the computed values of the discrete Fourier transform to produce a transparency (the hologram), which reconstructs the object wave when it is suitably illuminated.

Two approaches have been followed for this purpose. The first is analogous to off-axis optical holography (see section 2.2). The complex amplitudes of a plane reference wave and the object wave in the hologram plane are added, and the squared modulus of their sum is evaluated. These values are used to produce a transparency whose amplitude transmittance is real and positive everywhere. A variation [Burch, 1967] makes use of the fact that in the expression on the right-hand side of (2.7), the first two terms do not contribute to the reconstructed image  Hence, it is necessary to compute only the last term at regular intervals and add a constant bias to all the samples to make them positive

An alternative approach, which is possible only with a computer-generated hologram, is to produce a transparency that records both the amplitude and the phase of the object wave in the hologram plane. This can be thought of as the superimposition of two transparencies, one of constant thickness having a transmittance proportional to the amplitude of the object wave, and the other having thickness variations corresponding to the phase of the object wave, but no transmittance variations. Such a hologram has the advantage that it forms a single, on-axis image.

In either case, the computer is used to control a plotter that produces a large-scale version of the hologram. This master is photographically reduced to produce the required transparency. An optical system similar to that shown in fig. 2.7 is then used to reconstruct the object wavefront.

## 10.1 Binary detour-phase holograms

Although it is possible to use an output device with gray scale capabilities to produce the hologram, a considerable simplification results if the amplitude transmittance of the hologram has only two levels – either zero or one. Such a hologram is called a binary hologram.

The best known hologram of this type is the binary detour-phase hologram [Brown & Lohmann, 1966, 1969], which is made without explicit use of a reference wave or bias. To produce such a hologram, the output format covered by the plotter is divided into $N \times N$ cells, which correspond to the $N \times N$ coefficients of the discrete Fourier transform of the complex amplitude in the object plane. Each complex Fourier coefficient is then represented by a single transparent area within the corresponding cell, whose size is determined by the modulus of the Fourier coefficient, while its position within the cell represents the phase of the Fourier coefficient. The method derives its name from the fact that a shift of the transparent area in each cell results in the light transmitted by it travelling by a longer or shorter path to the reconstructed image. Figure 10.1(a) shows a typical binary detour-phase hologram of the letters ICO; fig. 10.1(b) shows the image produced by it. The first-order images are those above and below the central spot; in addition, higher-order images are seen due to nonlinear effects (see section 10.4).

To understand how this method of encoding the phase works, consider a rectangular opening $(a \times b)$ in an opaque sheet (the hologram) centred on the origin of coordinates, as shown in fig. 10.2, which is illuminated with a uniform coherent beam of light of unit amplitude. The complex amplitude $U(x_i, y_i)$ at

(a)

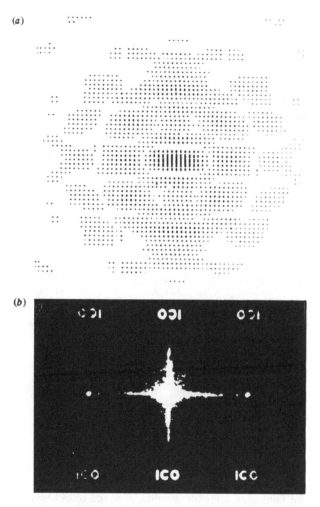

(b)

Fig. 10.1. Binary detour-phase hologram: (*a*) the hologram; (*b*) the reconstructed image [Lohmann & Paris, 1967].

a point $(x_i, y_i)$ in the diffraction pattern formed in the far field is given by the Fourier transform of the transmitted amplitude and is

$$U(x_i, y_i) = ab \text{ sinc } (by_i/\lambda z),  \tag{10.1}$$

where $\text{sinc } x = (\sin \pi x)/\pi x$.

We now assume that the centre of the rectangular opening is shifted to a point $(\Delta x_0, \Delta y_0)$, and the sheet is illuminated by a plane wave incident at an angle. If the complex amplitude of the incident wave at the sheet is $\exp[i(\alpha \Delta x_0 + \beta \Delta y_0)]$,

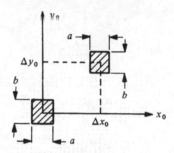

Fig. 10.2. Diffraction at a rectangular aperture.

the complex amplitude in the diffraction pattern becomes

$$U(x_i, y_i) = ab \operatorname{sinc}(ax_i/\lambda z) \operatorname{sinc}(by_i/\lambda z)$$
$$\times \exp\left[i\left(\alpha + \frac{2\pi x_i}{\lambda z}\right)\Delta x_0 + i\left(\beta + \frac{2\pi y_i}{\lambda z}\right)\Delta y_0\right],$$
$$= ab \operatorname{sinc}(ax_i/\lambda z) \operatorname{sinc}(by_i/\lambda z)$$
$$\times \exp[i(\alpha\Delta x_0 + \beta\Delta y_0)]$$
$$\times \exp\left[i\left(\frac{2\pi}{\lambda z}x_i\Delta x_0 + \frac{2\pi}{\lambda z}y_i\Delta y_0\right)\right]. \qquad (10.2)$$

If $ax_i \ll \lambda z$, $by_i \ll \lambda z$, (10.2) reduces to

$$U(x_i, y_i) = ab \exp[i(\alpha\Delta x_0 + \beta\Delta y_0)]$$
$$\times \exp\left[i\left(\frac{2\pi}{\lambda z}x_i\Delta x_0 + \frac{2\pi}{\lambda z}y_i\Delta y_0\right)\right]. \qquad (10.3)$$

If, then, the computed complex amplitude of the object wave at a point $(n\Delta x_0, m\Delta y_0)$ in the hologram plane is

$$o(n\Delta x_0, m\Delta y_0) = |o(n\Delta x_0, m\Delta y_0)| \exp[i\phi(n\Delta x_0, m\Delta y_0)], \qquad (10.4)$$

its modulus and phase at this point can be encoded, as shown in fig.10.3, by making the area of the opening located in this cell equal to the modulus so that

$$ab = |o(n\Delta x_0, m\Delta y_0|, \qquad (10.5)$$

and displacing the centre of the opening from the centre of the cell by an amount $\delta x_{nm}$ given by the relation

$$\delta x_{nm} = (\Delta x_0/2\pi)\phi(n\Delta x_0, m\Delta y_0). \qquad (10.6)$$

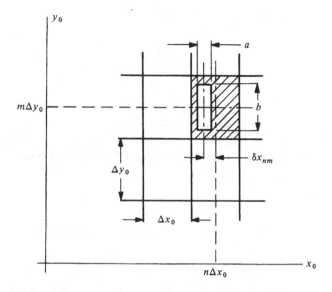

Fig. 10.3. Typical cell in a binary detour-phase hologram.

To show the validity of this method of encoding, we consider the complex amplitude in the far field due to this opening, which is, from (10.3),

$$U_{nm}(x_i, y_i) = |o(n\Delta x_0, m\Delta y_0)|$$
$$\times \exp[i\alpha(n\Delta x_0 + \delta x_{nm}) + i\beta m\Delta y_0]$$
$$\times \exp[(i2\pi/\lambda z)(nx_i\Delta x_0 + my_i\Delta y_0 + \delta x_{nm})]. \quad (10.7)$$

The total diffracted amplitude in the far field, which is obtained by summing the complex amplitudes due to all the $N \times N$ openings, is therefore

$$U(x_i, y_i) = \sum_{n=1}^{N} \sum_{m=1}^{N} |o(n\Delta x_0, m\Delta y_0)| \exp(i\alpha\delta x_{nm})$$
$$\times \exp[i(\alpha n\Delta x_0 + \beta m\Delta y_0)]$$
$$\times \exp[(i2\pi/\lambda z)(nx_i\Delta x_0 + my_i\Delta y_0)]$$
$$\times \exp[(i2\pi/\lambda z)\delta x_{nm}]. \quad (10.8)$$

If the dimensions of the cells and the angle of illumination are chosen so that

$$\alpha\Delta x_0 = 2\pi, \quad (10.9)$$
$$\alpha\Delta y_0 = 2\pi, \quad (10.10)$$
$$\delta x_{nm} \ll \lambda z, \quad (10.11)$$

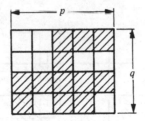

Fig. 10.4. Typical cell in a generalized binary detour-phase hologram.

(10.8) reduces to

$$U(x_i, y_i) = \sum_{n=1}^{N} \sum_{m=1}^{N} |o(n\Delta x_0 + m\Delta y_0)| \exp[i\phi(n\Delta x_0, m\Delta y_0)]$$

$$\times \exp[(i2\pi/\lambda z)(nx_i\Delta x_0 + my_i\Delta y_0)], \qquad (10.12)$$

which is the discrete Fourier transform of the computed complex amplitude in the hologram plane or, in other words, the desired reconstructed image.

Binary detour-phase holograms have several attractive features. It is possible to use a simple pen-and-ink plotter to prepare the binary master, and problems of linearity do not arise in the photographic reduction process. Their chief disadvantage is that they are very wasteful of plotter resolution, since the number of addressable plotter points in each cell must be large to minimize the noise due to quantization of the modulus and the phase of the Fourier coefficients. When the number of phase-quantization levels is large, this noise is effectively spread over the whole image field, independent of the form of the signal. However, when the number of phase-quantization levels is small, the noise terms become shifted and self-convolved versions of the signal, which are much more annoying [Goodman & Silvestri, 1970; Dallas, 1971a,b].

## 10.2 Generalized binary detour-phase holograms

In this method, as shown in fig. 10.4, rather than producing a single transparent area with variable size and position in the cell, corresponding to each Fourier coefficient, a combination of $p \times q$ transparent and opaque subcells is used [Haskell & Culver, 1972; Haskell, 1973]. This method permits finer quantization of both amplitude and phase, resulting in less noisy images. However, it is necessary for the computer to identify the proper binary pattern out of the $2^{pq}$ possible patterns that is the best approximation to the desired complex Fourier coefficient before plotting it.

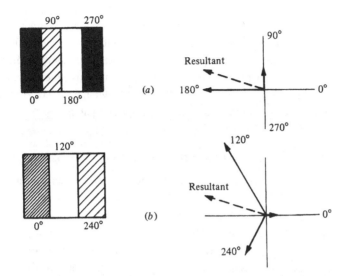

Fig. 10.5. Typical subcell arrangements and phasor diagrams for detour-phase holograms developed by (*a*) Lee [1970]; and (*b*) Burckhardt [1970].

### 10.2.1 Lee's method

An alternative method of recording a computer-generated hologram is due to Lee [1970]. In this method, each cell in the hologram is divided into four equal-sized subcells arranged side by side, as shown in fig. 10.5(*a*). The four subcells then contribute phasor components with relative phases of $0°$, $90°$, $180°$, and $270°$ because of their positions within the cell, the amplitude of each contribution being determined by the transmittance of that subcell. In actual use, two of the subcells in any cell are totally opaque, while the other two are partially transmitting. This gives complete control of both the amplitude and the phase of the resultant contribution.

A simplification of this technique is due to Burckhardt [1970], who pointed out that the same result could be achieved with only three subcells for each Fourier coefficient, as shown in fig. 10.5(*b*).

## 10.3 Phase randomization

The Fourier transforms of the wavefronts corresponding to most simple objects have very large dynamic ranges, because the coefficients of the dc and low-frequency terms have much larger moduli than those of the high-frequency terms. This results in nonlinearity because of the limited dynamic range of the recording medium (see section 6.5).

To minimize this problem, it is convenient, where the phase of the final reconstructed image is not important, to multiply the complex amplitudes at the original sampled object points by a random phase factor before calculating the Fourier transform. This procedure is optically equivalent to placing a diffuser in front of the object transparency and has the effect of making the magnitudes of the Fourier coefficients much more uniform, as shown in fig. 10.6. However, the reconstructed image is then modulated by a speckle pattern.

### 10.4  Error diffusion methods

Several coding techniques have been developed to reduce quantization errors and achieve a satisfactory compromise between smoothing the object spectrum and minimizing speckle (see the review by Bryngdahl & Wyrowski [1990]).

The methods studied include a binary-search approach [Seldowitz, Allebach & Sweeney, 1987], an iterative approach [Broja, Wyrowski & Bryngdahl, 1989], and procedures based on error diffusion (ED)[Hauck & Bryngdahl, 1984]. However, because binary search and iterative methods are more time consuming, ED methods have been gaining favour [Barnard, 1988; Weissbach, Wyrowski & Bryngdahl, 1989].

The basic procedure followed in ED algorithms involves comparing the computed value for the first pixel with a threshold value, say 1/2. If the computed value exceeds this threshold, the output is set to 1, otherwise to 0. This operation produces an error, defined as the difference of the two values, which is then distributed among the values for future pixels. Different ED algorithms differ mainly in the number and relative positions of the pixels and in the weighting array used to diffuse the error. A comparison of ED methods with respect to reconstruction errors and diffraction efficiency has been made by Eschbach [1991].

Another approach, known as the interlacing technique (IT), involves designing a number of subholograms that are interlaced to generate the actual hologram. The first subhologram is designed to reconstruct the desired image, while the succeeding subholograms are designed to correct the residual errors in the image. An extension is the iterative interlacing technique (IIT), in which the error remaining after the last subhologram is fed back to the first subhologram, and the process is continued until there is no further reduction in the reconstruction error [Ersoy, Zhuang & Brede, 1992]. The IIT method can be combined with the ED method to retain the advantages of both methods [Chang & Ersoy, 1993].

### 10.5  The kinoform

If the object is diffusely illuminated, the magnitudes of the Fourier coefficients are relatively unimportant, and the object can be reconstructed using only the

(a)

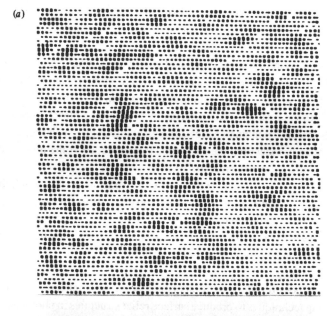

(b)

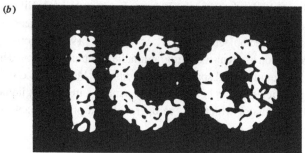

Fig. 10.6. Simulation of illumination through a diffuser (object with random phase: (a) the hologram; (b) the reconstructed image) [Lohmann & Paris, 1967].

values of their phases. This fact led to the concept of a completely different type of hologram called a kinoform [Lesem, Hirsch & Jordan, 1969].

The kinoform is a computer-generated hologram in which all the cells are completely transparent, so that the moduli of all the Fourier coefficients are arbitrarily set equal to unity, and only the phase of the transmitted light is controlled in accordance with the phase of the computed Fourier coefficients. Thus, the amplitude transmittance $\mathbf{t}_{nm}$ of the cell corresponding to a Fourier coefficient with modulus $|o_{nm}|$ and phase $\phi_{nm}$ would be

$$\mathbf{t}_{nm} = \exp(i\phi_{nm}). \tag{10.13}$$

To simplify recording, integral multiples of $2\pi$ radians are subtracted from the computed phases, so that they vary only between 0 and $2\pi$ over the entire kinoform.

To record the kinoform, the computed values of the phase $\phi_{nm}$ are encoded on a multilevel gray scale and used to control a photographic plotter, which exposes a piece of film. The resulting master is then photographed once again to reduce it to the final size, and bleached with a tanning bleach to convert the gray levels to corresponding changes in optical thickness. With proper control of exposure and processing, the amplitude transmittance of the final kinoform can be made to conform closely to (10.13).

Kinoforms have the advantage that they can diffract all the incident light into the final image. However, to achieve this result, care is necessary to ensure that the phase-matching condition expressed by (10.13) is satisfied accurately [Kermisch, 1970]. Any error in the recorded phase shift results in light diffracted into the zero order [see the spot in the centre of figs. 10.7 (*a*) and 10.7 (*c*)], which can spoil the image.

Interesting possibilities have been opened up by the application of electron-beam writing techniques to produce surface relief structures in photoresists that can generate multiple phase levels [Ekberg, Larsson, Hård & Nilsson, 1990]. These techniques make it practicable to write directly computer-generated kino-forms of complicated objects [Urquhart, Stein & Lee, 1993]. Another possibil-ity is the production of a real-time programmable computer-generated hologram by means of a liquid-crystal spatial light modulator [Mok, Diep, Liu & Psaltis, 1986]. High efficiency can be achieved by using a birefringent liquid-crystal spatial light modulator which can be operated as a programmable kinoform [Amako & Sonehara, 1991].

## 10.6 The referenceless on-axis complex hologram (ROACH)

This type of hologram makes use of multilayer colour film as a recording medium to obtain most of the advantages of the kinoform without its ma-jor disadvantages [Chu, Fienup & Goodman, 1973]. Different layers of the film are exposed selectively by light of different colours. When illuminated with light of a given colour, one layer of the film absorbs part of the light, while the other layers, which are effectively transparent, can cause phase shifts due to variations in the film thickness and refractive index. Thus, both the amplitude and phase of the transmitted beam can be controlled by a single element.

To record a ROACH that is to be illuminated finally with red light the com-puted values of the moduli $|o_{nm}|$ of the Fourier coefficients are used, in the

(a)

(b)

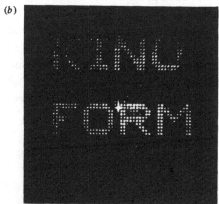

(c)

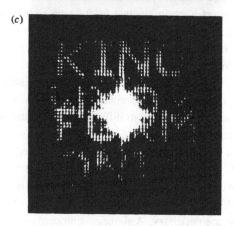

Fig. 10.7. Two-dimensional images produced by kinoforms: (*a*) undermodulation (too short an exposure); (*b*) correct exposure (good phase matching); (*c*) overmodulation (too long an exposure) [Lesem, Hirsch & Jordan, 1969; copyright 1969 by International Business Machines Corporation, reprinted with permission].

first instance, to control the brightness of a black-and-white CRT display. This display is then photographed on a reversal colour film, using a red filter. In the next step, the computed values of the phases $\phi_{nm}$ of the complex Fourier coefficients are displayed in the same manner and photographed on the same frame of colour film using a blue-green transmitting filter.

After processing, the red absorbing layer of the film controls the amplitude of the transmitted red light, so that it is proportional to the moduli $|o_{nm}|$ of the Fourier coefficients. The blue- and green-absorbing emulsions are transparent to red light, but they introduce phase shifts corresponding to the phases $\phi_{nm}$ of the Fourier coefficients, due to the variations in optical thickness introduced by the blue-green exposure. In practice, the red-absorbing layer also introduces a phase shift proportional to the attenuation due to it, but it is possible to compensate for this unwanted phase shift by subtracting from the blue-green exposure a component proportional to $|o_{nm}|$.

Since all the light is diffracted into a single image, the diffraction efficiency of the ROACH is very high. In addition, because both the amplitude and the phase of the object wave are encoded, the image quality is superior to that of the kinoform. The ROACH is also superior to the binary detour-phase hologram, because only one display spot is required for each Fourier coefficient, and quantization noise is negligible.

## 10.7 Three-dimensional objects

The concept of computer holography can be generalized to a three-dimensional object [Waters, 1968; Lesem, Hirsch & Jordan, 1969; Brown & Lohmann, 1969]. Such an object is approximated by the sum of a number of equally spaced cross-sections perpendicular to the $z$ axis. Problems can arise due to distant parts of the object, which are normally hidden by surfaces in front, appearing in the image. To avoid this, it is necessary, at each point on the hologram, to sum only contributions to the object wave arising from points on the object that can be seen from that point on the hologram. Figure 10.8 shows a set of views of such an image from different angles, showing the resulting changes in parallax.

Generation of a three-dimensional image involves a very large amount of computation for two basic reasons. One is that the size of the hologram must be large enough to display effects due to parallax; the other is that to produce an acceptable three-dimensional image, a very large number of two-dimensional Fourier transforms are necessary. A significant amount of work has therefore been done on methods to reduce the computing time required.

One method that results in a substantial reduction in computing time is to

(a)

(b)

(c)

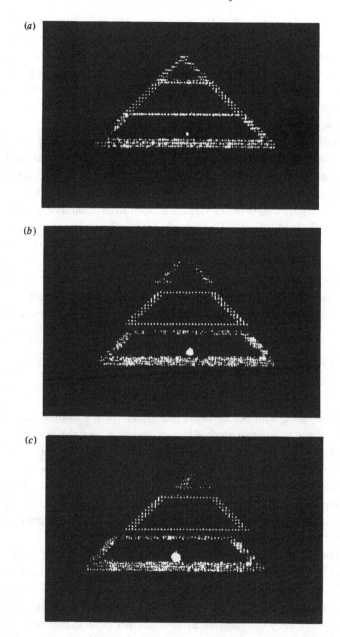

Fig. 10.8. Photographs of a three-dimensional image produced by a kinoform: (a) on-axis view; (b) and (c) off-axis views, showing changes in parallax [Lesem, Hirsch & Jordan, 1969; copyright 1969 by International Business Machines Corporation, reprinted with permission].

eliminate vertical parallax by performing a series of one-dimensional Fourier transforms corresponding to successive horizontal lines in the hologram [Leseberg & Bryngdahl, 1984; Leseberg, 1986]. Another method, which can be applied to objects, such as a cube, which can be decomposed into a series of tilted line segments or tilted planes, involves evaluating, from an analytic solution of the wave equation, the Fresnel transform of these lines or planes [Frère, Leseberg & Bryngdahl, 1986; Leseberg & Frère, 1988].

A completely different approach was followed by King, Noll, and Berry [1970]. Their technique is related to the techniques used in making holographic stereograms described in section 8.4 and has the advantage that it requires much less computer time. A computer is used to produce a series of perspective projections of the object, as seen from a number of angles in the horizontal plane. These views are then optically encoded as a series of vertical strip holograms on a single plate. The real image formed by this composite hologram, when it is illuminated by the conjugate reference beam, is then used to produce an image hologram. Since this real image is actually two-dimensional, it is located entirely in the plane of the final hologram, which can therefore be illuminated with white light to reconstruct a bright, almost achromatic image.

## 10.8 Computer-generated interferograms

Problems can arise with detour-phase holograms when encoding wavefronts with large phase variations, since a pair of apertures near the crossover may overlap when the phase of the wavefront moves through a multiple of $2\pi$ radians. This difficulty has been avoided in an alternative approach to the production of binary holograms based on the fact that an image hologram of a wavefront that has no amplitude variations is essentially similar to an interferogram, so that the exact locations of the transparent elements in the binary hologram can be determined by solving a grating equation [Lee, 1974].

Different methods can then be used to incorporate information on the amplitude variations in the object wavefront into the binary fringe pattern [Lee, 1979]. In one method, the two-dimensional nature of the Fourier transform hologram is used to record the phase information along the $x$ direction, and the fringe heights in the $y$ direction are adjusted to correspond to the amplitude. In another, the phase and the amplitude are recorded through the position and the width of the fringes along the direction of the carrier frequency, and in the third, the phase and amplitude of the object wave are encoded by the superimposition of two phase-only holograms.

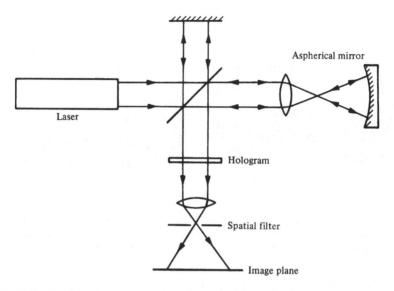

Fig. 10.9. Twyman–Green interferometer modified to use a computer-generated holo-gram to test an aspheric mirror [Wyant & Bennett, 1972].

## 10.9 Computer-generated holograms in optical testing

One of the main applications of computer-generated holograms is in interfer-ometric tests of aspheric optical surfaces. Normally, such tests would require either an aspheric reference surface or an additional optical element, commonly referred to as a null lens, which converts the wavefront produced by the ele-ment under test into a spherical or plane wavefront. However, Pastor [1969] and Snow and Vandewarker [1970] showed that a hologram of the reference surface or the null lens could be used instead, and it was not long before MacGovern and Wyant [1971] showed that a computer-generated hologram could be used instead of one recorded in the conventional manner.

An optical system using a Twyman–Green interferometer in conjunction with a computer-generated hologram to test an aspheric mirror is shown in fig. 10.9 [Wyant & Bennett, 1972]. The hologram is a binary representation of the interferogram that would be obtained if the wavefront from an ideal aspheric surface were to interfere with a tilted plane wavefront, and is placed in the plane in which the mirror under test is imaged. The superimposition of the actual interference fringes and the computer-generated hologram produces a moiré pattern which maps the deviation of the actual wavefront from the ideal computed wavefront.

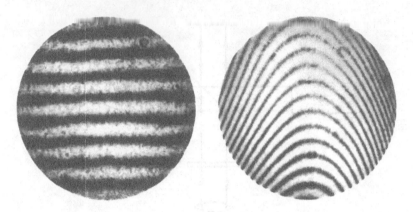

Fig. 10.10. Interference patterns obtained with an aspheric wavefront having a maximum slope of 35 waves per radius and a maximum departure of 19 waves from a reference sphere, (*a*) with, and (*b*) without a computer-generated hologram [Wyant & Bennett, 1972].

The contrast of the moiré pattern is improved by spatial filtering. This is done by reimaging the hologram through a small aperture placed in the focal plane of the reimaging lens. The position of this aperture is chosen so that it passes only the transmitted wavefront from the mirror under test and the diffracted wavefront produced by illuminating the hologram with the plane reference wavefront. These two wavefronts can be isolated if, in producing the computer-generated hologram, the slope of the plane reference wavefront is made greater than the maximum slope of the aspheric wavefront along the same direction. Typical fringe patterns obtained with an aspheric surface, with and without a computer-generated hologram, are shown in fig. 10.10.

To obtain good results, a number of potential sources of error must be kept in mind. The first of these is the effect of quantization. If there are $N$ resolvable points across the diameter of the hologram, any plotted point may be displaced from its proper position by a distance equal to $1/2N$ of the diameter. However, for the diffracted spectra not to overlap, the hologram must have a carrier frequency of $3s$ where $s$ is the highest spatial frequency (fringes per diameter) in the uncorrected interference pattern. The result is that the fringe frequency in the hologram can vary from a minimum of $2s$ to a maximum of $4s$. If there are $4s$ fringes across the hologram, an error in the fringe position of $1/2N$ would correspond to a wavefront error (expressed as a fraction of a fringe)

$$\Delta W = 2s/N. \tag{10.14}$$

Accordingly, if the wavefront error is not to exceed $\lambda/4$, the slope of the test wavefront should satisfy the condition

$$s < N/8, \qquad (10.15)$$

fringes per diameter.

Another source of error is plotter distortion. If $e$ is the maximum distortion introduced by the plotter over a plotting surface of diameter $D$, the corresponding wavefront error has a maximum value

$$\Delta W = (e/D)4s. \qquad (10.16)$$

For this error not to exceed $\lambda/4$,

$$e < D/16s. \qquad (10.17)$$

A method to minimize such errors [Fercher, 1976] is to use a hologram which contains two superimposed structures, one corresponding to the object wavefront, while the other is a linear grating. Plotter errors, being common to both, are eliminated.

Errors in the size of the hologram introduce a radial shear between the reference wavefront and the test wavefront. The wavefront error due to this cause is

$$\Delta W = s(\Delta d/d), \qquad (10.18)$$

where $d$ is the nominal diameter of the hologram. If this wavefront error is not to exceed $\lambda/4$,

$$\Delta d < d/4s. \qquad (10.19)$$

Improper positioning of the hologram also introduces a shear. A sideways translation $\Delta x$ results in a lateral shear causing a wavefront error

$$\Delta W = s(\Delta x/d), \qquad (10.20)$$

where $s$ is now the slope of the test wavefront along the direction of shear. The condition for this wavefront error to be less than $\lambda/4$ is, therefore,

$$\Delta x < d/4s. \qquad (10.21)$$

Different methods for plotting the fringes in such holograms have been discussed by Birch and Green [1972]. However, none of these methods are satisfactory when testing surfaces with large deviations from a sphere because of the relatively high fringe frequencies required. While Sirohi, Blume, and Rosenbruch [1976] have shown how the computation time involved can be reduced by merging two methods, a more common procedure is to use a combination of a null lens and a computer-generated hologram [Faulde, Fercher, Torge & Wilson, 1973; Wyant & O'Neill, 1974]. The design of the null lens can be quite simple, since it need only reduce the residual aberrations to a level that can be handled by the computer-generated hologram. Other approaches to minimize the number of fringes in the hologram have involved the use of on-axis holograms [Ichioka & Lohmann, 1972; Mercier & Lowenthal, 1980] and aberration-balancing techniques [Yatagai & Saito, 1978]. An entirely different technique is the dual computer-generated hologram described by Yatagai and Saito [1979]. This hologram consists of apertures whose positions are modulated along two different directions, allowing the simultaneous generation of two diffracted wavefronts whose sum and difference can be obtained by suitably positioned spatial filters.

Computer-generated holograms have been used widely to test aspheric surfaces: see for example the reviews by Lukin and Mustafin [1979] and Loomis [1980]. With the development of improved plotting routines and the application of techniques such as electron-beam recording and using layers of photoresist coated on optically worked substrates for the production of computer-generated holograms of very high quality [Biedermann & Holmgren, 1977; Leung, Lindquist & Shepherd, 1980; Arnold, 1985, 1988, 1989; Urquhart et al., 1989], recent work has focused on refinements of the technique to obtain the highest precision [Dörband & Tiziani, 1985; Arnold, 1989]. A preferred interferometer configuration is one in which both test and reference beams pass through the CGH so that aberrations of the substrate have no significant effect on the interferogram. It is also essential that the beams incident on the CGH should be collimated in order to reduce the effects of misalignment. Other factors to be considered are the use of null optics to simplify the requirements for the CGH, design of the optics to image the test surface on the CGH and design of the CGH, using ray-tracing software, to compensate for off-axis aberrations introduced by the imaging system.

# 11

# Special techniques

## 11.1 Polarization recording

With normal holographic techniques, the amplitude and phase of the object wavefront are recorded accurately, but information on its state of polarization is lost. The polarization of the reconstructed wave is determined by the polarization of the light used to illuminate the hologram.

### 11.1.1 Orthogonally polarized reference beams

Two basic methods for recording the state of polarization of the object wave have been described. The first method was proposed by Lohmann [1965a] and subsequently demonstrated by Bryngdahl [1967]. The experimental arrangement for this method, which is shown in fig. 11.1, uses two orthogonally polarized reference waves which interfere with the corresponding polarized component of the light from the object, so that two holograms are recorded on the same plate. After processing, when the plate is illuminated once again with the same reference beams, it yields two superimposed images that reproduce the polarization of the object wave. Care must be taken to adjust the angles of incidence of the beams so that the cross-talk images formed by diffraction of each of the reference beams at the hologram formed with the orthogonally polarized component do not overlap the desired image.

### 11.1.2 Coded reference beams

The other method, which is due to Kurtz [1969], uses a single reference beam in which, as shown in fig. 11.2, an opal glass diffuser is inserted. The light transmitted by this diffuser is depolarized, and the complex amplitudes of the two orthogonally polarized components of the reference beam at any point exhibit little or no correlation. As a result, the orthogonally polarized components of

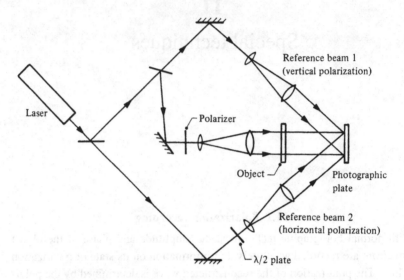

Fig. 11.1. Experimental arrangement for recording the state of polarization of the object wave using Lohmann's method [Gåsvik, 1975].

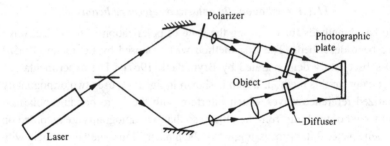

Fig. 11.2. Setup for recording the state of polarization of the object wave using Kurtz's method [Gåsvik, 1975].

the object wave are effectively encoded by different random wavefronts [see section 14.3].

### *11.1.3 Conditions for accurate recording*

Both these methods have been analysed in detail by Windischbauer *et al.* [1973] and by Gåsvik [1975]. The latter has shown that several conditions must be satisfied to ensure accurate reproduction of the state of polarization of the object wave.

A problem common to both methods is that the component of the electric vector parallel to the plane of incidence produces a hologram with lower modulation, and hence lower diffraction efficiency, than the component normal to the plane of incidence (see section 5.3). This difference can be brought within acceptable limits if the angle between the object beam and the reference beam(s) is made fairly small.

In addition, with Lohmann's method, inaccurate repositioning of the hologram after processing, as well as variations in the optical paths of the two reference beams due to temperature gradients, can cause changes in the polarization state of the reconstructed wave. These errors can be minimized by using a modified optical system in which the two reference beams run quite close to each other.

The latter problem is not present in Kurtz's method, but, in this case, the processed hologram must occupy exactly the same position in which it was exposed if it is to reconstruct an image (see section 14.3). Normally, this implies that the hologram must be processed *in situ*. In addition, the reconstructed image is quite noisy due to the presence of a speckled background due to the scattered light. With a transparent object, scattered light can be minimized by inserting an aperture in the back focal plane of the imaging lens, but there is always a considerable loss of light due to the diffuser.

Both methods require the hologram to be illuminated with the same wavelength as that used to record it. With Lohmann's method, a change in the wavelength results in a spatial variation in the state of polarization, whereas with Kurtz's method there is no reconstruction at all.

The main application of techniques for polarization recording has been in holographic photoelasticity (see section 16.2).

### *11.1.4 Polarization holography*

Kakichashvili [1972] was the first to point out that the absence of intensity modulation when orthogonally polarized object and reference waves interfere does not imply that no hologram is formed. In fact, in the region of overlap, we have standing waves with a spatially varying state of polarization. As a result, the polarization of the field in the hologram plane has a vectorial character which, however, cannot be recorded with normal photosensitive media.

The use of a photosensitive medium that develops anisotropy when exposed to polarized light makes it possible to record the polarization state of the standing wave. Such a recording, when illuminated by a coherent polarized wave, can reconstruct the polarization state of the original object wave, in addition to its amplitude and phase.

Two types of photoanisotropic materials are possible: one in which photoanisotropy is due to a reorientation of anisotropic absorbing centres and another in which it is due to variable bleaching depending on the polarization direction of the incident light. The former are suitable for recordings with a linearly polarized reference wave, while the latter can be used with a circularly polarized reference wave [Nikolova & Todorov, 1984; Todorov & Nikolova, 1989]. Gratings produced with the first type of material behave like a $\lambda/2$ plate with axes oriented at $\pm45°$ to the polarization directions of the recording waves. Gratings produced with two plane waves with orthogonal circular polarizations have only two diffraction orders ($\pm1$). The diffracted waves have orthogonal circular polarizations, and their intensities are proportional to the left- and right-hand circularly polarized components of the wave incident on the grating.

Some applications of polarization holography that have been studied include logical operations with images, determination of the sign of the displacement in double-exposure holographic interferometry [Todorov, Nikolova, Stoyanova & Tomova, 1985], and polarization-preserving phase-conjugation [Nikolova, Todorov, Tomova & Dragostinova, 1988] by four-wave mixing.

## 11.2 Holography with incoherent light

It is an interesting fact that image-forming holograms can be recorded even with incoherent light. While all the techniques developed for this purpose suffer from limitations, they suggest some interesting possibilities.

Mertz and Young [1962] were the first to show that the only necessary condition for recording a hologram was that each point on the object should produce a two-dimensional pattern that uniquely encoded its position and its intensity. For this, they used a mask in the form of a Fresnel zone plate, so that each object point cast a shadow of this form on the film. When illuminated, each Fresnel zone plate recorded on the film produced an image of the corresponding object point.

While this geometrical approach is applicable to very short wavelengths, such as x-rays, it breaks down at longer wavelengths. Accordingly, early methods of holography with spatially incoherent light were based on interference between two wavefronts derived from the object. This is possible with a shearing interferometer. With spatially incoherent illumination, interference takes place only between the waves derived from the same point on the object. As many independent superimposed interference patterns are recorded as there are independent points on the object. Paradoxically, this lack of spatial coherence between the light at different points on the object is an essential condition for this method to work. Of course, to obtain interference fringes of good contrast it is necessary for the coherence length of the light to be much greater than the optical path differences involved.

To encode each object point uniquely, the interferometer must produce a shear that is a function of the position of the object point. This is possible with either a rotational or a radial shear; in both cases the shear is directly proportional to the position vector of the point in the object plane.

### 11.2.1 Rotational shear systems

For plane objects it is possible to use an interferometer that introduces a rotational shear of 180° between the two waves [Stroke & Restrick, 1965; Worthington, 1966]. If we consider an object point $O$ with coordinates $(x_o, y_o)$, the optical system of the interferometer produces two images of this point located at $(x_o, y_o)$ and $(-x_o, -y_o)$ in the object plane. The interference pattern produced by these two virtual sources is a series of equally spaced, straight fringes at right angles to the line joining the two images. The intensity distribution in the hologram plane can then be written, apart from a constant factor, as

$$\Delta I(\xi, \eta) = I(x_o, y_o)\{1 + \cos[2\pi(x_o\xi + y_o\eta)]\}, \qquad (11.1)$$

where $I(x_o, y_o)$ is the intensity of the object point, $z_o$ is the distance from the object to the hologram, and $\xi$ and $\eta$ are coordinates in the hologram plane defined by the relations $\xi = 2x/\lambda z_o$ and $\eta = 2y/\lambda z_o$. The total intensity in the hologram plane is then merely the sum of these contributions, so that

$$I(\xi, \eta) = \int\int_{-\infty}^{\infty} I(x_o, y_o)\{1 + \cos[2\pi(x_o\xi + y_o\eta)]\}dx_o dy_o. \qquad (11.2)$$

The right-hand side of (11.2) is, apart from a constant background, the two-dimensional Fourier transform of the intensity distribution in the object. Accordingly, if the hologram is illuminated with a spatially coherent monochromatic wave, it will reconstruct, in the far field, two images of the object positioned symmetrically about the optical axis. Variation of the rotational shear permits varying the scale of the interference pattern and, hence, that of the reconstructed image [Lowenthal, Serres & Froehly, 1969].

### 11.2.2 Radial shear systems

Rotational shear systems cannot give a three-dimensional image, because the interference pattern produced by an object point is always a series of straight lines, irrespective of its distance from the hologram plane. To reproduce depth

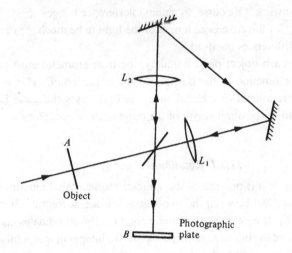

Fig. 11.3. Optical system used for holography with incoherent illumination, in which interference takes place between two waves with a radial shear [Cochran, 1966].

information, the shear introduced between the wavefronts must have a component parallel to the axis so that the fringe pattern is equivalent to a Fresnel transform of the scene [Lohmann, 1965b]. Cochran [1966] used the triangular-path interferometer shown in fig. 11.3 containing an afocal system which produces two wavefronts with a radial shear [Hariharan & Sen, 1961].

If, in this interferometer, the planes $A$ and $B$ are located at distances from $L_1$ and $L_2$ equal to their respective focal lengths, two images of $A$ are formed at $B$ with magnifications of $(1/\alpha) = -f_2/f_1$ and $(1/\beta) = -f_1/f_2$, respectively. An object point $O(x_o, y_o)$ gives rise to two images $O'(\alpha x_o, \alpha y_o)$ and $O''(\beta x_o, \beta y_o)$.

It can be shown (see Appendix 3) that the length of the optical path from an object point $(x_o, y_o, z_o)$ to a point $(x, y)$ in the reference plane is given by the relation

$$r(x, y) = r_o - (x_o x + y_o y)/r_o + (x^2 + y^2)/2r_o$$
$$- (x_o x + y_o y)^2/2r_o^3 - \ldots, \qquad (11.3)$$

where $r_o^2 = x_o^2 + y_o^2 + z_o^2$. Accordingly, the difference in the optical paths from the images $O'$ and $O''$ to a point $P(x, y)$ in the hologram plane, which we assume to be at a distance $z_o$, is

$$r'(x, y) - r''(x, y) = - (\alpha - \beta)(x_o x + y_o y)/r_o$$
$$+ (\alpha^2 - \beta^2)(x^2 + y^2)/2r_o$$
$$- (\alpha^2 - \beta^2)(x_o x + y_o y)^2/2r_o^3 - \ldots . \qquad (11.4)$$

The common factor $(\alpha - \beta)$ is merely a constant of proportionality which determines the scale of the interference pattern on the hologram, and we can equate it to unity. We can then simplify (11.4) further if we set

$$x_1 = x_o/(\alpha + \beta), \tag{11.5}$$
$$y_1 = y_o/(\alpha + \beta), \tag{11.6}$$
$$z_1 = z_o/(\alpha + \beta), \tag{11.7}$$
$$r_{1o} = r_o/(\alpha + \beta). \tag{11.8}$$

With these substitutions, the difference in the optical paths is

$$\begin{aligned} r'(x, y) - r''(x, y) = &- (x_1 x + y_1 y)/r_{1o} \\ &+ (x^2 + y^2)/2r_{1o} \\ &- (x_1 x + y_1 y)^2/2r_{1o}^3 - \dots. \end{aligned} \tag{11.9}$$

A comparison of (11.9) with (11.3) shows that (11.9) can be written as

$$r'(x, y) - r''(x, y) = r_1(x, y) - r_{1o}, \tag{11.10}$$

where $r_1(x, y)$ is the distance from a point $(x_1, y_1, z_1)$ to a point $(x, y)$ in the reference plane, and, from (11.5)–(11.8), $r_{1o}^2 = x_1^2 + y_1^2 + z_1^2$. For any given object point, $r_{1o}$ is a constant, and the optical path difference between the wavefronts defined by the right-hand side of (11.10) corresponds to that between a sphere of radius $r_{1o}$ with its centre at $(x_1, y_1, z_1)$ and the reference plane.

If the intensities of the two images are equalized by a proper choice of the reflectance and transmittance of the beam splitter, the intensity distribution in the resulting interference pattern is

$$I(x, y) = I_o\{1 + \cos(2\pi/\lambda)[r_1(x, y) - r_{1o}]\}. \tag{11.11}$$

The hologram is, therefore, a Fresnel zone pattern. When illuminated with collimated light, it reconstructs an image of the object point at $(x_1, y_1, z_1)$, as well as a conjugate image at $(-x_1, -y_1, -z_1)$.

With this arrangement, it is possible to make a hologram that reconstructs a three-dimensional image. In addition, the magnification of the image can be varied by varying the radial shear. However, the only magnification for which undistorted depth information can be obtained is $-1$ [Bryngdahl & Lohmann, 1970a].

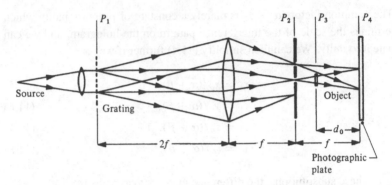

Fig. 11.4. Grating achromatic-fringe interferometer used to record holograms with light having an appreciable spectral bandwidth [Leith & Upatnieks, 1967].

### 11.2.3 Diffraction efficiency

The main drawback of holograms recorded with spatially incoherent light is that their diffraction efficiency drops off very rapidly as the complexity of the object increases. It is apparent from (11.11) that the intensity in the hologram plane due to a single object point is the sum of a uniform bias and a spatially varying term and, with a number of incoherently illuminated points, the intensity is the sum of the contributions due to the individual points. It can be shown then [Cochran, 1966] that, for an object consisting of $N$ independently radiating points, the contrast of the fringes in the hologram is proportional to $(1/N)$, while the diffraction efficiency is proportional to $(1/N^2)$. Various methods have been proposed to minimize this drop in diffraction efficiency. One proposal involves eliminating the bias term by time-modulation of one of the beams [Kozma & Massey, 1969]; another is to form an image in the vertical direction and a series of one-dimensional holograms in the horizontal direction [Bryngdahl & Lohmann, 1968$b$], each of which receives contributions from only a limited number of points.

### 11.2.4 Achromatic systems

The temporal coherence required to record a hologram can be reduced by using an interferometer that yields achromatic fringes [Leith & Upatnieks, 1967]. As shown in fig. 11.4, a grating located in the plane $P_1$ and illuminated by a collimated beam is imaged at unit magnification into the plane $P_4$. The diffracted waves are focused at $P_2$, the back focal plane of the lens, where an aperture passes only the zero-order and one of the first-order diffracted beams,

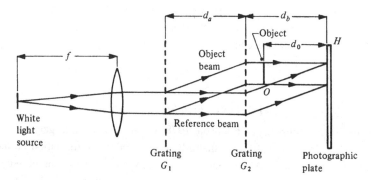

Fig. 11.5. Grating interferometer used to record holograms with a white-light source [Chang, 1973].

which then form an image of the grating at $P_4$. A scattering object, such as a transparency, is located at $P_3$ in the zero-order beam, and the diffracted beam serves as the reference beam.

It is obvious that the interference fringes formed at $P_4$ may also be regarded in this case as images of the grating rulings. Accordingly, the temporal coherence requirements are determined mainly by the characteristics of the object. Good holograms of transparencies can be recorded with a high-pressure mercury-vapour lamp illuminating a 500 $\mu$m pinhole.

For the system to be truly achromatic the diffraction pattern of the object in the plane of the hologram must be the same for all wavelengths. This is possible if the object is also imaged at, or near, the hologram plane. Bryngdahl and Lohmann [1970b] showed that it was then possible to record a hologram with a xenon arc lamp as the source.

A more efficient system is possible with an interferometric arrangement that uses diffraction gratings both to split and recombine the beams [Leith & Chang, 1973]. Such an interferometer can be made achromatic and can, in addition, use an extended source. The sizes of the object and the hologram can then be made relatively large without altering the coherence requirements.

In a holographic setup of this type shown in fig. 11.5 [Chang, 1973], two identical gratings $G_1$ and $G_2$ are used to split and recombine the beams. The object transparency $O$ is inserted in one of the beams at a distance $d_0$ from the hologram plane $H$. An interesting feature of this arrangement is that its modulation transfer function can be varied by changing the position of one of the gratings. White light from a broad source can be used both to record holograms and to view the reconstructed image.

A more general analysis of a three-grating interferometer used as a hologram-forming system has been presented by Chang and Leith [1979]. Such a system can reduce the coherence requirements for off-axis holography to less than those for in-line holography, and can be adjusted to produce low-pass, band-pass, or high-pass characteristics.

## 11.3  Time-gated holography

Light-in-flight holography [Abramson, 1978] is a method of time-gated imaging, which makes use of the fact that a hologram records information on the object wave only when it is illuminated simultaneously by a coherent reference wave. For a pulsed light source, this means that light must reach the hologram plate from the object, as well as directly from the source, during a certain time interval to form an interference pattern which can then reconstruct an image. Similarly, with a continuous source of light with a limited coherence length, only those parts of the object for which the difference in the optical paths for the object and reference beams is less than the coherence length will be reconstructed in the image. It follows that in holography, where the image is formed by interference, a short coherence length produces results similar to a short light pulse.

With a large object, the reconstructed image seen from any point on the hologram is crossed by a bright fringe covering points at which the optical path difference is close to zero. As the point of observation is moved along the holographic plate, this fringe moves over the object to satisfy the condition of near-zero optical path difference. The hologram is therefore equivalent to a time-gated viewing system.

If a flat object surface and a hologram plate are both illuminated at an oblique angle by continuous radiation with a short coherence length, the hologram will record the movement of the object wave across the surface. Scanning along the hologram produces a motion picture of the light in flight. Figure 11.6 shows what happens to a spherical wavefront from a point source when it passes through a converging lens [Abramson, 1983].

Light-in-flight recordings of a single pulse with a duration of 25 ps have also been made [Abramson & Spears, 1989]. With this method, it was possible to measure temporal and spatial pulse shape, to record a single pulse during the time it was reflected by a mirror, and to measure the shape of a human hand and a rotating propeller.

Another application of time-gated viewing has been to look through a scattering medium. By studying only the part of the pulse that arrived first, and thus travelled the shortest path, it became possible to see profiles of transparent objects hidden between two sheets of ground glass.

This technique opens up possibilities of imaging through living tissues, where

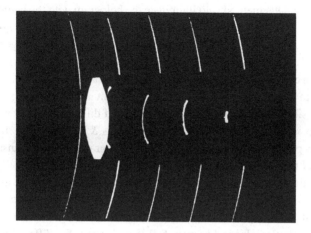

Fig. 11.6. A time series (left to right) of images of a 3 mm thick (equivalent duration, 10 ps) spherical wavefront focused by a converging lens [Abramson, 1983].

a difference in refractive index can cause a time shift in the arrival of the light pulse. However, a drawback is that only a small fraction of the light actually forms interference fringes at the hologram; the remainder forms an incoherent background, reducing the contrast of the fringes. This problem can be avoided if a short pulse is decomposed into its spectral components and a separate hologram is recorded for each component. These holograms can then be reconstructed with the appropriate phase delays and recombined in a computer to form an image at the desired gating. In this case, because each component is monochromatic, the fringe contrast of the individual holograms is much higher [Arons, Dilworth, Shih & Sun, 1993]. A detailed comparison of the two methods has been presented by Chen *et al.* [1993].

## 11.4 Hologram copying

There are quite a few situations in which many identical holograms are required. A considerable amount of time and labour can then be saved if copies are made from a single hologram of the original object. Several techniques have been used for this purpose [Vanin, 1978].

### 11.4.1 Optical methods

Optical methods have the advantage that they do not require special equipment and permit the use of a wide range of materials which can be handled easily in the laboratory. One method is to illuminate the hologram with the conjugate of the reference beam used to record it. The wave reconstructed by the hologram can then be used with another reference wave to record a second-generation hologram. This technique has the advantage of great flexibility. It is possible to produce a copy that reconstructs an orthoscopic real image [Rotz & Friesem, 1966] and to control the spatial frequency spectrum and the viewing angle. It is even possible to produce copies with improved diffraction efficiency from a hologram with low diffraction efficiency [Palais & Wise, 1971]. It is also possible to produce a number of reflection holograms from a transmission hologram of the object [Vanin, 1978; Růžek & Fiala, 1979].

The requirements on coherence and stability for this technique are almost as exacting as for recording the original hologram. Accordingly, a much simpler method that is widely used is to "contact print" the original on to another photosensitive layer [Harris, Sherman & Billings, 1966]. For diffraction effects to be negligible, the separation $\Delta z$ between the hologram and the copy must satisfy the condition

$$\Delta z < \Lambda^2/4\lambda, \tag{11.12}$$

where $\Lambda$ is the average spacing of the hologram fringes.


### 11.4.2 Coherence requirements for contact printing

In most cases, the spatial-carrier frequency of the hologram is high enough that (11.12) is not satisfied. It is then necessary to take into account diffraction effects at the hologram [Nassenstein, 1968a]. Consider a thin hologram and assume that, as shown in fig. 11.7, the hologram $H_1$ and the photographic plate $H_2$ used to copy it are separated by a small distance $\Delta z$ and illuminated by a uniform incoherent source of diameter $2\rho$ located at a comparatively large distance $z$. In this case, what is recorded on the photographic plate is actually the interference pattern formed by the light diffracted by the hologram and the light transmitted by it. Hence, for a satisfactory copy to be obtained, the coherence of the illumination must be adequate to produce interference fringes of high visibility at $H_2$.

If the hologram diffracts light at an angle $\theta$, interference takes place between rays originally separated by a distance $\Delta x$ where

$$\Delta x = \Delta z \tan\theta. \tag{11.13}$$

With an extended source, the degree of spatial coherence of the interfering beams is then given by the normalized Fourier transform of the intensity

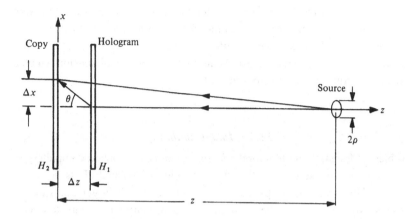

Fig. 11.7. Coherence requirements for copying a hologram by "contact printing."

distribution over the source (see Appendix A1.3), so that

$$|\gamma(v)| = 2J_1(v)/v, \qquad (11.14)$$

where $v = 2\pi\rho\Delta x/\lambda z$. If we assume that $\Delta z < 100\ \mu$m, and $\tan\theta < 0.3$, the degree of spatial coherence with a source of diameter $2\rho = 2$ mm at a distance of 1 m would be approximately 0.98, which is quite satisfactory.

It is also apparent from fig. 11.7 that the optical path difference between the beams is $\Delta x \sin\theta$. This must be less than $\Delta l$, the coherence length of the light, so that

$$\Delta l > \Delta x \sin\theta, \qquad (11.15)$$

or, from (11.13),

$$\Delta l > \Delta z \tan\theta \sin\theta. \qquad (11.16)$$

If, as before, $\Delta z < 100\ \mu$m and $\tan\theta < 0.3$, $\Delta l > 10\ \mu$m.

It is apparent from (11.13) and (11.16) that the coherence requirements for hologram copying by contact printing are much less stringent than for recording a hologram and that in many cases satisfactory copies of holograms can be made by this technique with a high-pressure mercury-vapour lamp, or even with a filtered white-light source [Phillips & Van der Werf, 1985].

To obtain a copy with good diffraction efficiency, the amplitudes of the transmitted wave and the diffracted wave from $H_1$ must be comparable. This condition can usually be met, when copying amplitude transmission holograms, by adjusting the exposure so that the amplitude transmittance of the master hologram is less than 0.1. An index-matching fluid is commonly used between the two plates to minimize spurious interference fringes formed by reflection between the emulsion surfaces. Uniform illumination can be obtained with a scanned laser beam, or with a laser beam expanded in one dimension which is moved at a uniform speed across the hologram.

### 11.4.3 Image doubling

Problems of spatial and temporal coherence can be eliminated completely by the use of a laser source, so that the separation of $H_1$ and $H_2$ is no longer critical. However, large separations result in a copy which generates double images [Brumm, 1967]. The separation of the primary and conjugate images is, in general, a function of the curvature of the reference and signal waves as well as the separation of $H_1$ and $H_2$ [Vanin, 1978]. For a plane reference wave, the separation of the twin images is twice the distance between $H_1$ and $H_2$, while for a Fourier hologram, all four images are at the same distance.

### 11.4.4 Volume holograms

The problem of image doubling is much less serious when copying volume transmission holograms, since the angular selectivity of the hologram can be made high enough to ensure that the amplitude of the conjugate reconstructed wave is negligible [Belvaux, 1967; Sherman, 1967]. However, with normal illumination, effects due to the thickness of the emulsion can be quite pronounced, resulting in low modulation in certain planes and copies of poor quality [Landry, 1967; Kaspar, 1974]. The MTF of the copying process has also been studied by Suhara, Nishihara, and Koyama [1975], who have shown that, for best results, the direction, curvature, and wavelength of the original reference wavefront must be duplicated in the copying system.

Volume reflection holograms can also be copied by interchanging the positions of $H_1$ and $H_2$ in fig. 11.7 [Belvaux, 1967; Kurtz, 1968]. In this case, the illuminating beam passing through the unexposed emulsion acts as the reference beam, while the light diffracted back from $H_1$ constitutes the object beam for the second hologram. If the amplitudes of the two beams are to be comparable, the diffraction efficiency of $H_1$ must be quite high. For this it is necessary to duplicate the direction, curvature, and wavelength of the reference beam used in recording $H_1$. In addition, it is necessary to see that $H_1$ is processed to avoid shrinkage of the photographic emulsion [Zemtsova & Lyakhovskaya, 1976]. With these precautions, good quality copies can be made.

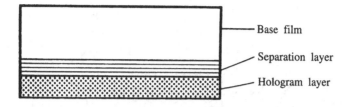

Fig. 11.8. Cross-section of a film used to make embossed copies of holograms.

### 11.4.5 Embossed holograms

Holograms recorded on a photoresist (see section 7.3) can be copied by mechanical methods making use of the surface relief of the original hologram. Copies can be made on a thermoplastic by pressing [Vagin & Shtan'ko, 1974] or with a cold-setting resin by casting [Trukhmanova & Denisyuk, 1977]. However, the method almost universally used now for large-scale replication of holograms is embossing [Bartolini, Hannan, Karlsons & Lurie, 1970; Iwata & Tsujiuchi, 1974].

The first step in the embossing technique is to make a stamper by electrodeposition of nickel on the relief image recorded on the photoresist [Iwata & Ohnuma, 1985]. When the nickel layer is thick enough, it is separated from the master hologram. The metal replica can be mounted on a metal backing plate and used directly as a stamper; alternatively it can be used, by repeating the same process, to produce additional stampers.

Figure 11.8 shows a cross section of the material used to make embossed copies. It consists essentially of a base film, a separation layer, and a hologram recording layer. Polyester film 10–25 $\mu$m thick is used as the base film because of its strength, dimensional stability, and resistance to heat. The separation layer is a resin which allows the base film to be peeled off finally. Since the separation layer then becomes the outermost layer, it has to be transparent and tough enough to protect the hologram. The hologram layer is the thermoplastic material in which the hologram is actually copied by embossing.

The embossing process can be carried out with a simple heated press, as shown in fig. 11.9. The bottom layer (the hologram layer) of the duplicating film is heated above its softening point and pressed against the stamper. The hologram layer then takes up the shape of the stamper and retains this shape when it is cooled and removed from the press. After the base film is stripped off, the hologram can be mounted suitably. A roll press can be used when a very large number of copies are required. Burns [1985] has given a detailed description of production methods for mass replication of large-format embossed holograms.

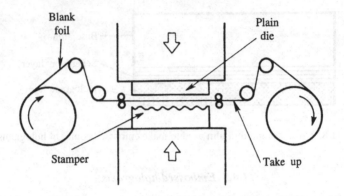

Fig. 11.9. Embossing press used to make copies of holograms.

For replicating some types of optical elements, such as holographic scanners (see section 13.3), which require a groove depth much greater than the groove width, the thermoplastic film can be replaced by a photopolymer which is cured by exposure to UV light before it is separated from the stamper [Shvartsman, 1991].

Embossed holograms can be transferred to an opaque surface, such as the cover of a book, by transcription. To permit viewing by reflected light, the transcription foil contains two more layers. The first is a reflecting layer of aluminium deposited in vacuum on the hologram recording layer; the second is an adhesive layer, usually a heat-sealing glue.

The transcription foil is placed on the substrate to which the hologram is to be transferred and pressed with a heated die. The bottom adhesive layer then melts and sticks to the substrate. When the adhesive layer has cooled, the base film can be lifted off from the separation layer leaving the other layers of the transcription foil, including the hologram, attached to the substrate.

Embossed holograms have been used in a range of items such as stickers, greetings cards, magazine covers, catalogues, and company reports. A major application has been as a security feature on credit cards and quality merchandise [Fagan, 1990].

# 12

# Applications in imaging

## 12.1 Holographic microscopy

As mentioned in Chapter 1, holographic imaging was originally developed in an attempt to obtain higher resolution in microscopy. Equations (3.20) and (3.21) show that it is possible to obtain a magnified image if different wavelengths are used to record a hologram and reconstruct the image, or if the hologram is illuminated with a wave having a different curvature from the reference wave used to record it. However, neither of these techniques has found much use, in the first instance because of the limited range of coherent laser wavelengths available, and, in the second, because of problems with image aberrations [Leith & Upatnieks, 1965; Leith, Upatnieks & Haines, 1965].

The most successful applications of holography to microscopy have been with systems in which holography is combined with conventional microscopy. In one approach, a hologram is recorded of the magnified real image of the specimen formed by the objective of a microscope, and the reconstructed image is viewed through the eyepiece [van Ligten & Osterberg, 1966]. While this technique offers no advantages for ordinary subjects, it is extremely useful for phase and interference microscopy [Snow & Vandewarker, 1968]. In another, a hologram is recorded of the object, and the reconstructed real image is examined with a conventional microscope. This technique is particularly well adapted to the study of dynamic three-dimensional particle fields, as described in the next section.

## 12.2 Particle-size analysis

Measurements on moving microscopic particles distributed throughout an appreciable volume are not possible with a conventional optical system, because a microscope which can resolve particles of diameter $d$ has a limited depth of field

$$\Delta z \approx d^2/2\lambda. \tag{12.1}$$

197

Holography permits storing a high-resolution, three-dimensional image of the whole field at any instant. The stationary image reconstructed by the hologram can then be examined in detail, throughout its volume, with a conventional microscope (see Thompson [1974], Trolinger [1975a], Cartwright, Dunn and Thompson [1980] and Vikram [1992]).

In-line holography can be used for such studies wherever a sufficient amount of light (> 80 per cent) is directly transmitted to serve as a reference beam. A very simple optical system, which is also economical of light, can then be used. However, a distinction must be made between such an in-line hologram of a particle field and a Gabor hologram. Because of the small diameter $d$ of the particles, the distance $z$ of the recording plane from the particles easily satisfies the far-field condition $z \gg d^2/\lambda$ (see Appendix 3), so that the diffracted field due to the particle is its Fraunhofer diffraction pattern. The hologram formed by the interference of the diffracted light and the directly transmitted light is, therefore, a Fraunhofer hologram (see section 2.6).

The permissible exposure time for recording a hologram of a moving particle field depends on the velocity of the particles. For size analysis, a useful criterion is that the particle should not move by more than a tenth of its diameter during the exposure. Typically, a 10 $\mu$m particle moving with a speed of 1 ms$^{-1}$ would require an exposure time less than $10^{-6}$ s. Suitable light sources are either a pulsed ruby laser or, where a higher repetition rate is necessary, a frequency-doubled Nd:YAG laser.

Again, to give a satisfactory reconstructed image of a spherical particle, the hologram must record the central maximum and at least three side-lobes of its diffraction pattern. This would correspond to waves travelling at a maximum angle

$$\theta_{max} = 4\lambda/d, \tag{12.2}$$

to the directly transmitted wave, and, hence, to a maximum spatial fringe frequency of $4/d$, which is independent of the values of $\lambda$ and $z$. Accordingly, for a 10 $\mu$m diameter particle, the recording material should have an MTF which extends beyond 400 mm$^{-1}$.

Films with lower resolution, which are faster and cheaper, can be used if a hologram is recorded of a magnified image of the particles. For this, the particle field is imaged near the hologram plane, as shown in fig. 12.1, with a telescopic system having a magnification of about 5. This does not affect the collimated reference beam and gives constant magnification over the whole depth of the field.

It is also apparent from this discussion that the depth of field is limited mainly by the dimensions of the recording material. From (12.2) it follows that $x$, the

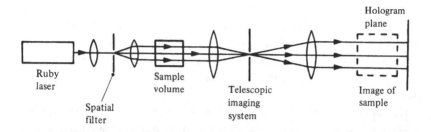

Fig. 12.1. In-line Fraunhofer holographic system for particle-size analysis.

half-width of the hologram, must be greater than $4z\lambda/d$. Hence, the maximum depth of the field over which the required resolution can be maintained is given by the relation

$$\Delta z_{max} = xd/4\lambda. \qquad (12.3)$$

For a 10 $\mu$m diameter particle and a film half-width of 1 cm, the depth of field is approximately 500 times that given by (12.1) for a conventional microscope.

To reconstruct the image, the hologram is illuminated with a collimated beam from a He-Ne laser. With normal processing, negative images are formed, but this is no problem for many technical applications. Two images are reconstructed, at equal distances $\pm z$ from the hologram and on opposite sides of it. However, with a Fraunhofer hologram, the contribution of one image in the plane of the other is essentially a constant and, therefore, does not degrade the image significantly.

Off-axis holography has also been used for particle-size analysis [Wuerker & Heflinger, 1970]. Apart from the fact that the two reconstructed images are completely separated and positive images are obtained, its major advantage is that the reference beam need not traverse the sample volume. This makes it possible to study samples with poor transmission in reflected light. Its disadvantages are the need for a separate reference-beam path and much more stringent requirements on the coherence of the illumination and the resolution of the recording medium.

A major problem in particle-field holography has been the drop in the contrast of the diffraction pattern of a particle with decreasing particle diameter, as well as with increasing distance to the hologram plane [Dunn & Thompson, 1982]. This drop in contrast can be reduced by several techniques, including the use of a Gaussian beam [Vikram & Billet, 1983], adjustment of the exposure to obtain a hologram with a low-amplitude transmittance [Vikram & Billet, 1984], and spatial filtering to attenuate the directly transmitted beam [Özkul,

Allano & Trinité, 1986], A significant improvement in contrast can also be obtained by using a divergent beam [Witherow, 1979]. A detailed analysis by Vikram and Billet [1988] has shown that the fringe contrast improves by a factor equal to $m_0$, the projected magnification. In addition, the fringe spacing increases by a factor $m_0$, so that high-speed films with lower resolution can be used.

The applications of holographic particle-size analysis include studies of fog droplets, dynamic aerosols, and marine plankton. Another significant area of application has been in bubble-chamber photography of short-lived particles, where high-resolution images of interactions in volumes up to 2.5 $m^3$ have been recorded [Haridas *et al.*, 1985; Brucker, 1991]. Double-exposure holography with a suitable time interval between the light pulses has also been used to measure the velocity spectrum of moving particles. A rapid method of analysis is based upon observations in the Fraunhofer plane on reconstruction [Ewan, 1979]. The diffraction pattern of the original particle field is modulated by fringes whose spacing can be shown to be related to the velocity of the particles. With unsteady flows, a small volume can be selected by an aperture placed at any desired point in the real image. The spatial autocorrelation of the field can then be processed to retrieve information on the three-dimensional velocity vector [Coupland & Halliwell, 1988, 1992].

## 12.3 Imaging through moving scatterers

Holography can be used quite effectively to produce an image of a stationary object masked by moving scatterers [Stetson, 1967*b*; Spitz, 1967]. This is possible because light scattered from a moving particle has its frequency shifted by the Doppler effect. Accordingly, if a hologram of the object is recorded with a reference beam which does not pass through the scattering medium, only the directly transmitted light contributes to the formation of the hologram. The light scattered by the moving particles cannot interfere with the reference beam and merely adds a constant exposure to the hologram plate. This has little effect on the reconstructed image as long as the recording is linear [Hamasaki, 1968]; however, there is a decrease in diffraction efficiency when the amount of scattered light is large.

The simplest way to eliminate this unwanted background is by using a dynamic recording medium, such as a photorefractive crystal [Tontchev & Zhivkova, 1992]. In this case, if the fringes formed by a scatterer in the field move at a rate that is greater than the rate of formation of the photorefractive grating, they produce only a weak recording that decays rapidly. As a result, a significant gain in the signal-to-noise ratio is obtained.

## 12.4 Imaging through distorting media

Another situation where holography permits forming an undistorted image of an object is where the reference wave can be made to undergo the same distortion as the object wave [Goodman, Huntley, Jackson & Lehmann, 1966]. This is possible if the distorting medium is very thin and lies close to the hologram plane, or if the angular separation of the two waves is very small.

Let the complex amplitudes of the object wave and the reference wave incident on the medium be $o(x, y)$ and $r(x, y)$, respectively. If the distorting medium modulates only the phase of an incident wave, its amplitude transmittance can be written as $\exp[-i\phi(x, y)]$. The complex amplitudes of the object wave and the reference wave at the hologram are then $o(x, y) \exp[-i\phi(x, y)]$ and $r(x, y) \exp[-i\phi(x, y)]$. If we assume linear recording, as defined by (2.2), the amplitude transmittance of the hologram is

$$\mathbf{t}(x, y) = \mathbf{t}_0 + \beta T \, |r(x, y) \exp[-i\phi(x, y)] + o(x, y) \exp[-i\phi(x, y)]|^2,$$
$$= \mathbf{t}_0 + \beta T \, |r(x, y) + o(x, y)|^2. \tag{12.4}$$

Since the phase variations $\phi(x, y)$ due to the distorting medium have been eliminated, an undistorted image is formed when the hologram is illuminated by the undistorted reference wave $r(x, y)$.

## 12.5 Correction of aberrated wavefronts

Holography can also be used, within certain limits, to recover an image of an object, unaffected by lens aberrations [Upatnieks, Vander Lugt & Leith, 1966] or by the presence of an aberrating or diffusing medium in the optical path [Kogelnik, 1965; Leith & Upatnieks, 1966].

As shown in fig. 12.2(*a*), the object is illuminated with coherent light, and a hologram is recorded of the aberrated image wave using a collimated reference beam. Let the complex amplitude of the aberrated object wave in the hologram plane be

$$o(x) = |o(x)| \exp[-i\phi(x)]. \tag{12.5}$$

In the reconstruction step, the hologram is illuminated, as shown in fig. 12.2(*b*), by the conjugate to the original reference wave. The hologram then reconstructs the conjugate to the original object wave, whose complex amplitude in the hologram plane can be written, apart from a constant factor, as

$$o^*(x) = |o(x)| \exp[i\phi(x)]. \tag{12.6}$$

This wave has exactly the same phase errors as the original object wave, except

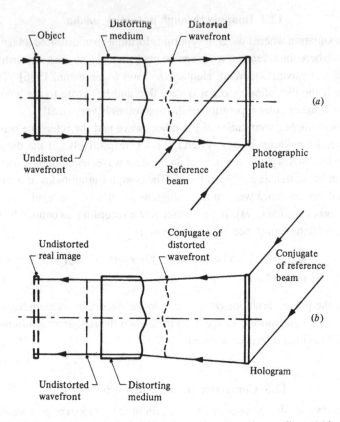

Fig. 12.2. Holographic system for imaging through an aberrating medium: (*a*) hologram recording, (*b*) image reconstruction.

that they are of the opposite sign. Hence, when this wave propagates back through the optical system, the phase errors cancel out exactly, so that the wave emerging from it is the undistorted object wave. As a result, a diffraction-limited real image of the object is formed in its original position.

Generation of an undistorted image in real time is possible with a hologram recording medium such as BSO (see section 7.7.2.). An optical system for this purpose is shown in fig. 12.3 [Huignard, Herriau, Aubourg & Spitz, 1979]. Continuous generation of the phase-conjugate wavefront is possible using an $Ar^+$ laser ($\lambda = 488$ nm) at power levels around $6W/m^2$. Generation of the phase-conjugate wavefront can also be looked at as an example of four-wave mixing in the BSO crystal.

A limitation of these methods is the need for the object beam to traverse the

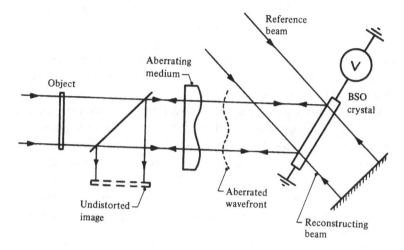

Fig. 12.3. Real-time imaging through an aberrating medium [Huignard *et al.*, 1979].

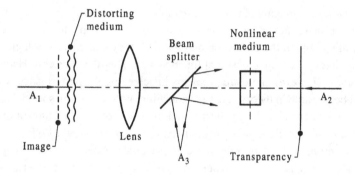

Fig. 12.4. Arrangement used to project an image of a transparency through an aberrating medium without distortion [Yariv & Koch, 1982].

aberrating medium twice, so that the image is formed on the same side of the aberrating medium as the object. A one-way technique described by Kogelnik and Pennington [1968] has been modified and made applicable to real-time imaging by Yariv and Koch [1982].

The scheme described by them is shown in fig. 12.4. In this arrangement, a plane wave $A_1$ passes through the aberrating medium, which is imaged on a nonlinear medium. At the same time, the object wave $A_2$ from an illuminated transparency is incident on the nonlinear medium from the opposite side. When the hologram created by these two beams is illuminated by a spherical wave $A_3$

propagating along the same axis as the object wave, but in the opposite direction, it produces an aberration-free image on the other side of the aberrating medium.

With this technique, perfect compensation is possible for only a thin aberrating medium. This limitation can be avoided with a more elaborate method of imaging through inhomogeneities involving phase-conjugation, which has been described by Cunha and Leith [1988].

It is also possible to use a hologram to correct the aberrations in a telescope objective. This technique does not rely on phase-conjugate reconstruction, but makes use of a correcting plate which is an image hologram of the aberrations of the objective [Munch & Wuerker, 1989]. To produce the hologram, the imperfect telescope is illuminated with a collimated beam of light, and an off-axis hologram is recorded in the plane in which the objective is imaged by the eyepiece. When this hologram is replaced in its original position, it reconstructs an aberration-free image.

## 12.6 High-resolution projection imaging

High-resolution imaging of photo-lithographic masks is an essential step in the production of semiconductor devices, but even the best photographic lenses have limitations on the resolution and the field they can cover. Holographic imaging has considerable potential for such work [Beesley, Foster & Hambleton, 1968]. However, there are problems in achieving the performance which should be possible in theory, due to degradation of the image by speckle, noise introduced by the recording material, and the need for exact alignment of the hologram with the illuminating beam [Champagne & Massey, 1969].

These problems can be minimized if the object plane and the hologram plane are very close to each other. A system for this purpose described by Stetson [1967a] is shown in fig. 12.5. In this system, a prism is cemented on to the back of the hologram plate. This prism makes it possible to bring in a collimated reference beam that is incident on the outer face of the emulsion at an angle greater than the critical angle, so that it is totally reflected. The object transparency, which is separated from the emulsion layer by a layer of tape, is illuminated from above, light transmitted through the emulsion being reflected out sideways by the prism.

In this arrangement, two holograms are recorded in the same emulsion. One is a reflection hologram due to interference of the object wave with the incident reference wave; the other is a transmission hologram formed by the object wave and the totally reflected reference wave.

When the hologram is illuminated with the conjugate of the original reference wave, this wave is diffracted by the transmission hologram and reconstructs the

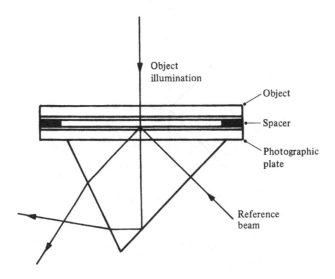

Fig. 12.5. Optical system used to record a hologram with a totally reflected reference wave [Stetson, 1967a].

conjugate of the original object wave. The component not diffracted by this hologram, which is totally reflected at the surface of the emulsion, is then diffracted by the reflection hologram and adds to this reconstructed wave. As a result, a real image is formed in the original object plane. Undiffracted light leaves the other face of the prism and does not affect the projected real image.

The proximity of the image plane to the hologram in this setup minimizes the effects of deviations from flatness of the hologram, and makes the image relatively insensitive to misalignment of the reconstruction beam and the spectral bandwidth of the source. With a proper choice of recording parameters, a resolution of approximately $600 \text{ mm}^{-1}$ was obtained [Stetson, 1968a].

## 12.7 Evanescent-wave holography

The experiments described above led to the study of holographic techniques using evanescent waves. Evanescent waves occur in diffraction and total reflection [Born & Wolf, 1980], but, because they are exponentially damped, exist only very close to the surface where they are formed. Holography can be used to record evanescent waves; the hologram can then reconstruct the information they carry in the form of homogeneous or propagating waves [Bryngdahl, 1973].

As shown schematically in fig. 12.6, an evanescent wave is formed by total

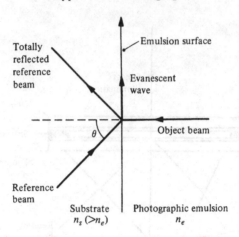

Fig. 12.6. Hologram recording with an evanescent wave.

internal reflection at the surface of the photographic emulsion in which the hologram is to be recorded. For this to happen, the emulsion must either be coated on a substrate having a higher refractive index [Nassenstein, 1968*b*] or immersed in a liquid with a relatively high refractive index [Bryngdahl, 1969*b*]. The evanescent wave then propagates along the interface. Its wavelength is

$$\lambda_e = \lambda_0/(n_s \sin \theta), \tag{12.7}$$

where $\lambda_0$ is the wavelength of the radiation in vacuum, $n_s$ is the refractive index of the medium adjacent to the photographic emulsion and $\theta$ is the angle of incidence. We assume that this wave interferes with a homogeneous wave incident normally on the interface. The wavelength of this homogeneous wave is

$$\lambda_h = \lambda_0/n_e, \tag{12.8}$$

where $n_e$ is the refractive index of the photographic emulsion.

The complex amplitude of the evanescent wave can be written as

$$a_e = a_r \exp(-2\pi n_s z d) \exp(-i2\pi x/\lambda_e), \tag{12.9}$$

where $a_r$ is the amplitude at the surface, and $d = \lambda_0/(n_s^2 \sin^2 \theta - n_e^2)^{1/2}$ is a measure of the depth to which the evanescent wave penetrates into the photographic emulsion. The complex amplitude of the homogeneous wave, which may be incident from either side, as shown in figs. 12.7(*a*) and (*b*), is

$$a_h = a_0 \exp(\pm i2\pi z/\lambda_h). \tag{12.10}$$

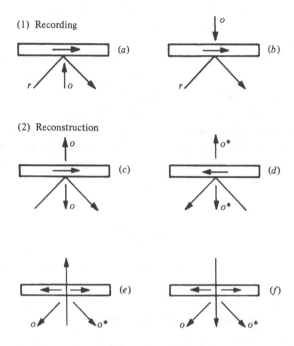

Fig. 12.7. Recording and reconstruction schemes for holograms formed by a homogeneous object wave $o$ and an evanescent reference wave $r$. Both the recording schemes shown, ($a$) as well as ($b$), give the same reconstructed waves with the reconstruction schemes shown in ($c$), ($d$), ($e$) and ($f$) [Bryngdahl, 1969$b$].

The intensity in the interference pattern formed by these two waves is then

$$I = |a_e + a_h|^2,$$
$$= a_0^2 + a_r^2 \exp(-4\pi n_s z/d)$$
$$+ 2a_0 a_r \exp(-2\pi n_s z/d) \cos\{2\pi[(x/\lambda_e) - (z/\lambda_h)]\}.$$

$$(12.11)$$

The interference pattern described by the last term in (12.11) is rapidly attenuated as the wave moves from the surface ($z = 0$) into the emulsion. Typically, the depth of penetration of the evanescent wave is only about 1 $\mu$m, so that a thin hologram is formed, with a period equal to the wavelength $\lambda_e$ of the evanescent wave.

The waves reconstructed by such a hologram differ significantly from those reconstructed by a conventional hologram. Thus, when it is illuminated, as shown in fig. 12.7($c$), with the same evanescent wave used to record it, the hologram reconstructs the original homogeneous wave used to record it, and, instead of its conjugate, a mirror-image copy of it propagating in the opposite direction. If, however, the direction of the evanescent wave is reversed, as shown in fig. 12.7(d), the conjugate of the homogeneous wave is reconstructed, along with its mirror image. An interesting feature is that in both of these cases only a single image, unaccompanied by any directly transmitted light, appears on one side of the hologram. Finally, if a homogeneous wave is used to illuminate the hologram, as shown in figs. 12.7(e) and (f), evanescent waves are created which propagate along the hologram. If the plate is immersed in a liquid with a high refractive index, these evanescent waves are converted into homogeneous waves which leave one side of the hologram at angles greater than the critical angle in air.

It is also possible to record a hologram using two evanescent waves. These waves can interfere at any angle, provided they have a common component of polarization. Accordingly, if $\lambda_{e1}$ and $\lambda_{e2}$ are the wavelengths of the two evanescent waves, the spacing of the hologram fringes can range from

$$\Lambda = \lambda_{e1}\lambda_{e2}/(\lambda_{e1} - \lambda_{e2}), \qquad (12.12)$$

when the waves propagate in the same direction, to

$$\Lambda = \lambda_{e1}\lambda_{e2}/(\lambda_{e1} + \lambda_{e2}), \qquad (12.13)$$

when they travel in opposite directions. Since from (12.13) the minimum spacing of the fringes is $\lambda_e/2$, and from (12.7) and (12.8), $\lambda_e < \lambda_h$, it is possible to produce finer fringe patterns than with homogeneous standing waves of the same frequency.

In general, an incident evanescent wave is diffracted by such a hologram into other evanescent waves, which, when the boundary conditions are met, are converted into propagating homogeneous waves. An interesting result is that the corresponding wave vectors need not lie in the plane of incidence.

A feature of holograms recorded with evanescent waves is that white light can be used to reconstruct the image [Bryngdahl, 1969b]. This is because, as shown in fig. 12.8, constructive interference occurs for a particular wavelength only when a condition equivalent to the Bragg condition in a volume hologram is satisfied. However, in this case, changes in the thickness of the photographic emulsion do not cause a wavelength shift, because the hologram is recorded only at the surface of the emulsion.

Theoretical studies of the diffraction efficiency of evanescent-wave holograms have been made by Lukosz and Wüthrich [1974], Lee and Streifer [1978a,b], and Woznicki [1980]. They show that if light of the same frequency polarized with the electric vector perpendicular to the plane of incidence (s-

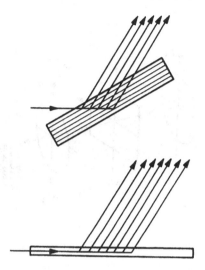

Fig. 12.8. Reconstruction by (*a*) a conventional volume hologram and (*b*) an evanescent-wave hologram. The wavelength selectivity of the hologram allows white light to be used in both cases [Bryngdahl, 1973].

polarization) is used for reconstruction as well as for recording, the diffraction efficiency exhibits sharp maxima when either the angle of incidence of the illuminating wave or the angle of diffraction of the reconstructed wave is equal to the critical angle. This phenomenon is confirmed by experimental measurements [Wüthrich & Lukosz, 1975]. The analysis also shows that the diffraction efficiency for waves polarized with the electric vector in the plane of incidence (*p*-polarization) is lower and exhibits dips not observed with the *s*-polarization. Quite high diffraction efficiency is possible at the critical angle. This tallies with earlier observations by Nassenstein [1969], who obtained a diffraction efficiency of 0.226 with an amplitude hologram. This is much higher than the theoretical maximum value (0.0625) for an amplitude hologram recorded with homogeneous waves.

An obvious application of evanescent-wave holography is in high-resolution imaging. An advantage here is that the reference and illuminating beams are confined entirely to one side of the recording medium, while the object and the image are on the other side, as shown in fig. 12.7(*b*) and (*c*). The object can therefore be located very close to the hologram, and an object field with a solid angle close to $2\pi$ can be recorded. Nassenstein [1970] has shown that information about details smaller than the normal resolution limit can be obtained if the object is illuminated with evanescent waves.

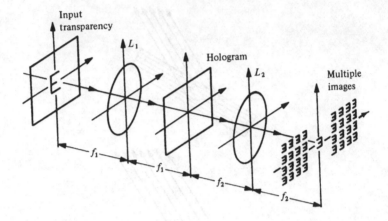

Fig. 12.9. Multiple imaging by a Fourier hologram [Lu, 1968].

Another promising application of holography with evanescent waves is in integrated optics. A hologram recorded in a layer over a planar dielectric waveguide, with a guided mode as the reference wave, can be used as a grating coupler with relatively high efficiency [Lukosz & Wüthrich, 1976; Wüthrich & Lukosz, 1980]. In addition, such holograms can provide information storage in a form compatible with integrated optics technology.

## 12.8 Multiple imaging

There are many applications in which it is necessary to produce an array of identical images. While this is normally done by making a series of exposures with a step-and-repeat camera, it is also possible to use holographic techniques.

### 12.8.1 Multiple imaging using Fourier holograms

To produce an $n \times m$ array of images separated by intervals $x_0$, $y_0$, a hologram is used with an amplitude transmittance

$$H(\xi, \eta) = \sum^{n} \sum^{m} \exp[-i2\pi(nx_0\xi + my_0\eta)]. \qquad (12.14)$$

This hologram represents a set of plane waves travelling in directions corresponding to the centres of the images in the array.

The hologram is then placed in the back focal plane of the lens $L_1$ in the optical system shown in fig. 12.9 [Lu, 1968]. If a transparency with an amplitude transmittance $f(x, y)$ located in the front focal plane of $L_1$ and illuminated by a

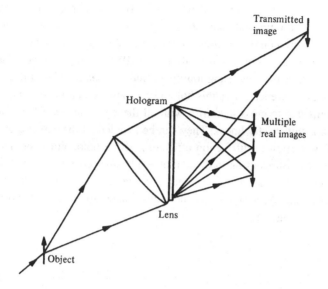

Fig. 12.10. Multiple imaging by means of a hologram of an array of point sources [Groh, 1968].

collimated beam is used as the input, its Fourier transform $F(\xi, \eta)$ is displayed in the back focal plane of $L_1$.

The wavefront emerging from the hologram can therefore be written as

$$G(\xi, \eta) = F(\xi, \eta)H(\xi, \eta), \tag{12.15}$$

and a second Fourier transform operation by the lens $L_2$ produces a set of multiple images

$$g(x, y) = f(x, y) * \sum^{n} \sum^{m} \delta(x - nx_0, y - my_0),$$

$$= \sum^{n} \sum^{m} f(x - nx_0, y - my_0). \tag{12.16}$$

### 12.8.2 Multiple imaging using lensless Fourier holograms

This technique [Groh, 1968] uses a much simpler optical arrangement. In the first step, a lensless Fourier hologram of an array of point sources $P_1 \ldots P_n$ is recorded with a point reference source. As shown in fig. 12.10, this hologram is illuminated with the conjugate to the original reference wave by means of a lens placed behind it.

With a point source, the hologram produces real images of the array of object points $P_1 \ldots P_n$ in their original positions. However, if the point source is replaced by an illuminated transparency, an array of images of the transparency is formed, centred on the positions of the original point sources $P_1 \ldots P_n$.

A variation of these methods [Kalestynski, 1973, 1976] is to record a hologram of the transparency using multiple reference beams. When illuminated with a single reference beam, this hologram produces an array of images.

Problems due to cross talk arise with all these techniques if the hologram recording is not strictly linear. They can be avoided, at the expense of a considerable reduction in diffraction efficiency, if the hologram is produced by successive exposures using individual object beams separately, rather than simultaneously. Another problem is that only the centres of the images are free from aberrations. Accordingly, the individual images must subtend only a small angle at the hologram.

# 13

# Holographic optical elements

## 13.1 Holographic diffraction gratings

Diffraction gratings formed by recording an interference pattern in a suitable light-sensitive medium (commonly called holographic diffraction gratings) have replaced conventional ruled gratings for many applications. While Burch and Palmer [1961] first showed that transmission gratings could be made by photographing interference fringes using silver halide emulsions, it was the use of photoresist layers coated on optically worked blanks which finally led to the production of spectrographic gratings of high quality [Rudolph & Schmahl, 1967; Labeyrie & Flamand, 1969]. After processing, the photoresist layer yields a relief image (see section 7.3) which can be coated with an evaporated metal layer and used as a reflection grating.

Holographic gratings have several advantages over ruled gratings. Besides being cheaper and simpler to produce, they are free from periodic and random errors and exhibit much less scattered light. In addition, it is possible to produce much larger gratings of finer pitch, as well as gratings on substrates of varying shapes, and gratings with curved grooves and varying pitch. This makes it possible to produce gratings with unique focusing properties and opens up the possibility of new designs of spectrometers [Namioka, Seya & Noda, 1976].

Against this, their main disadvantage is that the groove profile cannot be controlled as easily as in ruled gratings. While even a sinusoidal profile can give high diffraction efficiencies for small grating spacings ($\approx \lambda$) [Loewen, Maystre, McPhedran & Wilson, 1975], it is usually necessary to produce a triangular groove profile for maximum diffraction efficiency. Accordingly, a number of methods have been proposed for the production of blazed holographic gratings [Schmahl & Rudolph, 1976; Hutley, 1976, 1982].

One method of achieving this result is to expose the photoresist to a sawtooth irradiance distribution, which is built up by a process of Fourier synthesis,

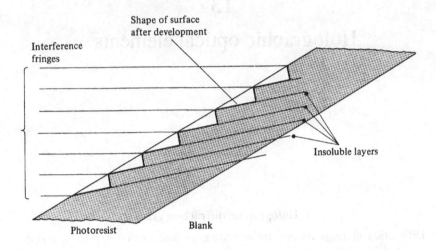

Fig. 13.1. Production of blazed gratings in a photoresist [Hutley, 1982].

either by using more than two beams to produce the fringes, or by making multiple exposures to fringe patterns of appropriate periodicities and phases [McPhedran, Wilson & Waterworth, 1973; Schmahl, 1975; Breidne, Johansson, Nilsson & Åhlen, 1979].

Another method is to start with a sinusoidal or partially blazed profile and modify it by ion-beam etching to produce a well-formed triangular profile [Aoyagi & Namba, 1976; Aoyagi, Sano & Namba, 1979].

In the most widely used method [Sheridon, 1968; Hutley, 1975], the photoresist layer is aligned obliquely to the fringe pattern, as shown in fig.13.1, to produce layers which are alternately soluble and insoluble within the thickness of the resist. After development, the surface profile is determined by the shape of the insoluble layers near the surface. The only disadvantage of this technique is that one of the beams is incident through the back of the blank which must, therefore, be of optical quality.

An optical system for this purpose is shown in fig. 13.2. In it, light from an Ar$^+$ laser ($\lambda$ = 458 nm) is split into two beams of equal intensity which are focused by microscope objectives on pinholes. Each pinhole is located at the focus of an off-axis parabolic mirror, so that a collimated beam is obtained upon reflection from the mirror. As shown in fig.13.1, the photoresist-coated blank is placed in the interference field at a small angle to the standing waves.

To produce gratings of good quality, the optics must produce wavefronts plane to $\lambda/10$. The liquid photoresist is applied to the optically worked blank, which is then spun rapidly to produce a uniform layer, about 0.5 $\mu$m thick.

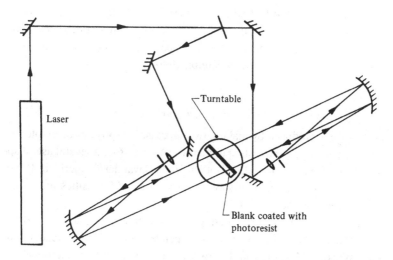

Fig. 13.2. Optical system used to produce blazed holographic gratings [courtesy I. G. Wilson, CSIRO Division of Chemical Physics, Melbourne, Australia].

In addition to the precautions normally taken to ensure stability of the fringes during the exposure, a closed-loop servo system is used to maintain the optical path difference in the interferometer stable to better than $\lambda/50$.

### 13.1.1 Zero-order gratings

As mentioned earlier, surface relief gratings with high spatial frequencies can be produced in photoresist layers by holographic techniques. If the groove spacing is smaller than the wavelength of the incident light, the power diffracted into the non-zero orders becomes negligible; such gratings are commonly called zero-order gratings (ZOGs).

Gratings with groove spacings comparable to the wavelength and groove depths equal to the groove spacing do not satisfy the criteria for thin gratings (see section 4.8). Theoretical studies by Moharam and Gaylord [1982] using the coupled-wave theory showed that diffraction efficiencies > 0.85 should be possible. Subsequently, work by Enger and Case [1983] confirmed that such high diffraction efficiencies could be achieved in practice.

A more detailed study by Moharam *et al.* [1984] showed that the diffraction efficiency varies much more rapidly with changes in the angle of incidence and the wavelength for the $TE$ polarization (electric field parallel to the grooves) than for the $TM$ polarization. This is because these gratings exhibit form birefringence [Born & Wolf, 1980], so that the effective refractive indices for the $TE$ and $TM$ polarizations are different. As a result, such gratings behave like homogeneous birefringent materials and can be used as $\lambda/4$ and $\lambda/2$ retarders [Flanders, 1983; Cescato, Gluch & Streibl, 1990].

### 13.1.2 Crossed gratings

Crossed gratings produced by making two successive exposures to an interference pattern have many interesting properties and have found several useful applications [McPhedran, Derrick & Botten, 1980; Savander & Sheridan, 1993], including the formation of light spot arrays with equal intensities for optical interconnections (see section 14.7).

Other applications for crossed gratings are as selective solar absorbers [Horwitz, 1974] as well as for polarization-independent antireflection treatment of surfaces [Wilson & Hutley, 1982]. Crossed ZOGs also offer an additional degree of flexibility when used as retarders, since the phase difference between the two polarizations can be controlled by varying the periods or filling factors of the component gratings.

## 13.2 Holographic filters

Volume reflection holograms recorded in dichromated gelatin can be used as narrow-band rejection filters (notch filters). To make such filters, a beam of laser light refracted at an angle $\theta$ within a layer of dichromated gelatin (refractive index $n$) is reflected off a mirror contacted to its back surface, to produce interference fringes parallel to the surface with a spacing

$$\Lambda = \lambda/2n\cos\theta. \tag{13.1}$$

The angle $\theta$ is chosen so that after processing [Chang & Leonard, 1979] the filter has its peak reflectance at the desired wavelength.

A major application of such notch filters has been for eye protection against laser radiation, while maintaining high visual transmittance [Magariños & Coleman, 1987; Tedesco, 1989]. If the filter has the configuration of a spherical mirror with its centre of curvature coincident with the centre of rotation of the eye, then all the rays incident towards the centre of the eye will be normal to the holographic filter. It is possible therefore, with proper design, to minimize shifts of the wavelength band rejected over a wide range of viewing angles.

Another application has been as a channel selector for wavelength-multiplexed optical fibre systems. In this case, a divergent beam from a point source is used, producing an increase in the angle of incidence along the length of the hologram and a corresponding variation of the peak reflection wavelength [Duncan, McQuoid & McCartney, 1985].

Holographic notch filters have revolutionized Raman spectroscopy, since they make it possible to suppress the Rayleigh line while allowing high transmission in the Stokes region [Carraba, Spencer, Rich & Rault, 1990]. Notch filters have been produced with a transmittance dropping from 0.7 to a few parts in $10^6$ within a wavelength range of 6 nm [Rich & Cook, 1991]. The rejection band can be tuned over a few nanometres by varying the angle of incidence.

Broad-band filters for reducing solar heating which reject the near infrared (0.7–1.2 $\mu$m), while transmitting more than 75 per cent of visible light, have also been produced in dichromated gelatin by the same technique [Rich & Petersen, 1992]. In this case, the gelatin layer is contacted to a prism with an index-matching fluid during recording, so that the angle of incidence within the gelatin layer can be greater than the critical angle, and the second interfering wave is produced by total reflection at the back surface. Filters produced by this technique typically have a transmittance of 0.84 over the visible region, and a rejection bandwidth of 360 nm.

Another interesting advance has been the production of high-finesse Fabry-Perot etalons using a pair of holographic mirrors recorded simultaneously with a collimated laser beam in dichromated gelatin coatings on the two faces of a glass substrate [Kuo et al., 1990]. Since the recording process ensures that the fringes in the second mirror are parallel to those in the first mirror, tolerances on the quality of the glass substrates are not stringent, and the peak playback wavelength is determined by the recording angle. Efficiencies of 99 per cent with a spectral bandwidth less than 20 nm are possible. If the cavity contains a thin, nematic liquid-crystal layer, the peak can be tuned over several free spectral ranges by applying an electrical field to the liquid crystal. A multi-channel narrow-band optical filter can also be realized by producing such a Fabry-Perot etalon with a wavelength-multiplexed reflection hologram as an integral part of the cavity [Lin, Chou, Strzelecki & Shellan, 1992]. The same technique can also be used to produce stratified volume holographic elements in which multiple layers of a holographic recording material are interleaved with optically homogeneous buffer layers. These structures exhibit a periodic angular sensitivity and have potential applications in optical array generation, wavelength notch filtering, and grating spatial-frequency filtering [Johnson & Tanguay, 1988; Nordin & Tanguay, 1992].

A narrow-band reflection filter for solar astronomy has also been developed by recording a volume reflection hologram produced by two counter-propagating beams in a photorefractive material [Rakuljic & Leyva, 1993].

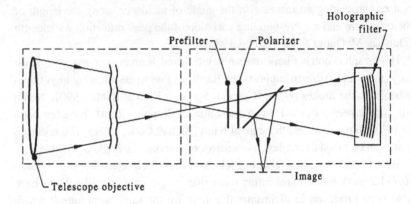

Fig. 13.3. Optical arrangement for using a narrow-band holographic filter with a telescope [Rakuljic & Leyva, 1993].

These filters are used with a telescope in the configuration shown in fig. 13.3. A bandwidth of 0.0125 nm, with a peak throughput of 10 per cent, can be obtained with an 8 mm thick filter produced in $LiNbO_3$.

### 13.3 Holographic scanners

Holographic scanners have replaced mirror scanners for many applications. They were first used commercially in point-of-sale terminals for bar code scanning [Dickson & Sincerbox, 1991], but are now incorporated in high-resolution imaging systems for applications such as photo typesetting and photocomposition [Kramer, 1991].

A simple disk hologram scanner [Cindrich, 1967; McMahon, Franklin & Thaxter, 1969] is shown in fig. 13.4. The disc has a number of holograms recorded on it, with a point source as the object and a collimated reference beam. Each of these holograms, when illuminated with the conjugate to the reference beam, forms an image of the point source. Rotating the deflector about an axis perpendicular to its surface causes the reconstructed image spot to scan the image plane.

The main problem with this simple system is that the scanning line is, in general, an arc of a circle with a radius

$$r = d \sin\theta \tan\theta, \qquad (13.2)$$

where $d$ is the distance from the scanning facet to the centre of the image plane, and $\theta$ is the angle between the diffracted principal ray and the normal to the hologram.

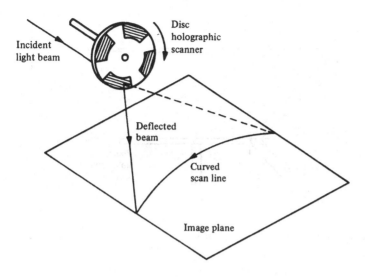

Fig. 13.4. Disc holographic scanner [Kramer, 1981].

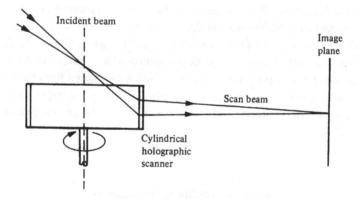

Fig. 13.5. Cylindrical holographic scanner [Kramer, 1981].

One method of obtaining a straight line scan is to use a cylindrical scanning element such as that shown in fig. 13.5 [Pole & Wolenmann, 1975; Pole, Werlick & Krusche, 1978]. However, such scanners are much more difficult to fabricate than disc scanners, which can be replicated easily.

Another method is to use an auxiliary reflector, in conjunction with a disc scanner, to make the principal diffracted ray incident normal to the imaging surface [Ih, 1977]. A better arrangement [Kramer, 1981] is shown in fig. 13.6.

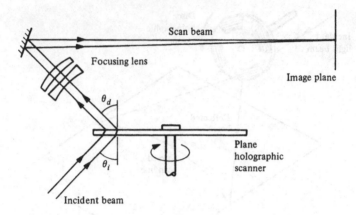

Fig. 13.6. Plane-grating holographic scanner producing a straight-line scan [Kramer, 1981].

In this system, a straight-line scan is obtained, which is self-compensated for wobble of the scanner when $\theta_i = \theta_d = 45°$. Aberrations due to the scanner [Herzig & Dändliker, 1988] are eliminated by using a separate focusing lens.

A compact straight-line scanner for laser printers using only a holographic disk and a holographic lens has been described by Hasegawa, Yamagishi, Ikeda, and Inagaki [1989]. A completely different approach has been followed in a very compact design for a point-of-sale laser scanner which uses a fixed multilayer hologram window, in conjunction with a rotating pentagonal mirror and three fixed mirrors, to generate horizontal, oblique, and vertical scan lines [Yamazaki *et al.*, 1990].

## 13.4 Holographic optical elements

A hologram can be used to transform an optical wavefront, in much the same manner as a lens [Schwar, Pandya & Weinberg, 1967]. Generalized holographic optical elements (HOEs) have been found very useful in several specialized applications, because they make possible unique system configurations and functions.

One of the main advantages of HOEs is the fact that, unlike conventional optical elements, their function is essentially independent of substrate geometry. In addition, since they can be produced on quite thin substrates, they are relatively light, even for large apertures. Another advantage is the possibility of spatially overlapping elements, since several holograms can be recorded in the same layer. Finally, HOEs provide the possibility of correcting system aberrations, so that separate corrector elements are not required.

A major consideration in the design of systems using HOEs is their optical efficiency, which can vary considerably with the angle and the wavelength. Because of this, a complete analysis of their imaging properties requires a wave-optical treatment [Forshaw, 1973]. This is not very easy in practice, and an approach based on ray tracing is commonly used in system design [Latta, 1971c], calculations being made of the amplitude as well as the direction of the diffracted ray. With high numerical aperture objectives, it is also necessary to take into account variations in the diffraction efficiency across the aperture due to the variations in the interbeam angle [Kostuk, 1991b].

Analysis of the aberrations (see section 3.4) is made more difficult by the fact that HOEs frequently have to be produced on curved substrates and are used in systems which do not have rotational symmetry. However, conditions for aberration-free imaging have been derived [Welford, 1973; Smith, 1977].

More general formulas for the aberrations of holograms recorded on a surface with any specified shape have been given by Peng and Frankena [1986] and by Verboven and Lagasse [1986]. The possibilities of compensating some of the third-order aberrations of holographic lenses recorded on quadrics of revolution for small fields of view have been studied in detail by Masajada and Nowak [1991], who have shown that a spherical geometry offers the simplest solutions.

Because the imaging characteristics of HOEs vary considerably with wavelength, they are most commonly used with quasi-monochromatic illumination, though achromatic systems have been proposed by Bennett [1976] and by Sweatt [1977] and studied experimentally by Weingärtner and Rosenbruch [1980a,b, 1982].

More recently, Faklis and Morris [1989] have made a Fresnel diffraction analysis of an imaging system that contains three lenses of arbitrary dispersion and have derived a general solution that simultaneously corrects the system for both longitudinal and lateral paraxial chromatic aberration. This general solution has been used to design a specific system consisting of real achromatic doublets and holographic lenses.

Production of HOEs involves recording a hologram in a suitable material using an exposing system which provides the appropriate wavefronts [Close, 1975]. However, once the recording system has been set up, HOEs can be produced at a rate limited only by the time taken to prepare, expose, and process the substrates. The recording material must have high resolution, good stability, high diffraction efficiency, and low scattering; photopolymers and dichromated gelatin are used most commonly. Electronic fringe stabilization is necessary because of the long exposure times.

Since these recording materials are sensitive only to blue and violet light, it is necessary to design the system so as to produce, with such wavelengths, an HOE that can give aberration-free performance at the wavelength at which it is to be used [Latta & Pole, 1979]. This can also be done by introducing aberration-compensating elements in the recording setup [Malin & Morrow, 1981].

An attractive possibility is the use of computer-generated holograms to produce the wavefronts [Bryngdahl, 1975; Fairchild & Fienup, 1982]. Since a CGH can produce a wavefront having any arbitrary shape, it is possible to realize a variety of unusual optical components and to achieve geometrical transformations as well as phase transformations.

A CGH which can transform rings to a line of points whose position corresponds to the radius of the ring has been described by Cederquist and Tai [1984]. Such an element makes it possible to replace a ring detector by a linear detector array in applications such as optical fibre data transmission and optical spatial frequency analysis.

A major area of application of HOEs has been in conjunction with laser diodes to correct the divergence and astigmatism of the beam [Hatakoshi & Goto, 1985; Chen, Hershey & Leith, 1987]. The problem of aberrations arising from a difference in the recording and readout wavelengths can be solved by recording the final hologram with an aberrated wavefront derived from an intermediate hologram [Amitai & Friesem, 1988; Amitai, Friesem & Weiss, 1990]. However, since the output frequency of laser diodes varies with the operating conditions as well as with aging, it is also necessary to compensate for chromatic dispersion, which can cause variations of the spot size and position. This can be done with an assembly of two holograms [Amitai & Goodman, 1991b]. In addition, by deliberately introducing a certain amount of spherical aberration, it is possible to increase the focal depth and, thereby, to relax the tolerance on the focal distance [Aharoni, Goodman & Amitai, 1993].

An interesting device described by Shariv, Amitai, and Friesem [1993] is a compact holographic beam-expander consisting of two holographic lenses on a single plate. The first lens converts the narrow input beam into a strongly diverging spherical wave making a large angle with the axis. This wave undergoes several total reflections as it propagates along the plate, until it reaches the second lens and emerges as an expanded plane wave.

Another useful application has been to convert a Gaussian beam from a laser into a uniform circular or rectangular beam to obtain greater efficiency. In this case, it is necessary to use two HOEs; the first redistributes the energy uniformly, and the second recollimates the beam [Han, Ishii & Murata, 1983; Roberts, 1989; Eismann, Tai & Cederquist, 1989].

An interesting application of holographic optical elements has been to generate beams with an amplitude profile described by a Bessel function [Vasara, Tu-runen & Friberg, 1989; Cox & Dibble, 1991]. Such a beam has the property that its intensity profile does not change as it propagates, making it very useful for precision alignment. MacDonald, Chrostowski, Boothroyd, and Syrett [1993] have described a binary-phase reflective HOE which, when illuminated with a diode laser, produces a beam with a $1/e$ amplitude radius for the central lobe of 100 $\mu$m whose intensity profile remains unchanged over a distance of 1 m.

A simpler method of building up a specific image light distribution, which is acceptable for many applications, is by means of a multifacet hologram [Case, Haugen & Løkberg, 1981]. The surface of the hologram is divided into a number of small elements (facets), each of which contains a grating of a specific spatial frequency and orientation. When this hologram is illuminated by a plane or spherical wave, the light incident on each facet is diffracted to a specific location in the image plane. The image is thus built up of small patches of light, with near 100 per cent efficiency.

The applications of HOEs are mainly in light-weight systems performing quite complex functions [Mosyakin & Skrotskii, 1972]. They have also found use as beam splitters and beam combiners [Case, 1975] and in multiple-wave-length systems. Since aberration correction is possible via the optical recording setup, elements which perform quite complex phase transformations can be duplicated at relatively low cost.

A major application has been in head-up displays [McCauley, Simpson & Murbach, 1973; Close, 1975; Withrington, 1978]. Fisher [1989] has described the design procedure for the curved combiner in a wide-field display for aircraft. While aspheric waves must be used to record the HOEs for optimum results, the need for specialized optics or computer-generated holograms to produce these aspheric waves can be avoided by a recursive technique using intermediate conventional holograms [Amitai & Friesem, 1989].

Another large-scale application has been in optical heads for compact disc players [Lee, 1989]. In a typical arrangement, the HOE diffracts the outgoing beam to produce three focused spots on the disc surface. The centre spot is used in conjunction with a quadrant detector to focus the beam and to read out information, and the two outer spots provide a tracking error signal.

A potential application of volume holograms recorded with visible light in dichromated gelatin and photopolymers is in the production of high-efficiency, grazing-incidence optical elements, such as gratings, lenses, mirrors, and beam splitters, for x-ray and XUV microscopy and astronomy. Peak diffraction efficiencies up to 28 per cent at the Bragg angle have been obtained [Jannson, Savant & Qiao, 1989].

# 14

# Information storage and processing

As mentioned in Chapter 1, the development of holography was greatly stimulated by the application of the concepts of communication theory to optics. Early work on these lines [Elias, Grey & Robinson, 1952; Marèchal & Croce, 1953] led to the development of spatial filtering techniques using coherent light [O'Neill, 1956; Cutrona, 1960; Tsujiuchi, 1963]. In turn, holographic techniques were found extremely useful in the synthesis of complex filters used in image enhancement and restoration, as well as matched spatial filters used for optical pattern recognition [Vander Lugt, 1964].

Optical information processing has been a major field of research (see Casasent [1978], Lee [1981] for surveys of early work). This chapter will therefore be limited to a brief discussion of some aspects having close links with holography as well as the use of holography for information storage.

## 14.1 Associative storage

A hologram is merely a record of the interference pattern formed by two waves. If the recording process is linear, as defined by (2.2), the transmittance of the hologram can be written as

$$t(x, y) = t_0 + \beta T [|o(x, y)|^2 + |r(x, y)|^2$$
$$+ o(x, y)r^*(x, y) + o^*(x, y)r(x, y)], \qquad (14.1)$$

where $t_0$ is a uniform background transmittance, $T$ is the exposure time, $\beta$ is a constant, and $o(x, y)$, and $r(x, y)$ are the complex amplitudes of the object wave and the reference wave, respectively.

Normally, the hologram is illuminated with the reference wave. As shown in Chapter 2, it then reconstructs the object wave. However, it is apparent that if both $|r(x, y)|^2$ and $|o(x, y)|^2$ are constants, (14.1) is symmetrical as far as the two waves are concerned. If, then, the hologram is illuminated with the object

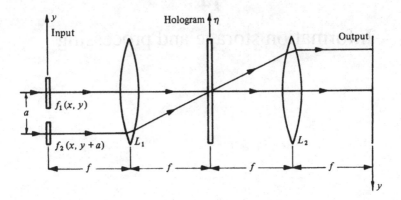

Fig. 14.1. Optical system used to analyse the conditions for associative storage of two waves.

wave, it reconstructs the reference wave. Such a situation, where illumination of the hologram with either one of a pair of waves results in reconstruction of the other, can be described by saying that the two complex amplitudes are stored in association.

To obtain the conditions for faithful imaging in such a case, consider the optical system shown in fig. 14.1. This setup is essentially the same as that used to record and reconstruct a Fourier transform hologram. If two transparencies whose amplitude transmittances are $f_1(x, y)$ and $f_2(x, y + a)$ are located in the front focal plane of the lens $L_1$ and illuminated by a plane wave of unit amplitude, the resultant complex amplitude in the hologram plane is

$$G(\xi, \eta) = F_1(\xi, \eta) + F_2(\xi, \eta) \exp(-i2\pi \eta a), \qquad (14.2)$$

where $F_1(\xi, \eta) \leftrightarrow f_1(x, y)$, and $F_2(\xi, \eta) \leftrightarrow f_2(x, y)$.

If we assume linear recording, the amplitude transmittance of the resulting hologram is

$$
\begin{aligned}
\mathbf{t}(\xi, \eta) &= \mathbf{t}_0 + \beta T \, |G(\xi, \eta)|^2, \\
&= \mathbf{t}_0 + \beta T \, [|F_1(\xi, \eta)|^2 + |F_2(\xi, \eta)|^2 \\
&\quad + F_1(\xi, \eta) F_2^*(\xi, \eta) \exp(i2\pi \eta a) \\
&\quad + F_1^*(\xi, \eta) F_2(\xi, \eta) \exp(-i2\pi \eta a)].
\end{aligned} \qquad (14.3)
$$

If this hologram is replaced in the same position in which it was recorded and illuminated by the wave due to $f_2(x, y + a)$ alone, the complex amplitude transmitted by the hologram is

$$
\begin{aligned}
H(\xi, \eta) = {}& \mathbf{t}(\xi, \eta) F_2(\xi, \eta) \exp(-\mathrm{i}2\pi\eta a), \\
= {}& \mathbf{t}_0 F_2(\xi, \eta) \exp(-\mathrm{i}2\pi\eta a) \\
& + \beta T [|F_1(\xi, \eta)|^2 + |F_2(\xi, \eta)|^2] F_2(\xi, \eta) \exp(-\mathrm{i}2\pi\eta a) \\
& + \beta T |F_2(\xi, \eta)|^2 F_1(\xi, \eta) \\
& + \beta T F_2^2(\xi, \eta) F_1^*(\xi, \eta) \exp(-\mathrm{i}4\pi\eta a).
\end{aligned}
\tag{14.4}
$$

The complex amplitude in the back focal plane of $L_2$ is then given by the relation

$$
h(x, y) \leftrightarrow H(\xi, \eta).
\tag{14.5}
$$

The only term of interest in $h(x, y)$ is the primary reconstructed image, which is given by the Fourier transform of the third term on the right-hand side of (14.4). If we neglect a constant factor, its complex amplitude is

$$
\begin{aligned}
h_3(x, y) = {}& \mathcal{F}\{|F_2(\xi, \eta)|^2 F_1(\xi, \eta)\}, \\
= {}& [f_2(x, y) \star f_2(x, y)] * f_1(x, y).
\end{aligned}
\tag{14.6}
$$

It follows from (14.6) that a perfect image of $f_1(x, y)$ will be reconstructed if the autocorrelation of $f_2(x, y)$ is a delta function. This is obviously the case if $f_2(x, y)$ is a point source.

However, it is also very nearly true if $f_2(x, y)$ is a randomly varying transmittance. In this case, the autocorrelation function of $f_2(x, y)$ has a narrow peak superimposed on a relatively weak background whose extent is twice that of $f_2(x, y)$. The finer the spatial structure of $f_2(x, y)$, and the greater its extent, the closer its autocorrelation approximates to a delta function (see Appendix A2.2). In this case, though, it is essential to preserve the geometry of the recording setup for the image to be reconstructed. This follows from (14.6), from which it is apparent that if the reference source moves by the width of the autocorrelation peak, the image will disappear.

The feasibility of using an extended diffusing object as a reference source has been demonstrated in experiments involving what is called "ghost imaging" [Collier & Pennington, 1966]. Two metal bars, one vertical and the other horizontal, were illuminated with light from a laser, and the scattered light was allowed to fall on a photographic plate. After processing, the photographic plate was replaced precisely in its original position, and the horizontal bar was removed. When the plate was illuminated with the scattered light from the vertical bar, it was found to reconstruct an image of the horizontal bar.

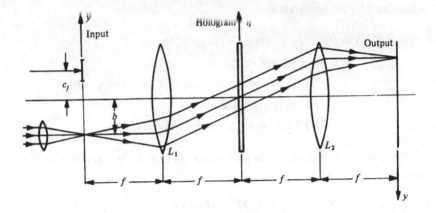

Fig. 14.2. Optical system used for experiments in pattern recognition.

## 14.2 Pattern recognition

The property of associative storage makes it possible to use a hologram for pattern recognition. This is essentially a spatial filtering operation in which the hologram functions as a matched filter [van Heerden, 1963a].

A typical optical system for this purpose is shown in fig. 14.2. To produce the matched filter, a transparency of the set of patterns to be identified is placed in the input plane, and a hologram of this transparency is recorded in the Fourier transform plane using a point reference source. For simplicity, we assume that the complex amplitude in the input plane due to the input transparency is a one-dimensional distribution

$$f(y) = \sum_{j=1}^{N} f_j(y - c_j), \qquad (14.7)$$

where $f_j(y - c_j)$ is the complex amplitude due to a typical pattern centred at $c_j$, while that due to the reference source is $\delta(y + b)$.

The transmittance of the resulting hologram is then

$$\mathbf{t}(\eta) = \mathbf{t}_0 + \beta T [1 + |F(\eta)|^2 + F^*(\eta) \exp(-i2\pi\eta b)$$
$$+ F(\eta) \exp(i2\pi\eta b)], \qquad (14.8)$$

where $F(\eta) \leftrightarrow f(y)$.

The hologram is replaced, after processing, in exactly the same position in which it was recorded and illuminated by a single pattern from the set, centred on the axis. If the amplitude due to this pattern in the input plane is $f_l(y)$, the transmitted amplitude at the hologram is

$$H(\eta) = F_l(\eta)\text{t}(\eta),$$
$$= (\text{t}_0 + \beta T)F_l(\eta) + \beta T F_l(\eta)|F(\eta)|^2$$
$$+ \beta T F_l(\eta)F^*(\eta)\exp(-\text{i}2\pi\eta b)$$
$$+ \beta T F_l(\eta)F(\eta)\exp(\text{i}2\pi\eta b). \qquad (14.9)$$

The complex amplitude in the output plane is then the Fourier transform of (14.9), which is

$$h(y) = (\text{t}_0 + \beta T)f_l(y) + \beta T f_l(y) * [f(y) \star f(y)]$$
$$+ \beta T f_l(y) \star f(y) * \delta(y + b)$$
$$+ \beta T f_l(y) \star f(y) * \delta(y - b). \qquad (14.10)$$

The only term of interest in (14.10) is the last but one on the right-hand side, which corresponds to the correlation of $f_l(y)$ with all the patterns of the set. If we ignore the constant factor $\beta T$, this term can be expanded as

$$f_l(y) \star f(y) * \delta(y + b) = [f_l(y) \star \sum_{j=1}^{N} f_j(y - c_j)] * \delta(y + b),$$
$$= [f_l(y) \star f_l(y)] * \delta(y + c_l + b)$$
$$+ \left[ f_l(y) \star \sum_{j=1, j \neq l}^{N} f_j(y - c_j) \right] * \delta(y + b).$$
$$(14.11)$$

If the autocorrelation function of the pattern presented is sharply peaked, the first term on the right-hand side of (14.11) represents a bright spot of light, which is the reconstructed image of the reference source, located at $y = -c_l - b$. The presence of this bright spot in the output plane corresponds to recognition of the pattern presented as one belonging to the original set. The fact that this image is reconstructed at a distance $-c_l$ from its correct position identifies the pattern presented as $f_l(y - c_l)$.

This basic pattern recognition technique has been extended by Vander Lugt, Rotz, and Klooster [1965] to permit simultaneous identification of all the characters on a page.

When real-time operation is not required, a more direct technique can be used [Rau, 1966; Weaver & Goodman, 1966]. This involves the use of two transparencies in the input plane. One of these $f_1(y)$ is a transparency of the character to be located, while the other $f_2(y + b)$ is a transparency of the page of characters to be searched. The transmittance of the Fourier hologram formed with these two inputs is then

$$\mathbf{t}(\eta) = \mathbf{t}_0 + \beta T [|F_1(\eta)|^2 + |F_2(\eta)|^2$$
$$+ F_1^*(\eta) F_2(\eta) \exp(-i2\pi \eta b)$$
$$+ F_1(\eta) F_2^*(\eta) \exp(i2\pi \eta b)], \qquad (14.12)$$

where $F_1(\eta) \leftrightarrow f_1(y)$ and $F_2(\eta) \leftrightarrow f_2(y)$.

If this hologram is illuminated with a plane wave, the complex amplitude in the output plane is proportional to the Fourier transform of $\mathbf{t}(\eta)$. As before, the only term of interest is the third term within the square brackets, which, if we neglect a constant factor, is

$$\mathcal{F}\{F_1^*(\eta) F_2(\eta) \exp(-i2\pi \eta b)\} = [f_1(y) \star f_2(y)] * \delta(y + b). \qquad (14.13)$$

If $f_2(y)$ is identical with $f_1(y)$, this term will result in a bright autocorrelation peak at $y = -b$. If, however, $f_2(y)$ contains more than one such character $f_j(y - c_j)$, identical with $f_1(y)$ but located at different positions, an equal number of autocorrelation peaks will be formed at locations $y = b - c_j$, corresponding to the centres of these characters.

Techniques and applications of holographic pattern recognition, including the use of computer-generated holograms, have formed the subject of several reviews [Horner, 1984; Casasent, 1985].

Interesting possibilities have been opened up by the development of electronically addressable spatial light modulators (SLMs) which make it possible to generate holographic matched filters in real time [Psaltis, Paek & Venkatesh, 1984; Mok, Diep, Liu & Psaltis, 1986]. However, a drawback of these devices is their limited space-bandwidth product. Schemes to minimize the effects of the resulting quantization errors have been presented by Mait and Himes [1989].

## 14.3 Coding and multiplexing

The technique of associated storage described in section 14.2 need not be restricted to the case of a single pair of wavefronts. It is possible to record holograms of a series of subjects on the same plate using a different coded reference wave for each hologram. Typically, this can be done by using an illuminated ground-glass screen as the reference source and moving the screen through a small distance ($\approx 20\mu$m) between exposures. If the processed hologram is replaced in exactly the same position in which it was recorded, any one of the stored images can be recovered separately by illuminating the multiplexed hologram with the appropriate coded reference wave, as can be seen from the following analysis [La Macchia & White, 1968].

Let $O_i(\eta)$ be the complex amplitude due to the $i$th object at the hologram and $R_i(\eta)$ be that due to the corresponding coded reference wave. If $N$ such exposures are recorded on the same photographic plate, the transmittance of the hologram is

$$\mathbf{t}(\eta) = \mathbf{t}_0 + \beta T \sum_{i=1}^{N} |O_i(\eta) + R_i(\eta)|^2. \tag{14.14}$$

If the processed hologram is replaced in exactly the same position in which it was recorded and illuminated by the $j$th reference wave, the transmitted amplitude is

$$H(\eta) = R_j(\eta)\mathbf{t}(\eta). \tag{14.15}$$

The only term in the expansion of (14.15) that is of interest is, as before, that corresponding to the primary reconstructed images, which is, apart from a constant factor

$$H_3(\eta) = R_j(\eta) \sum_{i=1}^{n} R_i^*(\eta)O_i(\eta),$$

$$= R_j(\eta)R_j^*(\eta)O_j(\eta) + R_j(\eta) \sum_{i=1, i\neq j}^{N} R_i^*(\eta)O_i(\eta).$$

$$\tag{14.16}$$

The complex amplitude in the image plane is then

$$h_3(y) = \mathcal{F}\{H_3(\eta)\},$$

$$= r_j(y) \star r_j(y) * o_j(y) + r_j(y) \star \sum_{i=1, i\neq j}^{N} r_i(y) * o_i(y).$$

$$\tag{14.17}$$

The first term on the right-hand side of (14.17) corresponds to the reconstructed image of the $j$th object (the signal). This image is superimposed on a diffuse halo due to the remaining cross-correlation terms (noise).

Since the diffracted power corresponding to each of the terms in the expansion of (14.17) is the same, it follows that the ratio of the signal power to the total noise power, measured in the hologram plane, is

$$I_S/I_N = 1/N. \tag{14.18}$$

This could result in a very poor signal-to-noise ratio (see section 6.8) when the number of holograms multiplexed on the same plate is large, but for the fact that we are concerned with the ratio of the intensities in the image plane. The most favourable case is where the image is that of a point source; the image power is then concentrated in a small, relatively bright spot, while the noise power is spread out over an area having twice the lateral extent of the coded source and, hence, is of relatively low intensity. In such a case, it is possible to superimpose more than 1000 coded reference holograms of individual points on a single plate, the practical limit being set mainly by the drop in the diffraction efficiency of the individual holograms, which is inversely proportional to the square of the number of exposures (see section 4.10).

## 14.4 Image processing

Holographic spatial filtering can also be used to improve a picture taken with an imperfectly corrected optical system, provided its impulse response is known [Stroke & Zech, 1967]. We assume that the picture is a positive transparency, which has been processed so that the product of the values of $\gamma$ (see Appendix 5) for the negative and positive materials is equal to 2. The amplitude transmittance of the transparency can then be written as

$$h(x, y) = f(x, y) * g(x, y),    \tag{14.19}$$

where $f(x, y)$ would have been its amplitude transmittance with a perfectly corrected optical system, and $g(x, y)$ is the impulse response of the system. The latter can be defined by another transparency, which is the image of a bright point recorded, under the same conditions, with the same optical system.

To retrieve $f(x, y)$, the transparency $h(xy)$ is placed in the front focal plane of the lens $L_1$, in an optical system such as that shown in fig. 14.1, and illuminated with a collimated beam. The complex amplitude in the back focal plane of $L_1$ is then the Fourier transform of $h(x, y)$ and is given by the relation

$$H(\xi, \eta) = F(\xi, \eta)G(\xi, \eta),    \tag{14.20}$$

where $F(\xi, \eta) \leftrightarrow f(x, y)$ and $G(\xi, \eta) \leftrightarrow g(x, y)$. This result can be rewritten as

$$\begin{aligned} F(\xi, \eta) &= H(\xi, \eta)/G(\xi, \eta), \\ &= H(\xi, \eta)[G^*(\xi, \eta)/|G(\xi, \eta|^2]. \end{aligned}    \tag{14.21}$$

It is apparent from (14.21) that if a filter with an amplitude transmittance $G^*(\xi, \eta)/|G(\xi, \eta|^2$ is inserted in the back focal plane of $L_1$, the complex amplitude in the back focal plane of $L_2$ will yield $f(x, y)$. Such a filter can be produced by superimposing two filters with amplitude transmittances of $|G(\xi, \eta|^{-2}$ and $G^*(\xi, \eta)$.

To produce the first filter, the transparency $g(x, y)$ is placed in the front focal plane of $L_1$, and the intensity distribution in its back focal plane is recorded on a photographic plate. If this plate is processed so that its $\gamma = 2$, its amplitude transmittance is

$$\mathbf{t}_1(\xi, \eta) = [|G(\xi, \eta)|^2]^{-\gamma/2},$$
$$= |G(\xi, \eta)|^{-2}. \qquad (14.22)$$

To produce the second filter, the optical system is modified, as shown in fig. 14.2, by the introduction of a collimated reference beam, and a Fourier hologram of $g(x, y)$ is recorded on a fresh photographic plate placed in the back focal plane of $L_1$. If we assume linear recording, as defined by (2.2), the transmittance of this hologram can be written as

$$\mathbf{t}_2(\xi, \eta) = \mathbf{t}_0 + \beta T[1 + |G(\xi, \eta)|^2 + G(\xi, \eta)\exp(\mathrm{i}2\pi\eta b)$$
$$+ G^*(\xi, \eta)\exp(-\mathrm{i}2\pi\eta b)]. \qquad (14.23)$$

The last term on the right-hand side of (14.23) contains the required transmittance. Hence, when the two filters $\mathbf{t}_1(\xi, \eta)$ and $\mathbf{t}_2(\xi, \eta)$ are superimposed and placed in the back focal plane of $L_1$, and the transparency $h(x, y)$ is illuminated with a plane wave, the corresponding term in the output plane yields $f(x, y)$.

Simplified techniques by which both the filters can be recorded on the same plate have been described by Zetsche [1982] and by Jo and Lee [1982].

## 14.5 Space-variant operations

A space-variant, linear operation can be defined as one whose effect on a point in the input field depends on the location of this point. Such operations are of considerable interest in data processing; their basic properties, and some of the optical techniques which can be used to perform them, have been reviewed by Walkup [1980], Goodman [1981], and Rhodes [1981].

To carry out a two-dimensional linear space-variant operation, it is necessary to have a system whose impulse response is a function of four independent variables (two more than a normal optical system). Two methods based on holographic techniques will be described in this section.

The first is a simple method developed by Bryngdahl [1974a,b] to perform the coordinate transformation

$$x = G_1(u, v), \qquad (14.24)$$
$$y = G_2(u, v). \qquad (14.25)$$

This transformation is performed with the optical system shown in fig. 14.3,

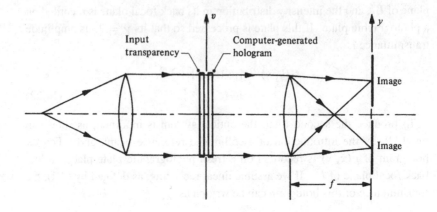

Fig. 14.3. Optical system used to produce a coordinate transformation.

which uses a computer-generated hologram whose spatial frequencies at any point $(u, v)$ are

$$s_u = G_1(u, v)/\lambda f,$$                    (14.26)

$$s_v = G_2(u, v)/\lambda f.$$                    (14.27)

Light from a point in the input plane having coordinates $(u, v)$ is then diffracted at an angle such that an image of this point is formed in the back focal plane of the lens $L_2$ at a point whose coordinates $(x, y)$ satisfy (14.24) and (14.25).

More general operations can be realized, in principle, by a hologram array. Each input pixel is backed by a hologram element which generates the desired impulse response for that pixel. However, there are serious limitations on the number of pixels which can be handled in this fashion, due to the limited resolution of the hologram elements when they are made very small.

Another method of obtaining a space-variant impulse response is to use a thick holographic element as a filter in the spatial frequency plane in an optical system such as that shown in fig. 14.2 [Deen, Walkup & Hagler, 1975]. This filter contains a number of superimposed holograms, each recorded with a plane reference wave incident at a different angle. Each point in the input plane gives rise to a plane wave whose angle of incidence on the filter depends on the coordinates of this point and, hence, generates an impulse response determined by the corresponding hologram. However, to avoid cross talk, the input field must contain only a small number of input points, since all points lying on a cone satisfy the Bragg condition.

Higher selectivity can be obtained by the use of coded reference beams

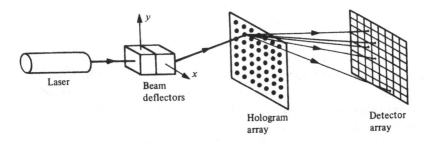

Fig. 14.4. Optical system for a read-only, page-organized holographic information store [Kogelnik, 1972].

[Krile, Marks, Walkup & Hagler, 1977; Jones, Walkup & Hagler, 1982]. For this, a diffuser is inserted in the input plane, and each of the holograms in the filter is recorded with a reference beam derived from a small area on this diffuser. Since the autocorrelation functions of the reference beams are sharply peaked, each point in the input produces an output from the corresponding filter. The diffuse background arising from the cross-correlation functions of the elementary diffusers is minimized by using a thick recording medium.

## 14.6 Information storage

Information can be stored in a compact form in microfilm. However, the maximum useful storage density with microfilm is set by the fact that, beyond a certain point, dust or scratches can result in total loss of significant parts of the stored information. Holographic information storage has the advantage that surface damage does not wipe out any particular item of information (see section 3.7), but only results in a drop in the overall signal-to-noise ratio. This makes it possible to use materials with much higher resolution and achieve much higher storage densities.

### *14.6.1 Holographic memories*

The virtual immunity of information stored on a hologram to degradation is particularly valuable where storage of information in a binary code for a computer is involved. This led at an early stage to the construction of page-organized holographic memories. As shown schematically in fig. 14.4, a typical arrangement [Anderson, 1968; Langdon, 1970] used a 32 × 32 array of small holograms, each containing 1024 bits of information. Two acousto-optic cells in tandem allowed $x$ and $y$ deflection of the laser beam so that any one of these holograms could be addressed. When illuminated by the laser beam, the information stored in the hologram was read out by a detector array in the image plane. This gave a capacity of about $10^6$ bits with an access time of a few microseconds.

These systems were initially developed as read-only stores. Subsequently, Stewart *et al.* [1973] showed that with a recording medium which could be recycled, such as a photothermoplastic, and a suitable page-composer, such a system could also be used as a read-write memory, though with a relatively long cycle time.

Since there is not much scope for reducing the cycle time of holographic memories, subsequent work has mainly been in the direction of increasing their capacity [Kiemle, 1974; Knight, 1975*b*].

The capacity of a simple holographic memory, such as that described above, is limited mainly by the resolution of the auxiliary optics to about $10^8$ bits [Vander Lugt, 1973]. One approach to increased capacity is to use a number of memory modules each consisting of a holographic storage plate with its own detector array, which are accessed by a common laser and beam deflector [Lang & Eschler, 1974]. Such a system can provide a storage capacity of approximately $2 \times 10^{10}$ bits and, in principle, can operate in a read-write configuration as well, using either a liquid-crystal page-composer or a multifrequency acousto-optic page-composer [Eschler, 1975].

An alternative approach has been to use the increased storage capacity made available by recording several holograms in the same thick recording medium [van Heerden, 1963*b*; Friesem & Walker, 1970]. Selective readout was achieved by means of electro-optic deflectors and beam translators which permit random access as well as Bragg-angle selection [D'Auria, Huignard, Slezak & Spitz, 1974]. Such a system can provide capacities of $10^{10}$–$10^{12}$ bits. A recent development has been the use of fractal-space multiplexing in combination with angular multiplexing to store as many as 5000 holograms in a single crystal of Fe:LiNbO$_3$ [Mok, 1993]. A detailed study of achievable bit-error rates (BER) has also been made by Hesselink and Bashaw [1993], who have shown that, with the use of error-correction codes, a BER of $10^{-6}$ can be achieved at readout rates of $6.3 \times 10^6$ pixels/s.

Several problems remain to be overcome with such large holographic memories. One is the need to combine the use of more sensitive recording materials with nondestructive readout of the stored information. A step towards a solution has been the use of a combination of two crystals, one for storage and the other for amplification, so that the stored information can be read out with a very low intensity beam [Rajbenbach, Bann & Huignard, 1992]. This technique also offers the possibility of increasing the storage density.

A potentially interesting line of development is the construction of a parallel-search holographic memory [Gabor, 1969]. This makes use of the associative properties of holographic storage and would be useful in applications involving rapid retrieval of data from a very large data base. In such a system [Knight, 1974, 1975*a*; Gerasimova & Zakharchenko, 1981] illumination of the memory plane with a search code produces an image in the detector plane for each hologram which contains a record of a logical match for the search code. This information is then used to steer the reference beam to each of these holograms, in turn, to read out the data stored in it.

### 14.6.2 Specialized information stores

Holographic information storage has had some success in meeting various specialized needs. One of these is in reducing the storage space required for archival copies of documents below what is possible with microfilms. This is done by recording holograms (1–2 mm diameter) of each microfilm frame [Vagin, Nazarova, Arseneva & Vanin, 1975]. The extent to which high-density storage and faithful grey-tone reproduction can be obtained with such systems has been studied by Killat [1977*b*]. Multicolour material such as maps and motion pictures can be stored as holograms since they have much better archival stability than colour film [Gale, Knop & Russell, 1975; Ih, 1975; Gale & Knop, 1976; Yu, Tai & Chen, 1978].

Holography has also been used in prototype videotape and videodisk systems [Hannan, Flory, Lurie & Ryan, 1973; Tsunoda, Tatsumo & Kataoka, 1976]. A simple colour-encoding technique for this purpose was proposed by Nishihara and Koyama [1979]. Digitized audio messages have been recorded as small one-dimensional Fourier-transform holograms on a disk in an audio response system with a 2000 word vocabulary and 250 Mbit/s transfer rate [Kubota *et al.*, 1980]. A holographic digital record/reproduce system has been described using a multi-channel acousto-optic modulator array and a mode-locked cavity-dumped laser to achieve a data rate of 500 Mbit/s [Roberts, Watkins & Johnson, 1974].

Holograms have also been used to provide additional machine-readable information on a conventional microfiche [Nelson, Vander Lugt & Zech, 1974]. An exploratory holographic information-processing system for libraries has been described by Tsukamoto *et al.* [1974] in which abstracts are stored in holographic arrays on 35 mm roll film. Relevant abstracts can be located with the appropriate keywords, using holographic correlation, after which the data are read out using a laser deflector. A large-capacity, high-speed holographic system for filing patents has been described by Sugaya, Ishikawa, Hoshino, and Iwamoto [1981]. This system can store up to 280,000 pages, any one of which can be retrieved within 1 second. A memory system using Fourier holograms has also been developed and used to store and reproduce 16,000 Kanji characters [Satoh, Kato, Fujito & Tateishi, 1989]. The memory consists of four groups of holograms of 63 pages, and can store 64 Kanji characters in each page. Digitized characters of 16 sizes can be generated at a rate of 330 characters per second.

Yet another interesting application of holographic information storage which has been studied is in credit- and identity-card verification. Systems for this purpose have been described by Sutherlin, Lauer, and Olenick [1974]; Abramson, Bjelkhagen, and Skande [1979]; and Greenaway [1980]. The use of a random phase mask for enhanced security has been proposed by Javidi and Horner [1994].

## 14.7 Holographic interconnections

The increasing data rates in modern digital computers have led to studies of the possibilities of replacing conventional electrical connections with optical interconnections using holographic optical elements [Goodman, Leonberger, Kung & Athale, 1984; Kostuk, Goodman & Hesselink, 1985; Bergman *et al.*, 1986; Kostuk, Goodman & Hesselink, 1987; Shamir, Caulfield & Johnson, 1989]. Optical interconnections minimize propagation delays; in addition, they reduce space requirements, since different optical signals can propagate through the same spatial volume without interference.

Computer-generated holographic optical elements for interconnections have to meet several specific requirements. They should have high diffraction efficiency and SNR, and produce well-focused output spots to match the dimensions of the detectors in an array. In addition, where required, they should provide space-variant imaging, so that each source in an array can have a different interconnection pattern [Feldman & Guest, 1987].

### 14.7.1 Fanouts

A specific area of interest has been the development of HOEs which can reconstruct arrays of equal intensity spots. Kinoforms with continuous phase profiles can reconstruct such arrays with a theoretical efficiency close to 100 per cent [Damman & Gortler, 1971; Streibl, 1989]. However, since the production of HOEs with continuous surface-relief or refractive-index profiles is an extremely demanding task, a considerable amount of research has gone into the design of kinoforms with only a few discrete, equally spaced phase levels [Damman,1970; Mait, 1990; Walker & Jahns, 1990].

An alternative approach involves producing HOEs with continuous phase profiles by a hybrid hologram technique [Bartelt & Case, 1982]. In this technique, a volume phase hologram is recorded optically using an object wave generated with the aid of optical filtering by a binary computer-generated hologram. The phase profile of the kinoform is encoded in the form of pulse width and pulse density variations on the CGH, which can be fabricated using electron-beam lithography [Robertson *et al.*, 1991*b*]. High efficiency and uniformity can be obtained if care is taken to optimize the relative phases of the object waves [Herzig, Ehbets, Prongué & Dändliker, 1992].

New possibilities have been opened up by the development of laser-beam writing systems which can be used to fabricate relief microstructures with smooth surfaces. The production of optimized kinoform structures for highly efficient fanout elements by such a system has been described by Prongué, Herzig, Dändliker and Gale [1992].

### 14.7.2 Space-variant interconnections

Another area of interest has been the development of two-dimensional optical interconnection networks, such as perfect-shuffle networks, which require the spatial permutation of a $P \times Q$ array of light beams [Jenkins *et al.*, 1984; Lohmann, 1986]. Such space-variant interconnections find applications in many operations such as a matrix transpose or fast Fourier transform and could be used in optical programmable logic arrays and optical cellular-logic image processors. A space-variant optical interconnection system can be fabricated, as shown in fig. 14.5, with a matched pair of space-variant HOEs, each consisting of an array of individual holographic lenses [Robertson *et al.*, 1991*a*]. The input signals from an array of optical switches are redirected by the elements of the first HOE ($H_1$) to produce the required spatial permutation at the elements of the second HOE ($H_2$) which focuses them on to an array of detectors.

### 14.7.3 Planar interconnects

While a volume reflection hologram can be used to distribute the light from an array of modulated light sources to an array of photo detectors, such a system is not space-efficient and is sensitive to variations in the wavelength of the source. A better arrangement uses two identical HOEs recorded in the same plate. The first HOE converts the light from a point source into a collimated beam which is reflected by a mirror [Brenner & Sauer, 1988], or totally reflected within the plate [Sauer, 1989; Kostuk, Huang, Hetherington & Kato, 1989] to the second HOE, which brings the beam to a focus on a photo detector. This system has the advantage that the chromatic dispersion of

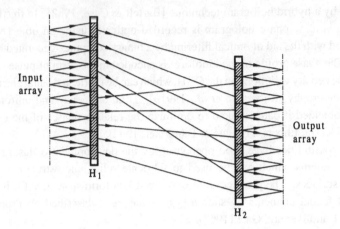

Fig. 14.5. Space-variant perfect-shuffle interconnection network using a pair of space-variant HOEs [Robertson *et al.*, 1991*a*].

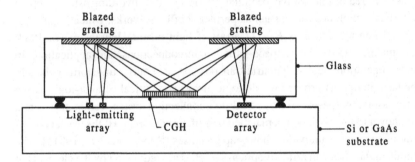

Fig. 14.6. Schematic of a perfect-shuffle interconnect using a planar optical configuration [Song *et al.*, 1993].

one hologram cancels that of the other, and opens up possibilities of various compact planar architectures with high interconnect density [Lin, Strzelecki & Jannson, 1990]. Problems arising from the difference in the recording and readout wavelengths can be overcome by recording the HOEs with aberrated beams derived from an intermediate hologram [Amitai & Goodman, 1991*a*]. A compact implementation of a perfect-shuffle interconnection, shown in fig. 14.6, uses a single computer-generated hologram in a planar optical configuration [Song *et al.*, 1993]. High packing densities can be achieved by using an aspheric field mirror to image the light sources on the receivers [Streibl *et al.*, 1993]. Substrate wave propagation and Bragg diffraction by multiplexed holographic gratings have also been used to produce a 1-to-30 fanout [Wang, Sonek, Chen & Jannson, 1992].

### *14.7.4 Reconfigurable interconnections*

The possibility of reconfigurable optical interconnections with volume reflection holograms using a tunable laser source has been studied by Wu *et al.* [1990]. An alternative scheme based on spatial division uses a set of pinhole holograms selected by a spatial light modulator. Volume holograms in photorefractive waveguides also offer a way to provide dynamically reconfigurable interconnections with integrated electronic devices [Wood, Cressman, Holman & Verber, 1989]. In a typical arrangement, beams from $N_1$ input channel waveguides coupled into a slab waveguide are collimated by a waveguide lens. The collimated beams are then diffracted by a volume hologram in the slab waveguide to a second integrated lens which focuses them on a set of $N_2$ output channel waveguides. The volume hologram is produced by two out-of-plane beams: a collimated reference beam and the Fourier transform of the signal from a spatial light modulator [Brady & Psaltis, 1991].

Another promising technique uses the polarization sensitivity of holograms recorded in photorefractive crystals [Song, Lee, Talbot & Tam, 1991]. If the grating vector and the wave vector of the reading beam are properly chosen, the intensity of the diffracted beam drops to zero when the polarization of the reading beam is changed. A crystal in which two holograms are recorded can therefore function as a $2 \times 2$ switch. A very large number of such switches can be built up in a single crystal and can be addressed either synchronously by a single beam of light, or asynchronously through a spatial light modulator.

## 14.8 Optical neural networks

Neural networks are useful for problems such as pattern recognition, in which the required transformation must be learned from examples. Optical neural networks are attractive because they offer large storage capacity as well as parallel access and processing capabilities during both the learning and reading phases [Psaltis, Brady & Wagner, 1988; Psaltis, Brady, Gu & Lin, 1990].

In a holographic neural network, neurons are represented by the pixels on a spatial light modulator (SLM), and the brightness of a pixel corresponds to the activation level of the neuron. If the SLM is placed in the input plane of the system shown in fig. 14.7, and a pair of neurons are illuminated with a coherent beam, a volume hologram is formed which corresponds to the weight between these neurons. If, subsequently, one of the original two beams is used

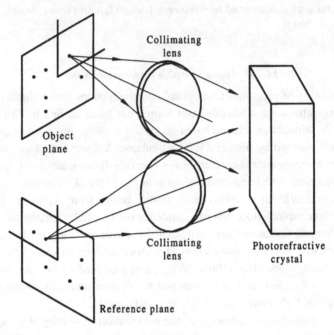

Fig. 14.7. Optical system used to record connection weights between neurons [Owechko, 1993].

to address the hologram, the other beam is reconstructed with an efficiency that represents the weight between these neurons. With a photorefractive recording material, a process of learning can be implemented by increasing or decreasing the weights selectively.

Two major factors that limit the useful storage capacity of photorefractive crystals are Bragg degeneracy and beam coupling. Partial solutions which have been proposed include subsampling of the SLMs, in which neurons are arranged in sparse nonredundant patterns, as well as similar sparse sampling of the output planes, so that false reconstructions occur on unused positions [Psaltis, Brady, Gu & Lin, 1990], and spatial multiplexing, in which the crystal is divided into separate volumes for each weight [Rastani & Hubbard, 1992]. An alternative is cascaded-grating holography in which, rather than a single grating, a set of angularly and spatially multiplexed gratings is used to store each weight. Two techniques have been studied to implement this method; the first, which was only partially successful, was the use of self-pumped phase-conjugate mirrors [Owechko & Soffer, 1991], while the second was based on the phenomenon of beam fanning.

In a high-gain photorefractive crystal, an incident beam is scattered by inhomogeneities in the crystal and generates a set of gratings which again scatter the beam, so that the beam literally fans out as it propagates in the crystal. With a fanned reference beam, the Bragg condition must be satisfied at a multitude of gratings at readout. As a result, Bragg degeneracy, cross talk, and beam coupling are virtually eliminated.

This technique has been applied in a holographic neural network using a single BaTiO$_3$ crystal, an SLM, and a CCD detector [Owechko, 1993]. The programmability of such an optical neurocomputer has been confirmed by demonstrating several networks that operate on the same hardware.

# 15

# Holographic interferometry

Holographic interferometry is an extension of interferometric measurement techniques in which at least one of the waves that interfere is reconstructed by a hologram.

The unique capabilities of holographic interferometry are due to the fact that holography permits storing a wavefront for reconstruction at a later time. Wavefronts which were originally separated in time or space or even wavefronts of different wavelengths can be compared by holographic interferometry. As a result, changes in the shape of objects with rough surfaces can be studied with interferometric precision.

One of the most important applications of holographic interferometry is in nondestructive testing (see Erf [1974], Vest [1981], and Rastogi [1994]). It can be used wherever the presence of a structural weakness results in a localized deformation of the surface when the specimen is stressed, either by the application of a load or by a change in pressure or temperature. Crack detection and the location of areas of poor bonding in composite structures are fields where holographic interferometry has been found very useful. An allied area of applications has been in medical and dental research, where it has been used to study the deformations of anatomical structures under stress, as well as for nondestructive tests on prostheses [Greguss, 1975, 1976; von Bally, 1979]; see also, Podbielska [1991, 1992].

Holographic interferometry has also proved its utility in aerodynamics, heat transfer, and plasma diagnostics. Yet another field of application has been in solid mechanics, where it has been used to evaluate the strains in complex structures, as well as to measure changes in shape due to corrosion or absorption of water. Since mechanical contacts are not involved, measurements can be carried out in hostile or corrosive environments [Vest, 1979].

## 15.1 Real-time holographic interferometry

In this technique, the hologram is replaced, after processing, in exactly the same position in which it was recorded. When it is illuminated with the original reference beam, it reconstructs the object wave, and the virtual image coincides with the object. If, however, the shape of the object changes very slightly, two sets of light waves reach the observer, one being the reconstructed wave (corresponding to the object before the change) and the other the directly transmitted wave from the object in its present state. The two wave amplitudes add at the points where the difference in optical paths is zero, or a whole number of wavelengths, and cancel at some other points in between. As a result, an observer viewing the reconstructed image sees it covered with a pattern of interference fringes, which is a contour map of the changes in shape of the object. These changes can be observed in real time.

If we consider a typical off-axis holographic recording system, as in section 2.2, the intensity at the photographic plate when the hologram is recorded is

$$I(x, y) = |r(x, y) + o(x, y)|^2, \qquad (15.1)$$

where $r(x, y)$ is the complex amplitude due to the reference beam and $o(x, y) = |o(x, y)| \exp[-i\phi(x, y)]$ is the complex amplitude due to the object in its normal state.

If we assume, as in (2.2), that the amplitude transmittance of the photographic plate after processing is linearly related to the exposure, the amplitude transmittance of the hologram is

$$t(x, y) = t_0 + \beta T I(x, y), \qquad (15.2)$$

where, as before, $\beta$ is the slope (negative) of the amplitude transmittance *versus* exposure characteristic of the photographic material, $T$ is the exposure time and $t_0$ is a uniform background transmittance.

When the processed hologram is replaced in the same position in which it was recorded, it is illuminated by the wave from the deformed object as well as the reference wave. Accordingly, the complex amplitude of the wave transmitted by the hologram is

$$u(x, y) = [o'(x, y) + r(x, y)]t(x, y), \qquad (15.3)$$

where $o'(x, y)$ is the complex amplitude of the wave from the deformed object. If the change in the shape of the object is very small, it can be assumed that only the phase distribution of the object wave is modified, so that

$$o'(x, y) = |o(x, y)| \exp[-i\phi'(x, y)]. \qquad (15.4)$$

The only terms of interest in the expansion of (15.3) are those corresponding to the reconstructed primary image and the directly transmitted object wave. If $|r(x, y)|^2 = r^2$, the complex amplitude due to these two terms is

$$u'(x, y) = \beta T r^2 o(x, y) + (t_0 + \beta T r^2) o'(x, y), \quad (15.5)$$

and the resultant intensity is

$$I'(x, y) = |o(x, y)|^2 \{\beta^2 T^2 r^4 + (t_0 + \beta T r^2)^2$$
$$+ 2\beta T r^2 (t_0 + \beta T r^2) \cos[\phi'(x, y) - \phi(x, y)]\}. \quad (15.6)$$

The reconstructed image is, therefore, covered with fringes. Since $\beta$ is negative, a dark fringe corresponds to the condition

$$\phi'(x, y) - \phi(x, y) = 2m\pi, \quad (15.7)$$

where $m$ is an integer.

The visibility of the interference pattern is a maximum when

$$|\beta T r^2| = |t_0 + \beta T r^2|, \quad (15.8)$$

which, since $\beta$ is negative, corresponds to the condition

$$|\beta T r^2| = t_0/2. \quad (15.9)$$

Precise repositioning of the hologram after processing is necessary to avoid the introduction of spurious fringes. In addition, with photographic emulsions, it is necessary to take precautions to avoid local deformations of the emulsion due to nonuniform drying. These problems can be avoided by exposing and processing the photographic plate *in situ* in a liquid gate [Van Deelen & Nisenson, 1969]. A typical arrangement is shown in fig. 15.1. In this system, processing is speeded up by using a monobath (see section 7.1.4); this also makes it possible to use a simple gravity system for handling the processing solutions [Hariharan & Ramprasad, 1973*b*]. A closed-circuit television camera is used to project the fringe pattern on the screen of a black-and-white television monitor. The fringes can be colour-coded to identify the sign of the displacement by photographing them with a Polaroid camera through red and green filters, with a small change in the load between the exposures [Hariharan, 1977*a*].

An alternative which completely eliminates the need for wet processing is to use thermoplastic recording [Thinh & Tanaka, 1973]. Automated systems using such materials make it possible to record a hologram and view the interference fringes in less than a minute. Since these materials produce a thin phase hologram, a high diffraction efficiency is obtained, and effects due to changes in the thickness of the recording material are eliminated. A possibility,

Fig. 15.1. System for real-time holographic interferometry. The hologram is processed *in situ*, and the interference fringes are viewed through a closed-circuit television system.

where maximum dimensional stability of the hologram is required, is to use solvent-vapour processing [Saito, Imamura, Honda & Tsujiuchi, 1980]. With such a phase hologram, the reconstructed image exhibits a phase shift of $\pi/2$ [Hariharan & Hegedus, 1975$b$] so that a dark zero-order fringe is no longer obtained.

A problem in real-time holographic interferometry is that while the light diffracted by the hologram is linearly polarized, the light scattered by a diffusely reflecting object is largely depolarized (see section 5.3), resulting in a significant drop in the visibility of the fringes. To avoid this, it is necessary to use a polarizer when viewing or photographing the fringes.

## 15.2 Double-exposure holographic interferometry

In double-exposure holographic interferometry, interference takes place between the wavefronts reconstructed by two holograms of the object recorded on the same photographic plate. Typically, the first exposure is made with the object in its initial, unstressed condition, and the second is made with a stress applied to the object. When the processed hologram is illuminated with the original reference beam, it reconstructs two images, one corresponding to the object in its unstressed state, and the other corresponding to the stressed object. The resulting interference pattern reveals the changes in shape of the object between the two exposures.

In this case, the intensity at the photographic plate during the first exposure is

$$I_1(x, y) = |r(x, y) + o(x, y)|^2, \tag{15.10}$$

and that during the second exposure is

$$I_2(x, y) = |r(x, y) + o'(x, y)|^2. \tag{15.11}$$

The amplitude transmittance of the resulting hologram is, therefore,

$$\mathbf{t}(x, y) = \mathbf{t}_0 + \beta T (I_1 + I_2). \tag{15.12}$$

When the hologram is illuminated once again with the same reference wave, the transmitted amplitude in the hologram plane is

$$u(x, y) = r(x, y)\mathbf{t}(x, y). \tag{15.13}$$

The only terms of interest in the expansion of (15.13) are those corresponding to the two superimposed primary images. The complex amplitude due to these images is

$$u_3(x, y) = \beta T r^2 |o(x, y)|$$
$$\times \{\exp[-i\phi(x, y)] + \exp[-i\phi'(x, y)]\}, \tag{15.14}$$

so that the resultant intensity is

$$I_3(x, y) \propto |o(x, y)|^2 \{1 + \cos[\phi(x, y) - \phi'(x, y)]\}. \tag{15.15}$$

In this case the hologram is a permanent record of the changes in the shape of the object.

Double-exposure holographic interferometry is much easier than real-time holographic interferometry, because the two interfering waves are always reconstructed in exact register. Distortions of the emulsion affect both images equally, and no special care need be taken in illuminating the hologram when viewing the image. In addition, since the two diffracted wavefronts are similarly polarized and have almost the same amplitude, the visibility of the fringes is good.

However, double-exposure holographic interferometry has certain limitations. The first is that where the object has not moved between the exposures, the reconstructed waves, both of which have experienced the same phase shift, add to give a bright image of the object. As a result it is difficult to observe small displacements. A dark field, and much higher sensitivity, can be obtained by holographic subtraction, which merely involves shifting the phase of the reference beam by $\pi$ between the two exposures [Collins, 1968; Hariharan & Ramprasad, 1972].

An alternative method, which also helps to resolve ambiguities in the sense of the displacement, is to shift the phase of the reference beam by $\pi/2$ between the two exposures, or, better, to introduce a very small tilt in the wavefront illuminating the object between the two exposures. In the latter technique, equally spaced reference fringes are obtained, whose position is modulated by the phase shifts being studied [Jahoda, Jeffries & Sawyer, 1967; Hariharan & Ramprasad, 1973a].

Another limitation of the double-exposure method is that information on intermediate states of the object is lost. This problem can be overcome to some extent by multiplexing techniques using spatial division of the hologram [Caulfield, 1972; Hariharan & Hegedus, 1973]. In the latter procedure, a series of masks are used in which the apertures overlap in a systematic fashion, and a sequence of holograms is recorded at different stages of loading. The images can then be reconstructed, two at a time, so that interference patterns between any two images can be studied. An alternative is a method described by Parker [1978], using a thermoplastic recording material, by which real-time fringes can be observed and the fringe pattern subsequently frozen to give a permanent holographic record.

An interesting possibility which has been studied by Yu and Chen [1978] is the use of rainbow holograms for double-exposure holographic interferometry. Fringe patterns corresponding to different effects can be displayed in different colours [Tai, Yu & Chen, 1979; Yu, Tai & Chen, 1979].

## 15.3 Sandwich holograms

Control of the fringes, to compensate for rigid body motion and eliminate ambiguities in interpretation, is not normally possible with a doubly exposed hologram. However, it is possible with two holograms recorded with different, angularly separated reference waves [Gates, 1968; Ballard, 1968; Tsuruta, Shiotake & Itoh, 1968]. These holograms may be either on the same plate or on different plates.

An elegant alternative is the sandwich hologram [Abramson, 1974, 1975, 1977; Abramson & Bjelkhagen, 1979]. In this technique, as shown in fig. 15.2, pairs of holographic plates (without any antihalation backing) are exposed in the same plate holder with their emulsion-coated surfaces facing the object. $B_1$, $F_1$ are exposed with the unstressed object, while $B_2$, $F_2$ and $B_3$, $F_3$, ...,

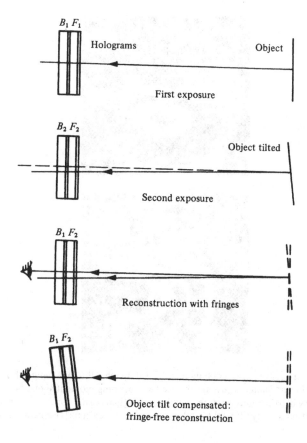

Fig. 15.2. Steps involved in sandwich hologram interferometry [Abramson, 1975].

are exposed with the object deformed by progressively increasing loads. After all the plates have been processed, $B_1$ is combined with, say, $F_2$ in the original plateholder and illuminated with the original reference beam to produce an interference pattern showing the surface deformations at the corresponding stage of loading, as shown in fig. 15.3($a$). A tilt of the sandwich then results, as shown in fig. 15.3($b$), in a change in the interference pattern exactly equivalent to a tilt of the original object. A detailed theoretical analysis of the changes in the fringes has been made by Dubas and Schumann [1977].

If $B_1$ is combined with $F_2$, $F_3$, ..., it is possible to study the total deformation at any stage, while combinations such as $B_1F_2$, $B_2F_3$, $B_3F_4$, ... will show the incremental deformations. A simplified version of this technique is also possible

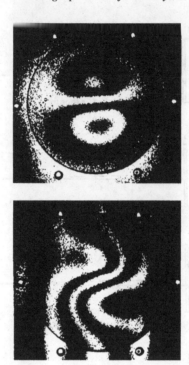

Fig. 15.3. Interference fringes obtained with a sandwich hologram using a thin, circular metal sheet clamped at its edge and subjected to a bending moment about the horizontal diameter, (*a*) when the sandwich is replaced in its original position and (*b*) when reference fringes are introduced by tilting the sandwich about a vertical axis [Hariharan & Hegedus, 1976].

using only two plates cemented together with a spacer [Hariharan & Hegedus, 1976].

## 15.4 Industrial environments

Holographic interferometry normally requires an extremely stable optical system. However, various techniques have been described which permit double-exposure holographic interferometry in an industrial environment.

The most common method is the use of a pulsed laser. Double-exposure holographic interferometry can then be used to study transient phenomena such as deformations due to impact loading, as shown in fig. 15.4 [Gates, Hall & Ross, 1972; Armstrong & Forman, 1977]. An electrical, optical, or acoustic signal is used to trigger the first pulse just before, or at the instant of impact, with the

Fig. 15.4. Holographic interferogram recorded with a double-pulsed ruby laser, showing the effect of a hammer blow on a crash helmet. Pulse duration 25 ns, pulse separation 25 $\mu$s [Gates, Hall & Ross, 1972; Crown copyright, National Physical Laboratory, reprinted with permission].

second pulse following after a predetermined delay. Techniques such as sandwich holography can be used to eliminate unwanted rigid body displacements and simplify interpretation of the fringes [Bjelkhagen, 1977b; Abramson & Bjelkhagen, 1978].

Even objects rotating at extremely high speeds can be studied with the aid of an optical derotator [Stetson, 1978] consisting of an inverting prism mounted in the hollow shaft of an electric motor. When this prism is aligned axially with the rotating object and rotated at half its speed, a stationary image is obtained. The residual movement is small enough to permit recording double-pulse holographic interferograms of quite large rotating objects (see fig. 15.5).

A technique which permits holographic interferometry in some situations with a continuous-wave laser source is object motion compensation. This is possible when the unwanted movements of the object are mainly out-of-plane translation or vibration, and is achieved by reflecting the reference wave also from a suitably chosen point on the surface, or from a mirror attached to it [Mottier, 1969; Waters, 1972]. Another method is to use reflection ("piggyback") holography (see section 5.1). Two holograms are recorded by illuminating the object through the hologram plate. Relative motion is virtually eliminated, since the hologram plate moves with the object [Neumann & Penn, 1972; Boone, 1975].

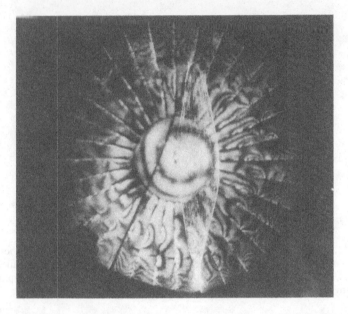

Fig. 15.5. Holographic interferogram of a turbine fan rotating at 4460 rpm recorded with an image derotator and a double-pulsed laser [courtesy K. A. Stetson, United Technology Research Center, East Hartford, USA].

Image holograms can also be used to minimize the effects of object movement, particularly when studying out-of-plane deformations [Klimenko, Matinyan & Dubitskii, 1975; Rowley, 1979, 1981].

## 15.5 Phase objects

Holographic interferometry has practical advantages even in applications, such as flow visualization and heat transfer studies, where conventional interferometry has been used for many years (see, for example, Tanner [1966] and Trolinger [1975*b*]).

In the first instance, mirrors and windows of relatively low optical quality can be used. Since the phase errors due to the optics contribute equally to both interfering wavefronts, they cancel out, and only the effects of changes in the optical path are seen. However, the most significant advantage is the possibility of incorporating a diffusing screen (a ground glass plate) in the interferometer, to obtain an interference pattern that is localized near the phase object and can be viewed and photographed over a range of angles. This makes it possible to study three-dimensional refractive index distributions.

If the refractive index gradients in the test section are assumed to be small, so that rays propagate through it along straight lines parallel to the $z$ axis, $\phi(x, y)$, the phase difference at any point in the interference pattern is given by the relation

$$\phi(x, y) = k_0 \int [n(x, y, z) - n_0]dz, \qquad (15.16)$$

where $n_0$ is the refractive index of the medium in the test section in its initial, unperturbed state and $n(x, y, z)$ is the final refractive index distribution.

The simplest case is that of a two-dimensional phase object with no variation of refractive index in the $z$ direction. In this case, the refractive index distribution can be calculated directly from (15.16). Fortunately, this is a valid approximation in many practical situations.

Another case which lends itself to analytic treatment is that of a refractive index distribution $f(r)$ which is radially symmetric about an axis normal to the line of sight (for convenience, say, the $y$ axis).

For a ray travelling in the $z$ direction at a distance $x$ from the centre, we then have

$$dz = (r^2 - x^2)^{-1/2}r\,dr, \qquad (15.17)$$

so that (15.16) becomes

$$\phi(x, y) = 2 \int_x^\infty f(r)(r^2 - x^2)^{-1/2}r\,dr. \qquad (15.18)$$

This is the Abel transform of $f(r)$, and it can be inverted to find $f(r)$ [Bracewell, 1978].

The evaluation of an asymmetric refractive index distribution $f(r, \theta)$ is more difficult and is possible only by recording a large number of interferograms from different directions [Sweeney & Vest, 1973]. The problem becomes even more complicated when the effects of ray curvature due to refraction cannot be neglected. An iterative technique which can be used under these conditions has been described by Cha and Vest [1979, 1981].

Holographic interferometry has been found extremely useful in plasma diagnostics [Zaidel, Ostrovskaya & Ostrovskii, 1969]. Since, unlike a neutral gas, a plasma is highly dispersive, measurements of the refractive index distribution at two wavelengths make it possible to determine the electron density directly. For this, two holograms are recorded simultaneously on the same plate with light from a ruby laser which has passed through a frequency doubler to produce two collinear beams with wavelengths $\lambda_1 = 694$ nm and $\lambda_2 = 347$ nm. If the plate is processed to give a nonlinear recording characteristic (see section 6.5), it is possible to make the second-order image reconstructed by one hologram interfere with the first-order image reconstructed by the other. Under these conditions it can be shown [Ostrovskaya & Ostrovskii, 1971] that the interference fringes are contours of constant dispersion and, hence, of constant electron density. Measurements carried out by this technique on a plasma produced in air by the beam from a $CO_2$ laser have been described by Radley [1975].

## 15.6 Phase-conjugate interferometry of phase objects

Many of the techniques of conventional interferometry such as multiple-beam and multiple-pass interferometry, as well as shearing interferometry, can be extended to holographic interferometry [see Vest, 1979]. A type of shear possible with holography is longitudinally reversed shear (phase-conjugate interferometry), in which the primary and conjugate images of the test wave front reconstructed by a hologram are made to interfere [Bryngdahl, 1969a; Fainman, Lenz & Shamir, 1981]. Interesting possibilities for phase-conjugate interferometry have been opened up by developments in dynamic holography which make it possible to generate a conjugate wave in real time [Hopf, 1980]. Optical systems for phase-conjugate interferometry have been described using degenerate four-wave mixing in a thin film of eosin [Bar-Joseph, Hardy, Katzir & Silberberg, 1981] and in a BGO crystal slice [Ja, 1982].

One way of obtaining increased sensitivity is by phase-difference amplification using the higher diffracted orders from a nonlinear hologram [Bryngdahl & Lohmann, 1968a; Matsumoto & Takashima, 1970]. A better method is the use of phase-conjugate interferometry [Matsuda, Freund & Hariharan, 1981].

## 15.7 Diffusely reflecting objects

One of the major advantages of holographic interferometry is that an interference pattern with high visibility can be obtained even with an object having a relatively rough surface. To understand how this is possible, and to obtain a relation between the interference order in the fringe pattern and the surface displacement at the corresponding point on the object, consider a small area of the surface which has suffered a simple translation, as shown in fig. 15.6, so that two points on it, $P$ and $Q$, undergo identical vector displacements $\mathbf{L}$ to $P'$ and $Q'$, respectively.

Let the complex amplitude of the wave reflected by the undeformed object be

$$o_1(x, y) = |o(x, y)| \exp[-i\phi(x, y)], \qquad (15.19)$$

where $\phi(x, y)$ varies in a random manner over the surface because of its microstructure. The complex amplitude of the wave reflected by the deformed

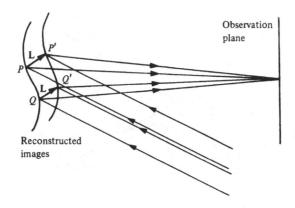

Fig. 15.6. Fringe formation in holographic interferometry with a diffusely reflecting object.

object can then be written as

$$o_2(x, y) = |o(x, y)| \exp\{-i[\phi(x, y) + \Delta\phi]\}, \qquad (15.20)$$

where $\Delta\phi$ is constant over the small area considered.

The intensity due to these two wavefronts is then

$$
\begin{aligned}
I(x, y) &= |o_1(x, y) + o_2(x, y)|^2, \\
&= o_1(x, y)o_1^*(x, y) + o_2(x, y)o_2^*(x, y) \\
&\quad + o_1(x, y)o_2^*(x, y) + o_1^*(x, y)o_2(x, y). \qquad (15.21)
\end{aligned}
$$

The resulting interference pattern exhibits a speckle structure (see Appendix 4) due to the rapid variation of $\phi(x, y)$ over the surface. However, the average intensity over a small area centred on $(x, y)$ containing a number of speckles is

$$
\begin{aligned}
I &= \langle |o_1 + o_2|^2 \rangle, \\
&= \langle o_1 o_1^* \rangle + \langle o_2 o_2^* \rangle + \langle o_1 o_2^* \rangle + \langle o_1^* o_2 \rangle, \\
&= I_1 + I_2 + (I_1 I_2)^{1/2}[\exp(i\Delta\phi) + \exp(-i\Delta\phi)], \\
&= I_1 + I_2 + 2(I_1 I_2)^{1/2} \cos \Delta\phi, \qquad (15.22)
\end{aligned}
$$

where $I_1$ and $I_2$ are the average intensities due to the individual wavefronts.

It is enough, therefore, to consider the path differences between corresponding points on the two wavefronts to evaluate the broader variations in intensity which constitute the fringes in the interference pattern.

The phase difference $\Delta\phi$ between the two wavefronts, which gives rise to the interference fringes seen by the observer, can now be found from the change in

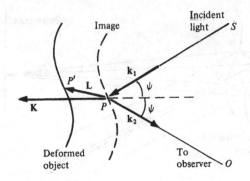

Fig. 15.7. Calculation of the phase difference in the interference pattern.

the total optical path from the source $S$ to the observer $O$. For small displacements ($|\mathbf{L}| \ll SP, PO$), as shown in fig. 15.7, $\Delta\phi$ is given by the relation

$$\Delta\phi = \mathbf{L} \cdot (\mathbf{k}_1 - \mathbf{k}_2),$$
$$= \mathbf{L} \cdot \mathbf{K}, \tag{15.23}$$

where $\mathbf{k}_1$ and $\mathbf{k}_2$ are the propagation vectors of the incident and scattered light. These are of magnitude $|\mathbf{k}_1| = |\mathbf{k}_2| = k_0 = 2\pi/\lambda$ and are taken along the directions of illumination and observation respectively [Aleksandrov & Bonch-Bruevich, 1967; Ennos, 1968; Sollid, 1969]. The sensitivity vector $\mathbf{K} = \mathbf{k}_1 - \mathbf{k}_2$ defined by (15.23) points along the bisector of the angle $2\psi$ between the illumination and viewing directions, and its magnitude is $|\mathbf{K}| = 2k_0 \cos\psi$.

## 15.8 Fringe localization

The visibility of the interference fringes formed with a diffusely reflecting object, as described in section 15.7, is a maximum for a particular position of the plane of observation. This position is commonly termed the plane of localization of the fringes. An early finding [Haines & Hildebrand, 1966] was that the position of the plane of localization depended on the type of displacement experienced by the object. Difficulties arise when the fringes are localized too far from the object for both the fringes and the object to be imaged sharply at the same time.

Fringe localization is due to the random phase variations across the object wavefront and is linked to the finite aperture of the viewing system. Detailed analyses of the phenomenon have been made by a number of authors including Stetson [1969, 1970b, 1974], Walles [1969], Steel [1970], and Dubas and

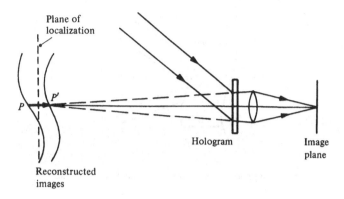

Fig. 15.8. Localization of fringes in holographic interferometry with a diffusely reflecting object.

Schumann [1974]. These studies showed that, actually, the fringes are not localized in a plane but along a line. Experiments have also been carried out by Molin and Stetson [1970*a,b*, 1971] on the localization of the fringes for different types of object displacements.

However, while fringe localization played a major part in early methods of interpretation of holographic interferograms, it can now be seen as a separate issue. Accordingly, this section is confined to a simple treatment of the phenomenon and a discussion of two limiting cases.

As shown in section 15.7, only waves from corresponding points on the two wavefronts contribute to the fringes. Hence, we need consider only the phase differences between a pair of such points, $P$ and $P'$, on the interfering wavefronts, over a range of viewing directions defined by the aperture of the viewing system, as shown in fig. 15.8.

At an arbitrarily chosen plane, this phase difference $\Delta\phi$ will, in general, vary over this cone due to the changes in the sensitivity vector in (15.23). However, a plane can be found at a distance from the object, determined by the geometry of the system and the magnitude and direction of the displacement, at which the value of $\Delta\phi$ is very nearly constant over this range of viewing directions. Accordingly, if the imaging system is focused on this plane, a fringe system with good visibility will be seen.

### 15.8.1 *Pure translation of the surface*

For pure translation, as shown in fig. 15.9, **L** is a constant over the whole object. If we assume that the object is illuminated with a plane wave, so that $\mathbf{k}_1$ is a

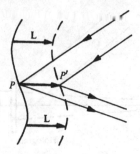

Fig. 15.9. Fringe localization for pure translation of the surface.

constant, the phase difference $\Delta\phi$ is

$$\Delta\phi = \mathbf{L} \cdot (\mathbf{k}_1 - \mathbf{k}_2),$$
$$= \text{a constant} - \mathbf{L} \cdot \mathbf{k}_2. \qquad (15.24)$$

The condition for localization ($\Delta\phi$ = a constant) is then simply,

$$\mathbf{k}_2 = \text{a constant}, \qquad (15.25)$$

over the detector, implying that the fringes are localized at infinity.

If a lens is placed at a distance equal to its focal length from the surface, plane waves corresponding to different values of $\mathbf{k}_2$ are brought to a focus at different points in its back focal plane. Fringes are then seen in this plane due to the corresponding variations of $(\mathbf{k}_1 - \mathbf{k}_2)$. Since the variation of $(\mathbf{k}_1 - \mathbf{k}_2)$ is a maximum in the plane containing $\mathbf{k}_1$ and $\mathbf{k}_2$ and is negligible at right angles to it, the interference pattern consists of parallel, straight fringes.

### 15.8.2 Pure rotation about an axis in the surface

Unlike pure translation, pure rotation of the object about an axis contained in its surface results in straight fringes localized very close to the surface.

As shown in fig. 15.10, the surface is considered to be initially in the $xy$ plane and to rotate about the $y$ axis through a small angle $\theta$. It is also assumed that the surface is illuminated by a collimated beam, and that the directions of illumination and observation lie in the $xz$ plane and make angles $\psi_1$ and $\psi_2$, respectively, with the $z$ axis. The phase difference between the waves scattered from a pair of corresponding points $P$ and $P'$ is then, from (15.23),

$$\phi = -(2\pi/\lambda)x\theta(\cos\psi_1 + \cos\psi_2). \qquad (15.26)$$

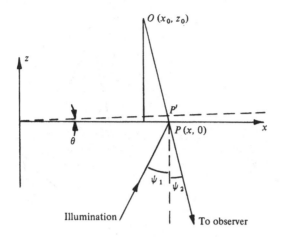

Fig. 15.10. Fringe localization for pure rotation about an axis in the surface.

This expression corresponds to a set of straight fringes running parallel to the $y$ axis with a spacing equal to $\lambda/\theta(\cos\psi_1 + \cos\psi_2)$.

To find the surface of localization of the fringes, we consider the variations of this phase difference over a narrow range of viewing directions centred on an arbitrary point $O(x_0, z_0)$ along the line of sight. We differentiate (15.26) to obtain the relation

$$\mathrm{d}\phi = -(2\pi/\lambda)[\theta(\cos\psi_1 + \cos\psi_2)\mathrm{d}x - \theta x \sin\psi_2\mathrm{d}\psi_2], \qquad (15.27)$$

since $\psi_1$ does not vary. Now,

$$\tan\psi_2 = (x - x_0)/z_0, \qquad (15.28)$$

so that

$$\mathrm{d}x = (z_0/\cos^2\psi_2)\mathrm{d}\psi_2. \qquad (15.29)$$

Hence,

$$\mathrm{d}\phi = -(2\pi/\lambda)\theta[(z_0/\cos^2\psi_2)(\cos\psi_1 + \cos\psi_2) - x\sin\psi_2]\mathrm{d}\psi_2. \qquad (15.30)$$

For the fringes to be localized at $O$, $(\mathrm{d}\phi/\mathrm{d}\psi_2)$ must be equal to zero, so that

$$z_0 = (x\sin\psi_2\cos^2\psi_2)/(\cos\psi_1 + \cos\psi_2). \qquad (15.31)$$

If the viewing direction is normal to the surface of the object, so that $\psi_2 = 0$, then $z_0 = 0$, and the fringes are localized on the surface of the object. If $\psi_2 \neq 0$, the surface of localization lies either in front of the object (when $\psi_2 > 0$), or

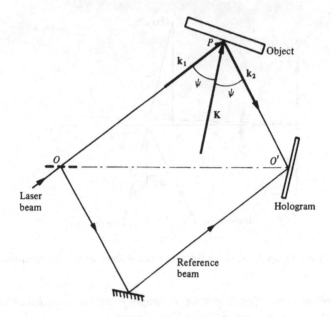

Fig. 15.11. Schematic of a holographic system.

behind it (when $\psi_2 < 0$). However, since $(\mathrm{d}\phi/\mathrm{d}\psi_2) = 0$ when $x = 0$, the
surface of localization always intersects the surface of the object at the axis of
rotation.

## 15.9 The holodiagram

The holodiagram [Abramson 1969, 1970*a*,*b*, 1971, 1972] provides a simple
geometrical representation of the relations discussed earlier. For the basic
hologram recording system shown in fig. 15.11, the holodiagram consists, as
shown in fig. 15.12, of a set of spheroids with their foci at $O$ and $O'$. Each
of these spheroids is the locus of points for which the distance $OPO'$ is a
constant, and this distance changes in steps of $\lambda$ from one spheroid to the
next. For simplicity, it is usually enough to consider the ellipses formed by the
intersection of these spheroids with the $xy$ plane.

A shift of one fringe in the interference pattern is caused by the movement
of $P$ from one ellipse to the next. The displacement of $P$ is a minimum when
its motion is normal to the ellipse, that is to say, along the sensitivity vector $\mathbf{K}$.
The holodiagram also shows that while a displacement of $P$ of $\lambda/2$ along the
normal to the ellipse, when $P$ lies on the $x$ axis, results in a shift of one fringe, a

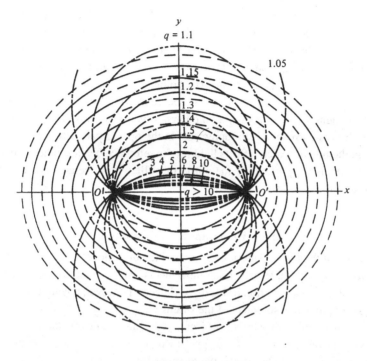

Fig. 15.12. The holodiagram showing (*a*) ellipses of constant optical path length, and (*b*) circles corresponding to constant values of the sensitivity vector [Abramson, 1969].

larger displacement $q\lambda/2$, where $q = 1/\cos\psi$, normal to the ellipse is required for the same fringe shift at any other location. The corresponding variations in the magnitude of the sensitivity vector are conveniently represented on the holodiagram by curves of constant $q$. These are the loci of points for which the angle $OPO' = 2\psi$ is a constant, and are, therefore, as shown in fig. 15.12, circles passing through $O$ and $O'$. The centres of these circles are located on the $y$ axis at points given by the relation

$$y = (l/2)(2 - q^2)(q^2 - 1)^{-1/2}, \qquad (15.32)$$

where $OO' = 2l$. The value of $q$ obtained from the holodiagram permits direct evaluation of the fringe pattern. If the fringe order at a point $P$ on the object is $N$, the component of the displacement normal to the ellipse passing through $P$ is $qN\lambda/2$.

The holodiagram is particularly valuable as an aid to understanding the fringe patterns which are obtained for different types of object displacements when the distances $OP$ and $PO'$ are comparable to $OO'$, since the sensitivity vector then varies considerably over the image field. It is also very useful to optimize a hologram recording system. Since the system is most sensitive to motion normal to the ellipses and least sensitive to tangential motion, the holodiagram can be used to design the system to maximize or minimize its sensitivity to a particular type of object motion. The holodiagram can also be useful when holograms of a large object are to be recorded with light having a limited coherence length $\Delta l$. For this, a holodiagram is drawn in which the spacing of the ellipses corresponds to increments in the optical path difference of $\Delta l$. If, then, the mirror which reflects light for the reference beam is placed on one ellipse, the visible portions of the object must lie within the area bounded by the adjacent two ellipses.

## 15.10 Holographic strain analysis

Inspection of the fringe pattern gives a considerable amount of information on the type and magnitude of surface movements, and is quite useful to detect areas of stress concentration or localized defects. However, quantitative strain analysis requires measurement of the strains.

If, at any point on the stressed object, $L_x$, $L_y$, and $L_z$ denote the $x$, $y$, and $z$ components, respectively, of the displacement, the three components of normal strain at this point are defined by the relations

$$\varepsilon_x = \partial L_x/\partial x, \tag{15.33}$$

$$\varepsilon_y = \partial L_y/\partial y, \tag{15.34}$$

$$\varepsilon_z = \partial L_z/\partial z, \tag{15.35}$$

and the three shear strains are

$$\gamma_{xy} = (\partial L_x/\partial y) + (\partial L_y/\partial x), \tag{15.36}$$

$$\gamma_{yz} = (\partial L_y/\partial z) + (\partial L_z/\partial y), \tag{15.37}$$

$$\gamma_{zx} = (\partial L_z/\partial x) + (\partial L_x/\partial z). \tag{15.38}$$

Several methods of analysis of the fringe pattern have been proposed to evaluate the surface displacements and, hence, the strains [Briers, 1976]. Early workers tended to favour methods using observations of fringe localization. They have the advantage (see Dubas and Schumann [1974, 1975]) that they can give direct measurements of the surface displacement and, in some cases, even permit direct evaluation of the strain [Stetson, 1976]. However, their accuracy is limited, and interpretation of the fringes becomes difficult when a combination of rotations and translations is involved.

Methods of differentiating the displacement data to obtain the local strain,

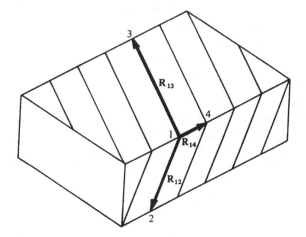

Fig. 15.13. Fringes due to homogeneous deformation of a three-dimensional object. The points 1, 2, and 3 define the plane of a fringe surface, while the point 4 is on the adjacent fringe [Stetson, 1975b].

making use of the moiré patterns obtained by superimposing two transparencies of the interferogram, have also been described [Boone & Verbiest, 1969; Stetson, 1970a]. These methods can give a quick picture of the surface strains.

Another direct method is the fringe vector method [Stetson, 1974, 1975a,b, 1979]. This method is based on the fact that any combination of homogeneous deformation and rotation of an object yields fringes that appear to be produced by the intersection of the object surface with a number of equally spaced surfaces which are contours of constant phase difference (the fringe locus function). The fringe vector $\mathbf{K}_f$ runs perpendicular to these surfaces, and its magnitude is inversely proportional to their separation.

If we consider a three-dimensional object such as that shown in fig. 15.13, the fringe vector must be normal to the plane defined by the points 1, 2, 3 which lie on the same fringe. Hence, $\hat{\mathbf{k}}_f$, the unit vector in the direction of the fringe vector, is given by the relation

$$\hat{\mathbf{k}}_f = (\mathbf{R}_{12} \times \mathbf{R}_{13})/|\mathbf{R}_{12} \times \mathbf{R}_{13}|, \qquad (15.39)$$

while the magnitude of the fringe vector is

$$|\mathbf{K}_f| = 2\pi/\hat{\mathbf{k}}_f \cdot \mathbf{R}_{14}. \qquad (15.40)$$

Accordingly, if the shape of the object is known, the fringe vector can be evaluated.

To apply this method, the fringe vectors $\mathbf{K}_{f1}$, $\mathbf{K}_{f2}$, $\mathbf{K}_{f3}$, corresponding to three different directions of viewing, and, hence, to three different values $\mathbf{K}_1$, $\mathbf{K}_2$, $\mathbf{K}_3$, of the sensitivity vector, are determined. Since it can be shown that the resolved components of the fringe vectors and the sensitivity vectors along the $x$, $y$ and $z$ axes are linked to the gradients of the displacements by the matrix relation

$$
\begin{bmatrix}
K_{f1x} & K_{f1y} & K_{f1z} \\
K_{f2x} & K_{f2y} & K_{f2z} \\
K_{f3x} & K_{f3y} & K_{f3z}
\end{bmatrix}
$$

$$
= \begin{bmatrix}
K_{1x} & K_{1y} & K_{1z} \\
K_{2x} & K_{2y} & K_{2z} \\
K_{3x} & K_{3y} & K_{3z}
\end{bmatrix}
\begin{bmatrix}
\partial L_x/\partial x & \partial L_y/\partial x & \partial L_z/\partial x \\
\partial L_x/\partial y & \partial L_y/\partial y & \partial L_z/\partial y \\
\partial L_x/\partial z & \partial L_y/\partial x & \partial L_z/\partial z
\end{bmatrix},
$$

$$(15.41)$$

the strains can be evaluated using (15.33) to (15.38).

This method can be extended to more general deformations, which can often be treated as approximately homogeneous over a limited region. Corrections can also be introduced for the variations in the sensitivity vector due to perspective over the region of the object which is being studied [Pryputniewicz & Stetson, 1976]. Experimental studies using this method have been described by Pryputniewicz and Bowley [1978], Pryputniewicz [1978, 1980], and Pryputniewicz and Stetson [1980].

The other approach which has been followed to calculate the strains involves evaluating and differentiating the actual surface displacements. The process is simplified by the fact that, in most cases, what is involved is not the absolute displacement of any point, but rather its displacement with respect to a point in the field of view which can be assumed to be stationary.

The latter can be evaluated if three observations of the fringe order are made with three different directions of observation [Shibayama & Uchiyama, 1971]. However, three separate holograms are then required to give adequate angular separation, and a better alternative is to use a single direction of observation and three different directions of object illumination [Hung, Hu, Henley & Taylor, 1973]. The measured phase differences $\Delta\phi_1$, $\Delta\phi_2$, $\Delta\phi_3$ are then linked to $L_x$, $L_y$, $L_z$, the three orthogonal components of the displacement by the matrix relation

$$
\begin{bmatrix}
K_{1x} & K_{1y} & K_{1z} \\
K_{2x} & K_{2y} & K_{2z} \\
K_{3x} & K_{3y} & K_{3z}
\end{bmatrix}
\begin{bmatrix}
L_x \\
L_y \\
L_z
\end{bmatrix}
=
\begin{bmatrix}
\Delta\phi_1 \\
\Delta\phi_2 \\
\Delta\phi_3
\end{bmatrix}.
\qquad (15.42)
$$

Data reduction can be simplified by illuminating the object from four different directions making equal angles with the viewing direction, two in the same vertical plane and two in the same horizontal plane [Goldberg, 1975]. This procedure also provides a check on the accuracy of the measurements.

The experimental errors in evaluating **L** from (15.42) fall into two categories: those related to the geometry of the measuring system and those associated with the measurement of the optical phase. The former have been analyzed by Matsumoto, Iwata, and Nagata [1973] and by Nobis and Vest [1978] and can be minimized by making the angular separation of the sensitivity vectors as large as possible.

## 15.11 Holographic moiré interferometry

In holographic moiré interferometry, images of the master object and a test object are superimposed by means of a suitable optical system, and a hologram is recorded of the resultant wave field. This hologram is processed and replaced in its original position.

If the test and master objects are illuminated along the same wave vector and viewed along the same wave vector, the intensity at any point in the reconstructed image, when both the objects are stressed, can be written as

$$I = I_0 \left[ 1 - \cos \left( \frac{\phi_1 - \phi_2}{2} \right) \cos \left( \frac{\phi_1 + \phi_2}{2} \right) \right]. \qquad (15.43)$$

In this equation, $\phi_1 = \mathbf{K} \cdot \mathbf{L}_1$, $\phi_2 = \mathbf{K} \cdot \mathbf{L}_2$, where **K** is the sensitivity vector and $\mathbf{L}_1$ and $\mathbf{L}_2$ are the vector displacements of the corresponding points on the two objects.

An observer sees moiré fringes generated by the term $\cos[(\phi_1 - \phi_2)/2]$ which contour the difference in the displacements of the two objects [Der, Holloway & Fourney, 1973; Hariharan & Hegedus, 1975a]. A moiré pattern of good contrast can be obtained if the individual interference patterns are dense and localized on, or very close to, the surface of the object.

Holographic moiré techniques can be used, in conjunction with an image-shearing system, to obtain information on the second derivatives of the surface displacements [Rastogi, 1984a, 1986].

## 15.12 Difference holographic interferometry

Difference holographic interferometry gives high-contrast fringes which contour the differences in the displacements or shapes of two macroscopically similar surfaces. It provides a practical method of detecting minor defects in a test piece, by comparison with a defect-free master [Füzessy & Gyimesi,

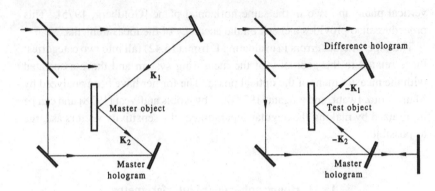

Fig. 15.14. Optical arrangement used for difference holographic interferometry: (a) for recording the master hologram; (b) for recording the difference hologram [Füzessy & Gyimesi, 1984].

1984; Rastogi, 1984b]. A typical experimental arrangement for difference holographic interferometry is shown in fig. 15.14.

In the first step, as shown in fig. 15.14(a), a pair of master holograms are recorded, one for the undeformed state of the master object and the other for the deformed state. The test object is then illuminated successively with the reconstructed waves obtained from the two holograms, using the conjugate reference beam in the setup shown in fig. 15.14(b), and a double-exposure hologram is recorded. The first exposure is made with the test object in the undeformed state, while it is illuminated by the reconstructed image of the undeformed master object, and the second exposure is made with the deformed test object, while it is illuminated by the reconstructed image of the master object in the deformed state.

The formation of the interference pattern can be explained as follows. We assume that the master object is illuminated in the direction $\mathbf{K}_1$ and observed along the direction $\mathbf{K}_2$. The phase difference at a point $P$ on the master object, due to a displacement $\mathbf{L}_1$, is then

$$\Delta\phi_m = (\mathbf{K}_1 - \mathbf{K}_2) \cdot \mathbf{L}_1. \tag{15.44}$$

In the second step, the master object is replaced by the test object and the master hologram is placed in the observation arm of the interferometer. The test object is now illuminated and observed along the directions $-\mathbf{K}_2$ and $-\mathbf{K}_1$, respectively. The phase difference introduced at the corresponding point on the test object by a displacement $\mathbf{L}_2$ is then

$$\Delta\phi_t = (\mathbf{K}_1 - \mathbf{K}_2) \cdot \mathbf{L}_2. \tag{15.45}$$

Accordingly, the phase difference in the final interference pattern is

$$\Delta\phi = (\mathbf{K}_1 - \mathbf{K}_2)(\mathbf{L}_2 - \mathbf{L}_1). \tag{15.46}$$

Alternatively, the two master holograms can be recorded on the same plate [Füzessy & Gyimesi, 1984*b*]. The master object is then replaced by the test object and a double-exposed difference hologram is recorded. The image reconstructed by this difference hologram then displays the interference pattern due to the four object waves. However, if the displacements of the master and test objects are sufficiently large, the only fringe pattern visible is that corresponding to their difference.

A real-time method in which the images of the master and test specimens are superimposed by a beam splitter has been described by Rastogi [1984(*b*)], while Simova and Sainov [1989] have studied additive and multiplicative moiré methods. A rigorous theoretical analysis of the formation of difference fringes has been made by Gyimesi and Füzessy [1988].

# 16

# Holographic interferometry: Further applications

## 16.1 Vibrating surfaces

Holographic interferometry can also be used to map the amplitude of vibration of a diffusely reflecting surface. The most commonly used technique for this purpose is called time-average holographic interferometry.

### 16.1.1 Time-average holographic interferometry

In this technique, a hologram is recorded of the vibrating surface with an exposure time which is long compared to the period of vibration [Powell & Stetson, 1965].

Consider a point $P(x, y)$ on the object whose displacement at time $t$ is given by the relation

$$\mathbf{L}(x, y, t) = \mathbf{L}(x, y) \sin \omega t. \qquad (16.1)$$

The phase of the light scattered from this point is then shifted by an amount which is a function of time and, from (15.23), can be written as

$$\Delta \phi(x, y, t) = \mathbf{K} \cdot \mathbf{L}(x, y) \sin \omega t, \qquad (16.2)$$

where $\mathbf{K}$ is the sensitivity vector.

Now, let $o(x, y) = |o(x, y)| \exp[-i\phi(x, y)]$ represent the complex amplitude of the scattered light from $P$ when the object is stationary in its equilibrium position. The complex amplitude of the scattered light from the vibrating object at any instant is then

$$o(x, y, t) = |o(x, y)| \exp\{-i[\phi(x, y) + \mathbf{K} \cdot \mathbf{L}(x, y) \sin \omega t]\}. \qquad (16.3)$$

Since the phase of the object wave changes very slowly with time compared to the electric field, the holographic recording process can be regarded as involving the recording of a very large number of superimposed holograms, one for

271

ԌuՆh slightly displaced position of the object. Accordingly, if the holographic recording process is assumed to be linear, the complex amplitude $u(x, y)$ of the wave reconstructed by the hologram will be proportional to the time average of $o(x, y, t)$ over the exposure interval $T$, so that we can write

$$
\begin{aligned}
u(x, y) &= \frac{1}{T} \int_0^T |o(x, y)| \exp\{-i[\phi(x, y) + \mathbf{K} \cdot \mathbf{L}(x, y) \sin \omega t]\} dt, \\
&= |o(x, y)| \exp[-i[\phi(x, y)] \\
&\quad \times \frac{1}{T} \int_0^T \exp[-i\mathbf{K} \cdot \mathbf{L}(x, y) \sin \omega t] dt, \\
&= o(x, y) M_T(x, y),
\end{aligned}
\tag{16.4}
$$

where $M_T(x, y)$ is known at the characteristic function. If the exposure time is long compared to the period of vibration ($T \gg 2\pi/\omega$), we have

$$
\begin{aligned}
M_T(x, y) &= \lim_{T \to \infty} \frac{1}{T} \int_0^T \exp[-i\mathbf{K} \cdot \mathbf{L}(x, y) \sin \omega t] dt, \\
&= J_0[\mathbf{K} \cdot \mathbf{L}(x, y)],
\end{aligned}
\tag{16.5}
$$

where $J_0$ is the zero-order Bessel function of the first kind. The intensity in the reconstructed image is then

$$
\begin{aligned}
I(x, y) &= |o(x, y) M_T(x, y)|^2, \\
&= I_0(x, y) J_0^2[\mathbf{K} \cdot \mathbf{L}(x, y)],
\end{aligned}
\tag{16.6}
$$

where $I_0(x, y)$ is the intensity when the object is at rest. The function $|M_T|^2$ is plotted against the parameter $\Omega = \mathbf{K} \cdot \mathbf{L}$ in fig. 16.1($a$). If the vibration amplitude varies across the object, (16.6) gives rise to fringes (contours of equal vibration amplitude) covering the reconstructed image. The dark fringes, at which the intensity drops to zero, correspond to the zeros of the function $J_0^2(\Omega)$, and the bright fringes to its maxima. The first maximum, which corresponds to the nodes, is the brightest, while successive maxima occurring at larger vibration amplitudes fall off progressively.

Time-average holography permits ready identification of the vibration modes as well as accurate measurements of the vibration amplitude in many cases. A typical series of interferograms is presented in fig. 16.2, showing the time-average fringes obtained with a model of an aircraft tail fin excited at frequencies corresponding to three of its principal vibration modes.

Apart from the fact that the fringes can be seen only after processing the hologram, the main limitation of this technique is that it does not give any information on the relative phases of the vibration at different points on the

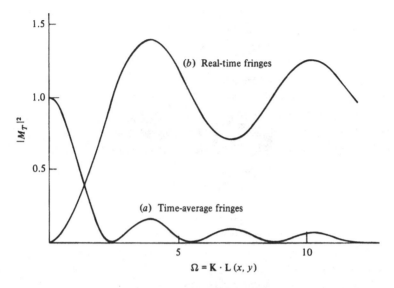

Fig. 16.1. Characteristic functions for a vibrating object whose displacement varies sinusoidally with time corresponding to (a) time-average fringes and (b) real-time fringes.

object. In addition, its sensitivity drops off at very small vibration amplitudes ($|\mathbf{L}| < \lambda/8$).

### 16.1.2 Real-time interferometry of vibrating surfaces

Real-time holographic interferometry (see section 15.1) can be used to identify the resonances of a test object by monitoring its response while varying the excitation frequency and the point of excitation [Stetson & Powell, 1965].

We assume that the hologram is recorded with the object stationary in its equilibrium position, and the system is adjusted so that the reconstructed image has the same intensity $I_0(x, y)$ as the object viewed through the hologram. The resultant intensity at any instant when the object is vibrating with an amplitude $\mathbf{L}(x, y)$ at a circular frequency $\omega$ is then

$$I(x, y, t) = I_0(x, y)\{1 - \cos[\mathbf{K} \cdot \mathbf{L}(x, y) \sin \omega t]\}. \qquad (16.7)$$

If the period of the vibration is much shorter than the time of response of the human eye ($\approx 0.04$ s), the observer sees the time-averaged intensity

$$\langle I(x, y)\rangle = I_0(x, y) \lim_{T \to \infty} \frac{1}{T} \int_0^T \{1 - \cos[\mathbf{K} \cdot \mathbf{L}(x, y) \sin \omega t]\}\mathrm{d}t,$$

$$= I_0(x, y)\{1 - J_0[\mathbf{K} \cdot \mathbf{L}(x, y)]\}, \qquad (16.8)$$

(a)

(b)

(c)

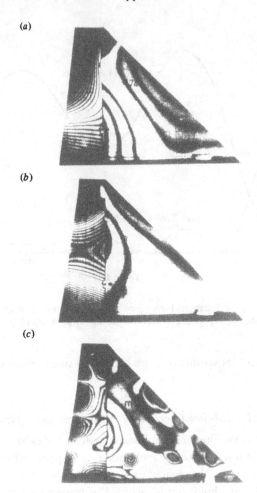

Fig. 16.2. Time-average interferograms showing the vibration modes of a model of an aircraft tail fin at vibration frequencies of (*a*) 670 Hz, (*b*) 894 Hz and (*c*) 3328 Hz [Abramson & Bjelkhagen, 1973].

and the characteristic function $M_T$ is defined by the relation

$$|M_T|^2 = \{1 - J_0[\mathbf{K} \cdot \mathbf{L}]\}. \tag{16.9}$$

Figure 16.1(*b*) shows the function defined by (16.9) plotted against $\Omega = \mathbf{K} \cdot \mathbf{L}$. Ideally, a dark-field interferogram is obtained with only half the number of fringes seen with the time-average technique. However, in practice, because of varying phase shifts introduced by the photographic emulsion, it is difficult to get a dark field even with *in situ* processing. Typical fringe patterns obtained

(a)

(b

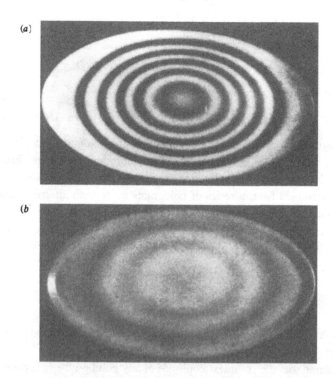

Fig. 16.3. Fringe patterns obtained with the same vibrating object (a tin can) using (a) the time-average technique, and (b) real-time interference [Biedermann & Molin, 1970].

with the same object using the time-average and real-time techniques are shown in fig. 16.3 [Biedermann & Molin, 1970].

Most studies on vibrating objects involve a combination of these techniques. Commonly, real-time holographic interferometry is used when searching for resonant modes, after which a time-average hologram can be made for more detailed measurements of the vibration amplitude. Typical applications have been in studies of musical instruments [Ågren and Stetson, 1972], loudspeakers [Chomat & Miler, 1973], turbine blades, and aircraft structures [Bjelkhagen, 1974].

### 16.1.3 Stroboscopic holographic interferometry

In stroboscopic holographic interferometry [Archbold & Ennos, 1968; Shajenko & Johnson, 1968; Watrasiewicz & Spicer, 1968] a hologram of a vibrating object is recorded using a sequence of light pulses that are triggered at times $\Delta t_1$ and $\Delta t_2$ during the vibration cycle. The characteristic function for an object displacement $\mathbf{L}(x, y) \sin \omega t$ is then

$$M_T = \exp[-i\mathbf{K} \cdot \mathbf{L}(x, y) \sin \omega \Delta t_1]$$
$$+ \exp[-i\mathbf{K} \cdot \mathbf{L}(x, y) \sin \omega \Delta t_2], \tag{16.10}$$

so that the intensity in the image is

$$I(x, y) = I_0(x, y)\{1 + \cos[\mathbf{K} \cdot \mathbf{L}(x, y)(\sin \omega \Delta t_1 - \sin \omega \Delta t_2)]\}. \tag{16.11}$$

The hologram is equivalent to a double-exposure hologram recorded while the object was in these two states of deformation, and the fringes have unit visibility, irrespective of the vibration amplitude. The phase of the vibration can be determined from a series of holograms made with different values of $\Delta t_2$, keeping $\Delta t_1$ fixed; alternatively, real-time observations can be made. More elaborate techniques employing pulsed illumination have been described by Vikram [1974a,b, 1975, 1976] for studying combinations of different types of motions.

### 16.1.4 Temporally modulated holography

Holography with a temporally modulated reference beam [Aleksoff, 1971] is an extremely powerful technique for the study of vibrations.

We assume that the complex amplitude at any point on the hologram plane due to a point on the vibrating object is $f_o(t)o$, where $f_o(t) = \exp(-i\mathbf{K} \cdot \mathbf{L} \sin \omega t)$, while that due to the reference beam is $f_r(t)r$, where $f_r(t)$ describes the temporal modulation imposed on the reference beam. The intensity at the hologram plane at time $t$ is then

$$I(t) = |f_o(t)o + f_r(t)r|^2. \tag{16.12}$$

As in section 16.1.1, we can consider this equivalent to recording a very large number of superimposed holograms, one for each displaced position of the object. If we assume linear recording, as in (2.2), the amplitude transmittance of the hologram is

$$\mathbf{t} = \mathbf{t}_0 + \beta \int_0^T |f_o(t)o + f_r(t)r|^2 dt. \tag{16.13}$$

To view the reconstructed image, the hologram is illuminated by the unmodulated reference wave. The complex amplitude of the transmitted wave is then $u = r\mathbf{t}$, and the characteristic function can be written, apart from a constant factor, as

$$M_T = \frac{1}{T} \int_0^T f_o(t) f_r(t) dt. \tag{16.14}$$

The modulation of the reference beam can take different forms; some of the most interesting cases are discussed in sections 16.1.5–16.1.7.

### 16.1.5 Frequency translation

If the circular frequency of the reference wave is shifted by an amount $n\omega$, where $n$ is an integer,

$$f_r(t) = \exp(-in\omega t), \tag{16.15}$$

and the characteristic function is

$$M_T = \lim_{T \to \infty} \frac{1}{T} \int_0^T \exp(-in\omega t) \exp(-i\mathbf{K} \cdot \mathbf{L} \sin \omega t) dt. \tag{16.16}$$

We then make use of the identity

$$\exp(iz \sin \phi) = \sum_{m=-\infty}^{\infty} J_m(z) \exp(im\phi), \tag{16.17}$$

and reverse the order of integration and summation to rewrite (16.16) as

$$M_T = \sum_{m=-\infty}^{\infty} J_m(\mathbf{K} \cdot \mathbf{L}) \lim_{T \to \infty} \frac{1}{T} \int_0^T \exp[-i(m-n)\omega t)] dt. \tag{16.18}$$

However, when $T \gg 2\pi/\omega$,

$$\frac{1}{T} \int_0^T \exp[-i(m-n)\omega t)] dt = \delta(m-n), \tag{16.19}$$

so that

$$M_T = J_n(\mathbf{K} \cdot \mathbf{L}), \tag{16.20}$$

and the intensity in the image is

$$I(x, y) = I_0(x, y) J_n^2[\mathbf{K} \cdot \mathbf{L}(x, y)]. \tag{16.21}$$

Frequency translation can be used to obtain greatly increased sensitivity for the detection of small vibration amplitudes [Zambuto & Fischer, 1973]. With time-average fringes, as pointed out in section 16.1.1, the intensity is proportional to $J_0^2(\mathbf{K} \cdot \mathbf{L})$ and exhibits very little change for small amplitudes of vibration. On the other hand, with frequency translation, when $M_T = J_1(\mathbf{K} \cdot \mathbf{L})$, a dark field with fringes whose intensity is proportional to $(\mathbf{K} \cdot \mathbf{L})^2$ is obtained.

Frequency translation can also be used to obtain reduced sensitivity for measurements of large vibration amplitudes since, from (16.20), the number of fringes formed decreases as $n$ is increased.

### 16.1.6 Amplitude modulation

In this case the amplitude of the reference wave is modulated at the same frequency as the vibrating object, so that $f_r(t) = \cos(\omega t - \psi)$, where $\psi$ is the phase difference with reference to the vibrating object point $P(x, y)$. When the hologram is illuminated with an unmodulated reference wave, the intensity in the image is

$$I(x, y) = I_0(x, y)J_1^2[\mathbf{K} \cdot \mathbf{L}(x, y)]\cos^2 \psi. \qquad (16.22)$$

A series of holograms recorded with different values of $\psi$ can then be used to map the variations in phase over the vibrating object [Takai, Yamada & Idogawa, 1976].

### 16.1.7 Phase modulation

Phase information can also be obtained by sinusoidal phase modulation of the reference beam at the frequency of the vibrating object [Neumann, Jacobson & Brown, 1970]. In this case $f_r(t) = \exp(if \sin \omega t)$, where $f$ is the amplitude of the cyclic phase shift. If the vibration at a point $P(x, y)$ on the object has a phase $\psi_0$, the characteristic function is

$$M_T = J_0\{[(\mathbf{K} \cdot \mathbf{L})^2 + f^2 - 2(\mathbf{K} \cdot \mathbf{L})f \cos \psi_0]^{1/2}\}. \qquad (16.23)$$

Techniques for obtaining useful phase information using a minimum number of interferograms have been described by Levitt and Stetson [1976] and by Vikram [1977].

### 16.1.8 Holographic subtraction

Holographic subtraction [Hariharan, 1973] involves making two equal exposures. During the first exposure the object is at rest, while during the second it is made to vibrate after introducing a phase shift of $\pi$ in one of the beams. The reconstructed amplitude due to the second exposure is subtracted from that due to the first, resulting in an interference pattern in which the intensity distribution is given by the relation

$$I(x, y) = I_0(x, y)\{1 - J_0[\mathbf{K} \cdot \mathbf{L}(x, y)]\}^2. \qquad (16.24)$$

The characteristic function is the square of that for real-time fringes defined by (16.9) and shown in fig. 16.1(b). The number and position of the fringes is the same, but their visibility is higher.

It is easy to get a perfectly dark field with holographic subtraction, giving a useful increase in sensitivity for small vibration amplitudes. This method has been extended by using weighted subtraction to generate contour lines of equal vibration amplitude at any given small level [Sato, Ogawa & Ueda, 1974]. Other applications, including measurements of large vibration amplitudes, separation of the effects of simultaneous uniform motion, and the study of periodic nonsinusoidal vibrations, have been listed by Hariharan [1976b].

### 16.1.9 Time-average holography of nonsinusoidal motions

We consider, in the first instance, separable motions for which the displacements of all the points on the surface of the object can be written as the product of a spatially varying vector amplitude $\mathbf{L}(x, y)$ and a single function of time $f(t)$. The characteristic function is then

$$M_T = \lim_{T \to \infty} \frac{1}{T} \int_0^T \exp[-i\mathbf{K} \cdot \mathbf{L} f(t)]\mathrm{d}t, \qquad (16.25)$$

where, as before, $\mathbf{K}$ is the sensitivity vector; this expression has been evaluated for various types of motion by Zambuto and Lurie [1970], Janta and Miler [1972], Stetson [1972a], and Gupta and Singh [1975a,b, 1976].

The analysis can be facilitated in some cases by an interpretation of the characteristic function suggested by Stetson [1971] according to which the reconstructed wavefront can be considered as the ensemble average of the object wavefronts recorded by the hologram. Accordingly, (16.25) can be written as

$$M_T = \int_{-\infty}^{\infty} p(f) \exp(i\Omega f)\mathrm{d}f, \qquad (16.26)$$

where $\Omega = \mathbf{K} \cdot \mathbf{L}$ and $p(f)$ is a probability density function.

More complex vibrations can be considered as the sum of several separable motions, so that the characteristic function can be written as

$$M_T = \lim_{T \to \infty} \frac{1}{T} \int_0^T \exp\left[-i\sum_{n=1}^{N} \Omega_n f_n(t)\right]\mathrm{d}t, \qquad (16.27)$$

where $\Omega_n = \mathbf{K} \cdot \mathbf{L}_n$.

In this case, if the motions are at irrationally related frequencies (temporally independent motions), the characteristic function is merely the product of the characteristic functions of the motions taken individually.

Vibrations at rationally related frequencies result in more complicated characteristic functions. Such vibrations have been studied by Molin and Stetson [1969], Stetson [1970c], Stetson and Taylor [1971], Wilson [1970, 1971], Wilson and Strope [1970], Dallas and Lohmann [1975], and Tonin and Bies [1978].

A useful approximation for the characteristic function can often be obtained
in such cases by using the method of stationary phase [Stetson, 1972*b*]. This
is equivalent to assuming that the time-average fringes are formed mainly by
interference between the reconstructed wavefronts corresponding to those positions of the object at which it dwells for a relatively long time, and is, in fact,
a simple physical explanation of the formation of time-average fringes.

### 16.1.10 *Localization of the fringes with vibrating objects*

Molin and Stetson [1970*a,b*] have shown experimentally that for separable
motions the localization of the fringes follows the same pattern as with a double-
exposure hologram for a displacement of the same nature. The situation is not
so simple for nonseparable motions, except for independent motions, where the
corresponding fringe systems localize independently [Molin & Stetson, 1971].
With mutually dependent vibrations, the fringes are not well localized and can
usually be observed only with a properly oriented slit aperture. However, with
temporally orthogonal vibrations, the fringes have high visibility where the
zero-order fringe of one component intersects the plane of localization of the
other.

## 16.2 Holographic photoelasticity

As is well known, information on the stresses in a model made of a material
which becomes birefringent when it is stressed can be obtained by studying
the state of polarization of the light transmitted by it. However, conventional
photoelastic measurements give only the isochromatics, the loci of points corresponding to constant values of $(\sigma_1 - \sigma_2)$, the difference in the principal
stresses.

Favre [1929] showed that interferometric measurements with unpolarized
light can give a set of fringes, called isopachics, corresponding to variations
in the thickness of the model. These fringes are the loci of points for which
$(\sigma_1 + \sigma_2)$, the sum of the principal stresses, is constant. Subsequently, Nisida
and Saito [1964] developed an interferometric method which gives $(\sigma_1 - \sigma_2)$
and $(\sigma_1 + \sigma_2)$ simultaneously. However, both these methods require a test
specimen of high optical quality.

Fourney [1968] and Hovanesian, Brcic, and Powell [1968] were able to
eliminate this requirement by applying holography to photoelastic measurements. They used a double-exposure technique to obtain a combined isopachic-
isochromatic fringe pattern which could be related to the stress distribution in
the model [Fourney & Mate, 1970]. Unfortunately, the analysis of this pattern

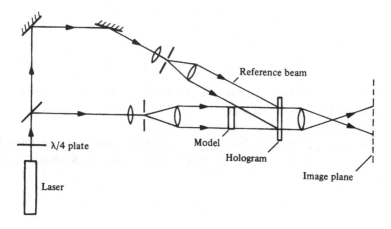

Fig. 16.4. Holographic system used to obtain the isochromatics of a photoelastic model [O'Regan & Dudderar, 1971].

is not very easy, and ambiguities can arise [Holloway & Johnson, 1971; Sanford & Durelli, 1971]. Because of these difficulties, methods have been developed to obtain separate isochromatic and isopachic patterns.

### 16.2.1 Isochromatics

A hologram of the stressed model recorded with coherent light which is either unpolarized or circularly polarized reconstructs the isochromatics. This occurs because the light incident at a point $P$ on the model, in the system shown in fig. 16.4, can be resolved into two orthogonally polarized components, with electric vectors parallel to the principal stresses at this point, whose complex amplitudes can be written as $o_1 = \exp(-i\phi_1)$ and $o_2 = \exp(-i\phi_2)$, respectively. The reference wave can also be resolved into two similar orthogonal components, $r_1$ and $r_2$. Interference takes place only between $o_1$ and $r_1$, and $o_2$ and $r_2$, respectively, so that two independent holograms are recorded on the same plate.

When the processed plate is illuminated once again by the reference beam, it reconstructs two waves whose complex amplitudes are proportional to $o_1$ and $o_2$, respectively. However, these two waves are now able to interfere, exactly as in double-exposure holographic interferometry, since their states of polarization are the same. The intensity distribution in the image is then

$$I = |o_1 + o_2|^2,$$

$$= 2[1 + \cos(\phi_1 - \phi_2)]. \tag{16.28}$$

The phase difference between the two waves is

$$\phi_1 - \phi_2 = (2\pi/\lambda)(n_1 - n_2)(d + \Delta d), \tag{16.29}$$

where $n_1$ and $n_2$ are the principal refractive indices of the material when it is stressed, $d$ is the thickness of the unstressed model, and $\Delta d$ is the change in its thickness.

In addition, $n_1$ and $n_2$ are related to $n_0$, the refractive index of the unstressed material, and $\sigma_1$ and $\sigma_2$, the principal stresses, by the Maxwell–Neumann equations

$$n_1 - n_0 = A\sigma_1 + B\sigma_2, \tag{16.30}$$

$$n_2 - n_0 = B\sigma_1 + A\sigma_2, \tag{16.31}$$

where $A$ and $B$ are the stress-optical coefficients of the material, so that

$$n_1 - n_2 = (A - B)(\sigma_1 - \sigma_2). \tag{16.32}$$

Since $\Delta d \ll d$, we then have from (16.28), (16.29), and (16.32),

$$I = 2\{1 + \cos[(2\pi/\lambda)(A - B)(\sigma_1 - \sigma_2)d]\}. \tag{16.33}$$

### 16.2.2 Isopachics

In order to obtain the isopachics, it is necessary to make two exposures on the hologram plate, one with a stress applied to the model and the other without any stress applied to it. The change in the thickness of the model at any point is then linearly proportional to the sum of the principal stresses at this point and is given by the relation

$$\Delta d = -d(\nu/E)(\sigma_1 + \sigma_2), \tag{16.34}$$

where $\nu$ and $E$ are the Poisson's ratio and the Young's modulus, respectively. In this case, if the model did not exhibit any birefringence, the image reconstructed by the hologram would exhibit fringes corresponding to the isopachics. As it is, because of the birefringence of the model, a complicated fringe system is observed arising from the interaction of the two fringe systems.

The effects of the stress-induced birefringence can be eliminated if, as shown in fig. 16.5, the object beam is made to traverse the model once again after passing through an optical rotator (a Faraday cell) which rotates the axes of polarization by 90° [Chau, 1968; O'Regan & Dudderar, 1971; Chatelain, 1973]. As a result, the vibration which propagates along the fast axis on the outward

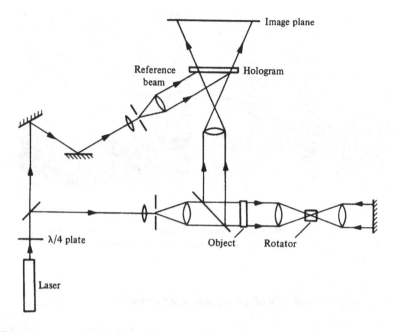

Fig. 16.5. Holographic system used to obtain the isopachics of a photoelastic model [O'Regan & Dudderar, 1971].

journey returns along the slow axis and *vice versa*, with the result that the phase difference between them due to the stress-induced birefringence is cancelled.

In this case, the effects of the change in thickness of the model can be neglected since they are relatively small compared to the effects due to the changes in the refractive index. The optical path through the stressed model can then be taken as $(n_1 + n_2)d$, while that through the unstressed object is $2n_0d$. Accordingly, the intensity in the image reconstructed by the double-exposed hologram can be written as

$$I = 2I_0[1 + \cos(\phi - \phi_0)], \qquad (16.35)$$

where $(\phi - \phi_0)$, the phase difference between the two reconstructed waves, is given by the relation

$$\phi - \phi_0 \approx (2\pi/\lambda)(n_1 + n_2 - 2n_0)d. \qquad (16.36)$$

Now, from (16.30) and (16.31),

$$(n_1 + n_2 - 2n_0) = (A + B)(\sigma_1 + \sigma_2). \qquad (16.37)$$

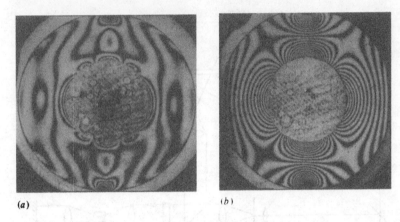

(a)                                                    (b)

Fig. 16.6. Holographic interference fringes obtained with a photoelastic model (an Araldite disc in diametral compression) showing (a) the isochromatics, (b) the isopachics [Chatelain, 1973].

Accordingly, the intensity of the reconstructed image is

$$I = 2\{1 + \cos[(2\pi/\lambda)(A + B)(\sigma_1 + \sigma_2)d]\}, \qquad (16.38)$$

which corresponds to the isopachic fringe pattern.

Isochromatic and isopachic fringes obtained with a typical test object (an Araldite disc in diametral compression) using these techniques are shown in fig. 16.6 [Chatelain, 1973].

### 16.2.3 Combined displays

While the isochromatic and isopachic patterns can be obtained separately in this fashion, it is also possible to combine the two by making use of the multiplexing capabilities of holograms. An optical system using two reference beams has been described by Assa and Betser [1964] with which it is possible to obtain either of the patterns without interference from the other.

Another method of separating the isochromatic and isopachic fringes using a simple real-time holographic interferometer has been described by Hovanesian [1974]. A combined fringe pattern identical to that obtained by Nisida and Saito [1964] with an interferometer can also be obtained in a holographic system if a volume diffuser is used to depolarize the object beam [Kubo & Nagata, 1976a,b; Ebbeni, Coenen & Hermanne, 1976]. Sandwich holography can also be used for this purpose [Uozato & Nagata, 1977].

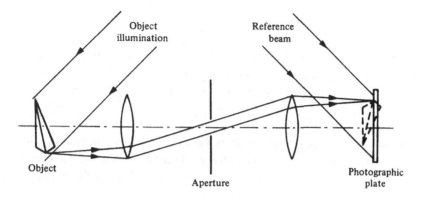

Fig. 16.7. Optical system for two-wavelength holographic contouring [Zelenka & Varner, 1968].

## 16.3 Holographic contouring

Holographic interferometry can produce an image of a three-dimensional object modulated by a fringe pattern corresponding to contours of constant elevation with respect to a reference plane.

### 16.3.1 Two-wavelength holographic contouring

In this method, two holograms of the object are recorded using light of two different wavelengths [Haines & Hildebrand, 1965; Hildebrand & Haines, 1966, 1967].

To minimize the transverse displacement between the reconstructed images and obtain plane contouring surfaces, a telecentric system is used, as shown in fig. 16.7, to image the object on the hologram, and a plane wave is used to illuminate the object [Zelenka & Varner, 1968]. Another plane wave making an equal but opposite angle with the axis of the optical system is used as the reference wave. Two exposures are made with light of two different wavelengths, $\lambda_1$ and $\lambda_2$. After processing, the hologram is replaced in its original position and illuminated with one of the wavelengths, say, $\lambda_2$.

Since the holograms are recorded and reconstructed with a collimated reference beam, it follows from (3.17) and (3.20) that the two reconstructed images have the same magnification (unity). In addition, since they are formed close to the hologram plane, (3.38) shows that any lateral displacement of the images is negligible. However, interference fringes are seen due to the axial displacement of one image with respect to the other which, from (3.39), is given by the relation

$$\Delta z = z(\lambda_1 - \lambda_2)/\lambda_1. \qquad (16.39)$$

Since the telecentric system limits the rays forming the image to a small angle with the axis, successive fringes correspond to changes in $\Delta z$ of $\lambda_2/2$, or, when $\lambda_1$ and $\lambda_2$ are not very different, to increments of $z$ given by the relation

$$\delta z \approx \lambda^2/2\Delta\lambda. \qquad (16.40)$$

A fairly wide range of contour intervals can be obtained with pairs of lines from an $Ar^+$ laser; thus, the two lines at $\lambda = 514$ nm and 488 nm give a contour interval $\delta z$ of approximately 5 $\mu$m, while those at $\lambda = 488$ nm and 477 nm give a contour interval of approximately 10 $\mu$m.

With a dye laser, the contour interval can be varied continuously. It is also possible to make the wavelength difference very small. In the latter case, the criteria for good fringe visibility can be satisfied more easily and simpler optical systems can be used [Friesem & Levy, 1976]. Photothermoplastics as well as BSO have been used as recording media for two-wavelength contouring [Leung, Lee, Bernal & Wyant, 1979; Küchel & Tiziani, 1981]; the latter permits real-time viewing of the contours.

### 16.3.2 Two-refractive-index contouring

The optical arrangement for this method [Tsuruta, Shiotake, Tsujiuchi & Matsuda, 1967; Zelenka & Varner, 1969], which requires only a single laser wavelength, is shown in fig. 16.8. The object is placed in a cell with a plane glass window and viewed through a telecentric system. A beam splitter is used to illuminate the object with a plane wave along the axis of the optical system. The hologram plate is located near the stop of the telecentric system.

Two holograms are recorded on the same plate, with the cell filled with fluids having refractive indices $n_1$ and $n_2$, respectively. Since the line of sight is normal to the window, there is no lateral displacement of the images; however, one of the images is longitudinally displaced with respect to the other by an amount

$$\Delta z = (n_1 - n_2)z, \qquad (16.41)$$

where $z$ is the distance from the window to the surface of the object.

Accordingly, when the hologram is replaced in the same position and illuminated once again by the same reference beam, successive fringes in the reconstructed image correspond to increments of $z$ given by the relation

$$\delta z = \lambda/2|n_1 - n_2|. \qquad (16.42)$$

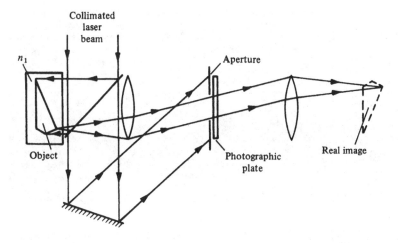

Fig. 16.8. Optical system for two-index holographic contouring [Zelenka & Varner, 1969].

The contouring interval $\delta z$ can be varied from about 1 $\mu$m to 300 $\mu$m by using air and a liquid, or a combination of liquids, while combinations of gases can be used to obtain even larger contour intervals [Marrone & Ribbens, 1975].

In practice, two-refractive-index contouring is the simpler and more flexible of the two techniques. Its only disadvantage is the need to immerse the object in a suitable liquid when a contour separation less than 300 $\mu$m is needed.

Typical contours obtained with both these techniques using the same object are presented in fig. 16.9.

### 16.3.3 Contouring by changing the angle of illumination

A simple method of holographic contouring involves making a double-exposure hologram with a small lateral displacement of the point source illuminating the object between the two exposures. In this case, the contouring surfaces consist of a set of hyperboloids of revolution with the two positions of the source as their common foci. Plane surfaces can be obtained if the dimensions of the object are small compared to the distance of the source, or if collimated illumination is used along with a telecentric imaging system.

To obtain contouring surfaces normal to the line of sight with this technique, the beam illuminating the object must also be normal to the line of sight [Menzel, 1974]. However, contouring surfaces at any desired angle can be generated by translating the object between the two exposures in a properly chosen direction

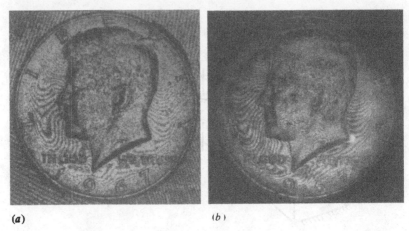

(a)                                                    (b)

Fig. 16.9. Depth contours obtained by (a) the two-wavelength method (contour interval 9.25 $\mu$m); (b) the two-index method (contour interval 11.8 $\mu$m) [Zelenka & Varner, 1969].

[Abramson, 1976b]. This displacement can be combined with a displacement of the illuminating source [De Mattia & Fossati-Bellani, 1978]. A convenient alternative is the use of a sandwich hologram [Abramson, 1976a]; the contouring surfaces can then be made to assume any desired orientation by tilting the sandwich through a small angle.

Rastogi and Pflug [1991a] have combined these techniques to obtain a method capable of contouring surfaces having a wide range of depths with good sensitivity. The first step involves producing a set of modulated interference fringes similar to those produced when a grid is projected at an angle on the object surface. For this, the object is illuminated at an angle with a collimated beam, as shown in fig. 16.10, and a hologram is recorded.

We can regard the modulation of the phase produced in this case as consisting of two parts, one ($\phi_1$) corresponding to a linear phase variation along the $x$ axis, which would have been produced by a flat surface, and another ($\phi_2$) due to the variations in depth of the object. The first term can be eliminated by generating an additional linearly varying phase term which has the same magnitude as $\phi_1$, but the opposite sign. This can be done by a number of methods, one being to tilt the reference beam through a small angle. The elimination of the unwanted phase term $\phi_1$ can be monitored by observing the fringes formed with a flat plate mounted next to the object. The contour interval per fringe is then

$$\Delta z = \frac{\lambda}{\sin \theta \sin \Delta\theta}. \tag{16.43}$$

Increased sensitivity can be obtained by repeating the process, so as to in-

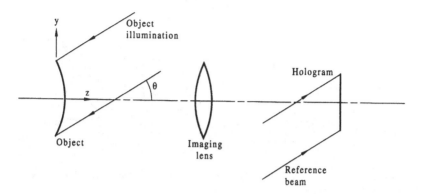

Fig. 16.10. Schematic of an optical system for multiple-source contouring [Rastogi & Pflug, 1991*a*].

crease the value of $\Delta\theta$. Any decrease in the visibility of the fringes can be overcome by imaging the object on the hologram plate and, where necessary, translating the hologram laterally. This procedure also makes it possible to change the orientation of the contouring planes with respect to the $xy$ plane.

Similar results can also be obtained by tilting the object around an axis in its plane parallel to the $y$ axis [Rastogi & Pflug, 1991*b*].

### 16.3.4 Contouring with reflection holograms

A simple contouring technique using reflection holograms has been described by Henshaw and Ezekiel [1974]. This technique also has the advantage that the contoured image can be viewed with white light.

The optical arrangement is shown in fig. 16.11 and is similar to that used for "piggyback" holograms (see section 5.1). The hologram plate is clamped to the object and illuminated by a collimated beam. Two exposures are made with the light incident on the hologram plate at angles $\theta_1$ and $\theta_2$, respectively. The contour spacing in the reconstructed image is then given by the relation

$$\delta z = \lambda \cos\theta_0/2 \sin(\langle\theta\rangle - \theta_0) \sin(\Delta\theta/2), \qquad (16.44)$$

where $\theta_0$ is the angle of viewing, $\langle\theta\rangle = (\theta_1 + \theta_2)/2$ and $\Delta\theta = \theta_1 - \theta_2$.

### 16.3.5 Photogrammetric contouring

In contouring techniques using holographic interference, the contour fringes formed are localized on or near the surface of the virtual image. If the object

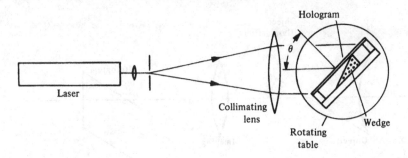

Fig. 16.11. Optical system for contouring using reflection holography [Henshaw & Ezekiel, 1974].

has appreciable depth, a picture of the fringes taken with a camera gives a perspective projection of the fringes instead of a true orthoscopic contour map. In such a case, there are advantages in using techniques in which the contours are plotted by observations on the image with an optical system having a very limited depth of focus [Stetson, 1968*b*; Gara, Majkowski & Stapleton, 1973].

A more convenient way is to use a stereo plotter [Balasubramanian, 1975], of the type developed for photogrammetry, in which a luminous dot (the tip of an optical fiber) attached to a three-coordinate measuring device is kept in apparent contact with the image while it is moved over a plane; it is then possible to plot contours (or profiles) directly. The multiplicity of perspectives available with an image hologram, or a holographic stereogram, makes precise settings easier than with a stereo model.

# 17
# Holographic interferometry: Advanced techniques

## 17.1 Photorefractive crystals

As described in section 7.7, photorefractive crystals can be used as recording materials for holography. In such crystals, exposure to light frees trapped electrons and induces a redistribution of charge within the crystal. The spatially varying electric field produced by this charge pattern modulates the refractive index of the crystal through the electro-optic effect, resulting in the formation of a volume phase hologram. If the incident light distribution changes, a new charge pattern and a new hologram are formed in a characteristic time $\tau_H$. The dynamic and adaptive properties of holograms recorded in photorefractive crystals have opened up new and interesting possibilities in holographic interferometry.

### 17.1.1 Time-average interferometry of vibrating surfaces

Photorefractive crystals of the sillenite family (BGO, BSO, and BTO) have been used most commonly for holographic interferometry because of their relatively high sensitivity to light. One of the earliest applications of such photorefractive crystals to holographic interferometry made use of the ability of a BSO crystal to generate a phase-conjugate image in real time [Huignard, Herriau & Valentin, 1977]. A typical optical system used to study the vibration modes of a loudspeaker is shown in fig. 17.1.

In this case, time-average fringes are formed because the crystal integrates over the characteristic time $\tau_H$ required to build up the hologram, which is usually much longer than the period of the vibration. With such an arrangement, it is possible to observe the changes in real time, as the excitation frequency is changed.

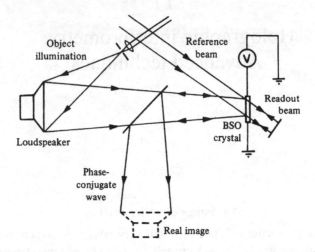

Fig. 17.1. Application of phase conjugation in BSO to the study of the modes of a vibrating loudspeaker [Marrakchi, Huignard & Herriau, 1980].

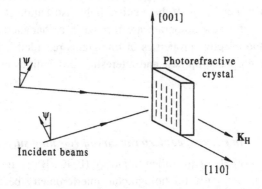

Fig. 17.2. Two-beam geometry used with photorefractive crystals of the sillenite family.

### 17.1.2 *Self-diffraction in photorefractive crystals*

Current techniques of holographic interferometry with sillenite crystals make use of the phenomenon of self-diffraction [Huignard & Marrakchi, 1981] using the two-beam geometry shown in fig. 17.2.

In this arrangement, the input and output faces of the crystal are the (110) crystallographic planes, and the grating vector $\mathbf{K}_H$ is parallel to the [110] axis. If, in the two incident beams, the electric vector makes an angle $\psi = \pm45°$ with the [001] axis, energy is transferred from one beam to the other. This transfer of energy is a maximum when the phase difference $\phi_H$ between the interference fringes produced by the two beams and the hologram formed in the crystal is 90°. This condition is satisfied automatically in the absence of an external field when the hologram is formed through the diffusion of free charge carriers [Huignard & Micheron, 1976]. However, the maximum diffraction efficiency that can be attained with no external field is low ($< 0.05$).

A higher diffraction efficiency can be obtained with an external electric field, but the phase shift $\phi_H$ then tends to either 0° or 180°, making energy transfer difficult. This problem can be partly overcome by continuously varying the phase of one of the beams [Marrakchi, Huignard & Günter, 1981; Stepanov, Kulikov & Petrov, 1982], but a more effective method is to use an alternating electric field with a frequency $\Omega \geq 2\pi/\tau_H$ [Stepanov & Petrov, 1985]. In this case, the phase difference between the interference pattern and the hologram is always 90°, and the diffraction efficiency of the hologram increases as the square of the field strength.

If, on the other hand, the electric vectors in the two incident beams are parallel to the [001] axis of the crystal, or at right angles to it, there is no change in the energy of the beams emerging from the crystal, but the light diffracted by the hologram has its polarization rotated by 90° [Petrov, Miridonov, Stepanov & Kulikov, 1979]. This transfer of polarization makes it possible to block the directly transmitted beam completely by means of a polarizer on the output side of the crystal and isolate the self-diffracted beam.

In practice, because of the optical activity of the crystal, the maximum separation of the self-diffracted and transmitted polarizations is obtained when the polarization of the reference/readout beam is parallel to the [001] axis of the crystal at its centre [Marrakchi, Johnson & Tanguay, 1986]. This condition can be satisfied by using a polarizer on the input side of the crystal, which can be set at the appropriate angle. When they emerge from the crystal, both the transmitted and diffracted beams have their planes of polarization rotated, because of the optical activity of the crystal. However, since the diffracted beam has its plane of polarization rotated by an additional 90° with respect to that of the transmitted beam, it can still be isolated by means of a polarizer on the output face.

### 17.1.3 Continuous readout techniques

Continuous readout is possible if the beam diffracted by the hologram can be separated from that directly transmitted by the crystal.

In a scheme using energy transfer ($\psi = \pm 45°$), the diffracted beam is actually the object beam amplified by the hologram. The directly transmitted beam now becomes a source of noise; another is the beam scattered by surface

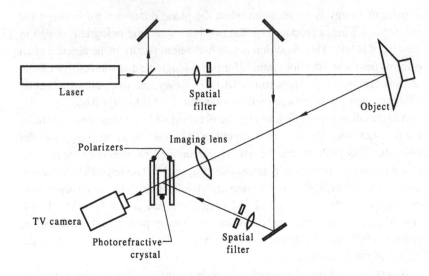

Fig. 17.3. Schematic of a two-beam holographic interferometer using polarization transfer for continuous readout [Kamshilin, Mokrushina & Petrov, 1989].

imperfections and defects in the crystal. The quality of the reconstructed image therefore improves with the coefficient of holographic amplification, which is higher with an alternating field. However, the coefficient of holographic amplification then peaks at fairly low spatial frequencies, typically around 200 mm$^{-1}$.

If polarization transfer is used ($\psi = 0°$ or $90°$), the diffracted beam and the directly transmitted object beam have orthogonal polarizations, so that a polarizer placed after the crystal, as shown in fig. 17.3, can be set to transmit the diffracted beam and suppress the original object beam [Kamshilin & Petrov, 1985]. An advantage of this technique is that holograms can be recorded with higher spatial carrier frequencies ($> 1000$ mm$^{-1}$), and the amount of scattered light is less than with energy transfer. However, a drawback is that, normally, this technique can be used only with no external field applied to the crystal, so that the diffraction efficiency is low.

Improved diffraction efficiency can be obtained by using a strong external alternating electric field, so that the phase difference between the polarization eigenstates of the beams emerging from the crystal is an integral multiple of $2\pi$ [Kamshilin & Mokrushina, 1989]. The beam emerging from the crystal is then always linearly polarized, and a polarizer can be used to suppress the directly transmitted object beam.

A detailed study of the optimum conditions for recording and readout using polarization separation in BSO has been carried out by Troth and Dainty [1991].

### 17.1.4 Time-average interferometry

An obvious application of continuous readout techniques is in time-average interferometry. In this case, the averaging period is always $\tau_H$, and does not depend on the exposure time. With polarization transfer, the intensity in the image of a vibrating object is

$$I(x, y) = I_0(x, y)J_0^2[\mathbf{K} \cdot \mathbf{L}(x, y)], \tag{17.1}$$

where $I_0(x, y)$ is the intensity when the object is at rest, $J_0$ is the zero-order Bessel function of the first kind, $\mathbf{K}$ is the sensitivity vector, and $\mathbf{L}(x, y)$ is the vibration amplitude. This is the same expression as that for a conventional time-average hologram.

With the energy-transfer technique, the intensity in the image of a vibrating object is

$$I(x, y) = I_0(x, y) \exp\{\Gamma J_0[\mathbf{K} \cdot \mathbf{L}(x, y)] - \alpha\}d, \tag{17.2}$$

where $\Gamma$ is the coefficient of holographic amplification, $\alpha$ is the coefficient of absorption of light in the crystal, and $d$ is the thickness of the crystal. While the intensity distribution in the image differs from that in a conventional time-average interferogram, the positions of the dark fringes are the same since, in both cases, they correspond to the zeros of the Bessel function.

### 17.1.5 Double-exposure interferometry

With continuous readout, a displacement of the object at any time $t_0$ results in the initiation of two concurrent processes. One is the decay of the hologram of the object in its initial state, the other is the formation of a hologram of the object in its new state. The waves reconstructed by the two holograms interfere to yield fringes contouring the displacement of the surface of the object.

In this case, the visibility of the fringes is a function of time, because the diffraction efficiency of the first hologram is decreasing with time while the diffraction efficiency of the second is increasing. The intensity in the image at any time $t$ is given by the relation [Kamshilin, Mokrushina & Petrov, 1989]

$$I(x, y, t) = I_0(x, y)\left[1 - 2\{1 - \cos[\mathbf{K} \cdot \mathbf{L}(x, y)]\}\right.$$
$$\left.\times \left(1 - \exp\frac{t_0 - t}{\tau_H}\right)\exp\frac{t_0 - t}{\tau_H}\right], \tag{17.3}$$

where $I_0(x, y)$ is the intensity in the reconstructed image of the undisturbed object, $\mathbf{K}$ is the sensitivity vector, and $\mathbf{L}(x, y)$ is the object displacement. The visibility of the fringes is a maximum when $t = t_0 + \tau_H \log_e 2$. The characteristic time $\tau_H$ depends on the crystal and the intensity and wavelength of the light, and is typically around 5 seconds for BTO and a light intensity of 4 W/m$^2$ at a wavelength of 633 nm.

Since a hologram recorded in a crystal such as BTO can be stored in darkness for several tens of minutes, the second exposure can be made after a time delay, during which the crystal is shielded from light and a mechanical or thermal stress is applied to the test object.

With BTO, a hologram recording can also be fixed by leaving the crystal in darkness [Zhivkova & Miteva, 1990], so that the positive charge carriers in the crystal move to compensate the field produced by the trapped photoelectrons. After fixing, the hologram can be read out over a period of tens of minutes, so that it is possible to compare the initial state of the object with a number of subsequent states. This technique has been applied in a holographic interference microscope by Tontchev, Zhivkova, and Miteva [1990].

### 17.1.6  Real-time interferometry

Another material which has been used for real-time interferometry is iron-doped lithium niobate (Fe : LiNbO$_3$). It has a lower sensitivity but permits storage of holograms over a much longer time. It has been used for real-time studies of heat-flow patterns from electronic chips over periods up to 10 minutes with minimal degradation of the reconstructed image [Wang, Magnusson & Haji-Sheikh, 1993].

Real-time measurements of vibrations are possible with sillenite crystals using polarization transfer. For this, the output polarizer is set so that the amplitudes of the beam diffracted by the hologram and the beam directly transmitted by the crystal are approximately equal. Any change in the phase difference between the object wave and the reconstructed wave then results in a change in the intensity. If the object vibrates, the time-varying component of the intensity at the vibration frequency can be processed to yield information on the amplitude and phase of the vibration at any point [Kamshilin & Mokrushina, 1986].

The optical path difference in such a system can be stabilized by modulating the phase of the reference beam at a frequency $\omega_0$ with a very small amplitude [Kamshilin, Frejlich & Cescato, 1986]. Under stable conditions, the output intensity then exhibits a modulation at a frequency $2\omega_0$. However, any additional change in the phase difference between the beams produces a signal at a frequency $\omega_0$, whose amplitude is proportional to this change, which can be applied to a mirror mounted on a PZT to compensate for this change. Active stabilization using this technique makes it possible to measure vibration amplitudes as small as 0.1 nm, at frequencies greater than 100 kHz [Frejlich, Kamshilin, Kulikov & Mokrushina, 1989].

## 17.2 Computer-aided evaluation

Manual methods of analysing the fringes in holographic interferograms are based on the assumption that the maxima and minima of the intensity distribution correspond, respectively, to values of the phase difference between the beams equal to an even or odd integral multiple of $\pi$. Because of the sinusoidal variation of the intensity with the phase difference, the accuracy with which the maxima and minima can be located from measurements on photographs is usually no better than 0.1 of the fringe spacing. In addition, when the number of fringes is small and they are unequally spaced, errors are introduced by the need for nonlinear interpolation to determine the fractional fringe order at any point.

This problem is particularly serious when the data are to be used to evaluate the local strains. In this case, it is necessary to fit a polynomial to the data, which can then be used for interpolation or differentiated analytically. Another problem is ambiguities which can arise where the sign of the displacement is not known.

One way to obtain higher accuracy, which was explored even at an early stage, was the use of television techniques to acquire and store the intensity distribution in the interference pattern [Kreis & Kreitlow, 1980; Schlüter, 1980; Nakadate, Magome, Honda & Tsujiuchi, 1981]. Conventional television cameras have now been replaced by charge-coupled detectors (CCDs) which sample the intensity at arrays containing from $100 \times 100$ to $1024 \times 1024$ discrete picture elements (pixels). A digitizer quantizes the intensity at each pixel into a number of gray levels (typically, 256 gray levels corresponding to 8 bits of data).

The distribution of the phase difference can then be determined by a number of techniques [Pryputniewicz, 1987; Robinson & Reid, 1993]. Most of these techniques require preprocessing to reduce speckle noise (usually by low-pass filtering) as well as to correct for local variations in image brightness.

### *17.2.1 Fringe skeletonizing*

Methods for fringe skeletonizing fall into two categories – those based on fringe tracking and those based on segmentation.

Fringe tracking methods use an algorithm which looks for pixels corresponding to local maxima or minima of the intensity [Funnell, 1981; Ennos, Robinson

& Williams, 1985]. The starting points for the algorithm are usually defined interactively, and the search direction is determined by the intensity values of the adjacent pixels. The operator may intervene to clear up ambiguities or correct apparent mistakes.

In segmentation techniques, the interference pattern is divided into regions corresponding to ridges, valleys, and slopes [Yatagai, Nakadate, Idesawa & Saito, 1982; Becker & Yu, 1985; Budzinski, 1992]. These regions are extended to eliminate any isolated points, and then thinned to line structures to obtain the fringe skeleton. Finally, the skeleton is completed (with operator assistance, if necessary) by linking interrupted lines, removing line crossings, and adding missing points.

### 17.2.2 Fourier-transform techniques

An additional tilt introduced in one of the beams (say, along the $x$ direction) generates background fringes corresponding to a spatial carrier frequency $\xi_0$ [Takeda, Ina & Kobayashi, 1982]. This spatial carrier is modulated by the phase difference between the beams, so that the intensity in the interference pattern can be written as

$$I(x, y) = a(x, y) + b(x, y) \cos[2\pi \xi_0 x + \Delta\phi(x, y)],$$
$$= a(x, y) + c(x, y)$$
$$\times \exp(2\pi i \xi_0 x) + c^*(x, y) \exp(-2\pi i \xi_0 x), \quad (17.4)$$

where $c(x, y) = (1/2)b(x, y) \exp[i\Delta\phi(x, y)]$.

The Fourier transform of the intensity with respect to $x$ is then [Macy, 1983],

$$\mathcal{I}(\xi, y) = A(\xi, y) + C(\xi - \xi_0, y) + C^*(\xi + \xi_0, y), \quad (17.5)$$

where $\xi$ is the spatial frequency in the $x$ direction. If the spatial carrier frequency $\xi_0$ is sufficiently high, the term $C(\xi - \xi_0, y)$ can be isolated and processed to obtain $C(\xi, y)$. The inverse Fourier transform of $C(\xi, y)$ then yields $c(x, y)$, from which the phase difference can be calculated, using the relation

$$\Delta\phi(x, y) = \arctan\left[\frac{\mathrm{Im}\{c(x, y)\}}{\mathrm{Re}\{c(x, y)\}}\right]. \quad (17.6)$$

Fourier transform techniques are most useful in studies of phase objects, where the sensitivity vector does not vary over the field and it is possible to produce straight, equally spaced carrier fringes [Nugent, 1985; Bone, Bachor & Sandeman, 1986; Toyooka, Nishida & Takezaki, 1989].

### 17.2.3 Phase unwrapping

The values of the phase difference obtained at any point from measurements on the interference fringes are ambiguous, because they are known only to modulo $2\pi$. The process of determining the number of $2\pi$ steps to be added to these raw values is called phase unwrapping.

Phase unwrapping requires a knowledge of the sign as well as the magnitude of the raw phase [Bone, 1991]. It is then possible, by choosing a starting point at which the phase difference is known to be zero and checking the values of the raw phase at adjacent pixels along a line, to decide whether to add or subtract $2\pi$ when crossing successive fringes. The same procedure can be extended to two dimensions by using pixels along the first line as new starting points.

In another method for phase unwrapping, a simple operation for local unwrapping is repeated many times to generate a consistent phase distribution [Ghiglia, Mastin & Romero, 1987]. Alternatively, phase unwrapping can proceed along paths chosen so that the change in the phase difference between adjacent data points is a minimum [Judge, Quan & Bryanston-Cross, 1992], or following an edge [Stetson, 1992].

## 17.3 Heterodyne holographic interferometry

In this technique [Dändliker, Ineichen & Mottier, 1973], two holograms are recorded of the object using the setup shown in fig. 17.4. The two holograms are recorded on the same plate with two angularly separated reference beams having the same frequency as the object beam.

At the reconstruction stage, a small frequency difference is introduced between the two beams illuminating the hologram, either by means of a rotating grating or by means of two acousto-optic modulators operated at slightly different frequencies. The electric fields corresponding to the two reconstructed waves can then be written as

$$E_1(x, y, t) = |a_1(x, y)| \cos[2\pi \nu_1 t - \phi_1(x, y)], \qquad (17.7)$$

and

$$E_2(x, y, t) = |a_2(x, y)| \cos[2\pi \nu_2 t - \phi_2(x, y)], \qquad (17.8)$$

where $\nu_1$ and $\nu_2$ are the frequencies of the two reference beams, $|a_1(x, y)|$ and $|a_2(x, y)|$ are the real parts of the amplitudes, and $\phi_1(x, y)$ and $\phi_2(x, y)$ are the phases of the two reconstructed waves.

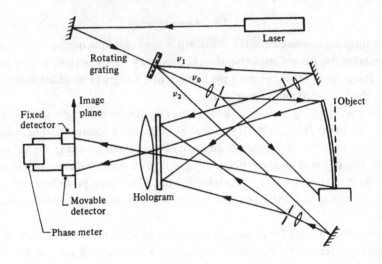

Fig. 17.4. Experimental setup for heterodyne holographic interferometry [Dändliker, 1980].

When these two waves are superposed at a photo detector with a limited frequency response, the output is

$$I(x, y, t) = |a_1(x, y)|^2 + |a_2(x, y)|^2 + 2|a_1(x, y)||a_2(x, y)|$$
$$\times \cos[2\pi(\nu_1 - \nu_2)t - \Delta\phi(x, y)], \qquad (17.9)$$

where $\Delta\phi(x, y) = \phi_1(x, y) - \phi_2(x, y)$. The output from the photo detector is modulated at the beat frequency $\nu_1 - \nu_2$, and the phase of this modulation corresponds to the optical phase difference between the reconstructed waves. This modulation can be separated by a band-pass filter, and its phase can be measured electronically with respect to a reference signal at the same frequency, derived from a second fixed detector, while the first detector is moved in steps across the interference field. Alternatively, an array of three, four, or five photo detectors can be scanned over the real image to obtain the differences in the values of $\Delta\phi$ between pairs of points with known separations in the $x$ and $y$ directions [Pryputniewicz, 1985].

This technique can measure phase differences with an accuracy of $2\pi/500$ and has found a number of applications, including measurements of vector displacements and strains [Dändliker, 1980; Thalmann & Dändliker, 1987] and vibration amplitudes [Stetson, 1982]. However, since it involves point-by-point measurements, it is slow and requires a very stable environment.

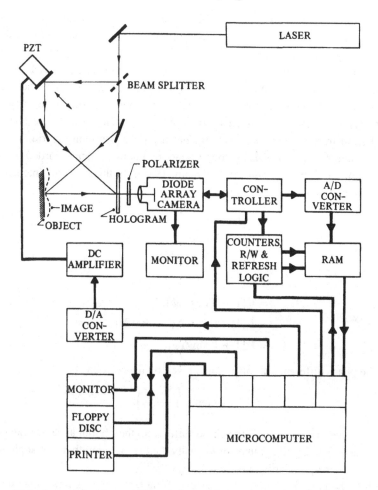

Fig. 17.5. System for phase-shifting holographic interferometry [Hariharan, Oreb & Brown, 1982].

## 17.4 Phase-shifting

Phase-shifting holographic interferometry permits rapid measurements on real-time fringes at a large number of points [Hariharan, Oreb & Brown, 1982; Dändliker, Thalmann & Willemin, 1982].

In this technique, as shown in fig. 17.5, an image of the real-time interference fringes is formed on a CCD array. During a single scan of the array, the values of the intensity at each of the pixels are read out and stored in the memory of a computer. Between successive scans of the array, the phase of the reference beam is shifted relative to that of the object beam by means of a mirror mounted on a piezoelectric translator to which appropriate voltages are applied by a dc amplifier controlled through a digital-to-analog converter by the computer. At the end of a measurement cycle the memory of the computer contains a set of readings of the intensity at each of the pixels in the array, which can then be processed to obtain the phase difference at the corresponding point in the image.

At least three frames of intensity data are required to evaluate the phase distribution in the interference pattern, but the most common algorithm for phase calculations is one using four frames of intensity data recorded with phase shifts of $0$, $\pi/2$, $\pi$ and $3\pi/2$ [Creath, 1988]. In this case, the intensity values at any point can be written as

$$I_1 = I_0[1 + \mathcal{V}\cos\Delta\phi], \tag{17.10}$$

$$I_2 = I_0[1 + \mathcal{V}\cos(\Delta\phi + \pi/2)],$$

$$= I_0[1 - \mathcal{V}\sin\Delta\phi], \tag{17.11}$$

$$I_3 = I_0[1 + \mathcal{V}\cos(\Delta\phi + \pi)],$$

$$= I_0[1 - \mathcal{V}\cos\Delta\phi], \tag{17.12}$$

$$I_4 = I_0[1 + \mathcal{V}\cos(\Delta\phi + 3\pi/2)],$$

$$= I_0[1 + \mathcal{V}\sin\Delta\phi], \tag{17.13}$$

and the phase difference at this point is given by the relation

$$\Delta\phi = \arctan\left(\frac{I_4 - I_2}{I_1 - I_3}\right). \tag{17.14}$$

Errors due to deviations of the phase shifts from their nominal values can be minimized by using five frames of intensity data recorded with nominal phase increments of $\pi/2$ [Hariharan, Oreb & Eiju, 1987].

While the precision offered by phase-shifting techniques is lower than that possible with heterodyne interferometry, phase-shifting has the advantage that measurements are made simultaneously over the entire array of points, so that the results are less sensitive to environmental effects. Typically, the intensity data can be processed in a microcomputer to obtain values of the phase difference with an accuracy of $2\pi/200$, over a $512 \times 512$ array of points, in a few seconds.

### 17.4.1 Measurements of vector displacements

One of the earliest applications of phase-shifting interferometry was for the measurement of vector displacements [Hariharan, Oreb & Brown, 1983]. They used an optical system which permitted four holograms to be recorded in quick

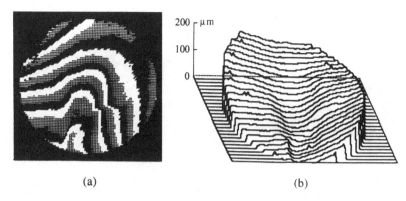

Fig. 17.6. (*a*) Contour map and (*b*) isometric plot of a wear mark on a flat surface obtained by phase-stepping holographic interferometry using the two-refractive-index technique [Hariharan & Oreb, 1984]

succession on a photothermoplastic material with the object illuminated from two different directions in the vertical and horizontal planes. Phase data from these holograms were then processed to obtain the components of the displacement vector at an array of points covering the surface of the test object.

The extension of this method to evaluate the in-plane and normal components of the strain in a pressure vessel has been described by Zarrabi, Oreb, and Hariharan [1990].

### 17.4.2 Contouring

Phase-shifting has been used with two-wavelength holography for measurements of the shape of aspheric optical surfaces [Wyant, Oreb & Hariharan, 1984]. Phase-shifting has also been used with the two-refractive-index technique [Hariharan & Oreb, 1984; Thalmann & Dändliker, 1985], a typical application being to study the shape of a wear pattern on a surface. Figure 17.6(*a*) shows a contour map of the surface. The contour interval was 200 $\mu$m, and readings could be repeated to $\pm 1$ $\mu$m. Figure 17.6(*b*) is an isometric plot from the same data.

Multiple-source contouring using phase-shifting has also been demonstrated by Rastogi and Pflug [1991*a*].

### 17.4.3 Difference interferometry

Phase-shifting and computer image processing have been combined by Rastogi, Barillot, and Kaufmann [1991] to map the difference in the displacements of two

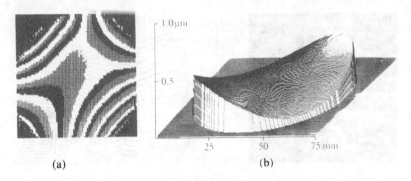

(a)                                              (b)

Fig. 17.7. (a) Contour plot of the surface displacements of a square metal plate vibrating at 231 Hz, obtained by the phase-shifting technique with stroboscopic illumination, and (b) a three-dimensional plot of the displacements over the centre of the plate [Hariharan & Oreb, 1986].

nominally identical specimens when loaded. The differences in the mechanical responses of the test specimen and a defect-free master reveal minor defects.

### 17.4.4 Analysis of vibrations

Stroboscopic holographic interferometry yields fringes with a cosinusoidal intensity distribution, so that accurate measurements of the surface displacements at any instant can be made by phase-shifting techniques. However, the conventional procedure for stroboscopic holographic interferometry involves recording a hologram of the stationary object and observing the real-time interference fringes with stroboscopic illumination. The pulse width then has to be reduced to a small fraction of the period of the vibration to ensure accuracy [Nakadate, Saito & Nakajima, 1986], resulting in a serious drop in the brightness of the image. This problem was solved by Hariharan and Oreb [1986] by using stroboscopic illumination to record a hologram of the vibrating object. The same interference fringes are then obtained with the stationary object and continuous illumination.

Figure 17.7(a) is a contour plot of the surface displacements of a square metal plate vibrating at 231 Hz, obtained by the phase-shifting technique using stroboscopic illumination, and fig. 17.7(b) is a three-dimensional plot of the displacements over the centre of the plate.

Phase-shifting can also be used when the direction of the motion is not known, or the motion is in two or three dimensions [Hariharan, Oreb & Freund, 1987]. Three or more holograms are recorded with the object illuminated from different directions to obtain different sensitivity vectors, so that the vector components of the vibration can be evaluated.

Phase-shifting has also been applied to the study of real-time time-averaged fringes by Nakadate [1986]. The intensity in the fringe pattern in this case can be written as

$$I = \alpha + \beta \cos(\zeta + \psi + \phi)J_0(\Omega), \tag{17.15}$$

where $\alpha$ is the bias intensity, $\beta$ is the maximum intensity modulation in the pattern, and $\zeta$, $\psi$, and $\phi$ are, respectively, the bias deformation, the phase change due to shrinkage of the hologram, and the phase shift introduced in the reference beam. Measurements of the intensity $I_i$ are made as the phase shift takes the values $\phi_i = 2\pi i/N$ ($i = 1, 2, 3 \ldots N$) and used to calculate the parameters

$$a_0 = (1/N) \sum_{i=1}^{N} I_i = \alpha, \tag{17.16}$$

$$a_1 = (2/N) \sum_{i=1}^{N} I_i \cos \phi_i = \beta \cos(\zeta + \psi)J_0(\Omega), \tag{17.17}$$

$$b_1 = (2/N) \sum_{i=1}^{N} I_i \sin \phi_i = \beta \sin(\zeta + \psi)J_0(\Omega), \tag{17.18}$$

where $\Omega = \mathbf{K} \cdot \mathbf{L}$ (see section 16.1.1). These parameters are then used in the relation

$$(a_1^2 + b_1^2)^{1/2}/a_0 = (\beta/\alpha)|J_0(\Omega)|. \tag{17.19}$$

Since the factor $(\beta/\alpha)$ corresponds to the intensity at the point where the vibration amplitude is zero, the right-hand side of this expression can be normalized to obtain $|J_0(\Omega)|$. As can be seen from fig. 17.8, high-contrast fringes are obtained. In addition, the discontinuities at the zeros of the function $|J_0(\Omega)|$ permit accurate measurements of the positions of the dark fringes.

### 17.4.5 Moiré (four-wave) interferometry

Phase-shifting can be applied to holographic moiré (four-wave) interferometry to evaluate specific components of the deformation of a surface, such as in-plane displacements and out-of-plane displacements, as well as the differences in the displacements of two surfaces.

A typical experimental arrangement uses two beams to illuminate the object simultaneously at equal and opposite angles of incidence. The intensity in the

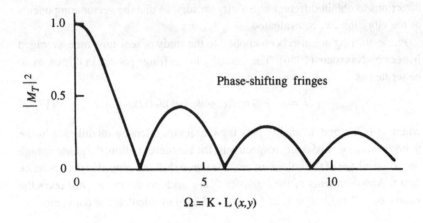

Fig. 17.8. Characteristic function for a vibrating object using phase-shifting.

real-time interferogram produced with such an arrangement can be written as

$$I = I_0[1 + (\mathcal{V}/2)(\cos \Delta\phi_1 - \cos \Delta\phi_2)], \qquad (17.20)$$

where $\Delta\phi_1$ and $\Delta\phi_2$ are the phase differences corresponding to the two directions of illumination produced by the object displacement. Three successive pairs of phase shifts with values of $(-2\pi/3, -4\pi/3)$, $(0, 0)$, and $(2\pi/3, 4\pi/3)$, respectively, introduced in the illumination beams then yield the intensity values [Rastogi, 1992a],

$$I_1 = 2I_0 \left\{ 1 + (\mathcal{V}/2) \left[ \cos \left( \Delta\phi_1 - \frac{2\pi}{3} \right) + \cos \left( \Delta\phi_2 - \frac{4\pi}{3} \right) \right] \right\},$$
$$(17.21)$$

$$I_2 = 2I_0[1 + (\mathcal{V}/2)(\cos \Delta\phi_1 + \cos \Delta\phi_2)], \qquad (17.22)$$

$$I_3 = 2I_0 \left\{ 1 + (\mathcal{V}/2) \left[ \cos \left( \Delta\phi_1 + \frac{2\pi}{3} \right) + \cos \left( \Delta\phi_2 + \frac{4\pi}{3} \right) \right] \right\},$$
$$(17.23)$$

from which it follows that

$$\Delta\phi_1 - \Delta\phi_2 = 2 \arctan \left[ \frac{3^{1/2}(I_1 - I_3)}{2I_2 - I_1 - I_3} \right]. \qquad (17.24)$$

Phase-shifting procedures have also been described which can be used for simultaneous determinations of the sum and difference of the two phases [Rastogi, 1992b, 1993a,b; Simova & Stoev, 1992a,b].

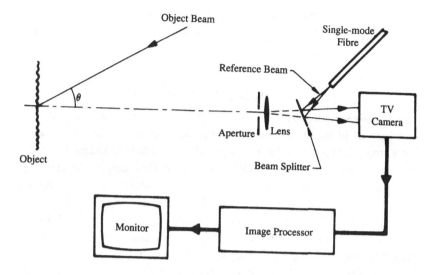

Fig. 17.9. Typical system for electronic speckle-pattern interferometry.

## 17.5 Electronic holographic interferometry

Electronic holographic interferometry can be regarded as having evolved from electronic speckle-pattern interferometry (ESPI). Early techniques were referred to as phase-shifting speckle-pattern interferometry, but recent, more advanced methods have been called electro-optic holography.

### 17.5.1 Electronic speckle-pattern interferometry

A typical system used for ESPI is shown in fig. 17.9. The object is imaged on the target of a television camera along with a coaxial reference beam. The resulting image hologram has a coarse speckle structure which can be resolved by the television camera. Any change in the shape of the object results in a change in the intensity distribution in the speckle pattern in the corresponding part of the image.

To measure displacements of the object, an image of the object in its original state is stored and subtracted from the signal from the television camera. Regions in which the speckle pattern has not changed, corresponding to the condition

$$\mathbf{K} \cdot \mathbf{L}(x, y) = 2m\pi, \qquad (17.25)$$

where $m$ is an integer, appear dark, while regions where the pattern has changed are covered with bright speckles [Butters & Leendertz, 1971; Macovski, Ramsey & Schaefer, 1971].

### 17.5.2 Phase-shifting speckle interferometry

Phase-shifting speckle interferometry uses phase-shifting techniques for direct measurements of the phase difference at an array of data points before and after a stress is applied to the test object [Creath, 1985; Nakadate & Saito, 1985; Robinson & Williams, 1986]. Each speckle, as seen by the camera, can be regarded as an individual interference pattern, and the phase difference between the beams at this point is measured by phase-stepping before and after the object is stressed. The difference of these two values is then calculated. Even though any two speckles may have different initial intensities, corresponding to different values of the amplitude and phase of the object wavefront, the change in phase will be the same for the same surface displacement. Accordingly, the result of subtracting the second set of phase values from the first is a contour map of the object deformation.

The phase change produced by the object displacement can also be calculated directly from the intensity data before and after the displacement using the relation

$$\Delta\phi = \arctan\left[\frac{\sin(\phi - \phi')}{\cos(\phi - \phi')}\right],$$

$$= \arctan\left[\frac{\sin\phi\cos\phi' - \cos\phi\sin\phi'}{\cos\phi\cos\phi' + \sin\phi\sin\phi'}\right],$$

$$= \arctan\left[\frac{(I_4 - I_2)(I_1' - I_3') - (I_1 - I_3)(I_4' - I_2')}{(I_1 - I_3)(I_1' - I_3') - (I_4 - I_2)(I_4' - I_2')}\right].$$

$$(17.26)$$

A simpler procedure which requires only three frames of data (one before, and two after the displacement) has been described by Kerr, Mendoza-Santoyo, and Tyrer [1990].

A drawback of these techniques is that the phase data obtained are noisy, due to the presence of terms arising from the speckle in the object beam. The noise can be reduced by averaging several sets of phase data obtained with slightly different directions of illumination, so that the speckle patterns in the images are uncorrelated. Spatial low-pass filtering can also be used to reduce the noise and improve the phase accuracy at some sacrifice of spatial resolution [Vikhagen, 1991].

### 17.5.3 Electro-optic holography

Stetson and Brohinsky [1986] showed that the speckle terms could be removed by processing the intensity data in a different way. They used the same eight frames of data acquired with phase increments of $\pi/2$ to evaluate the phase change $\Delta\phi$, due to the displacement of the object, from the relation

$$\Delta\phi = \arctan\left(\frac{C_1 - C_2}{C_3 - C_4}\right), \tag{17.27}$$

where the parameters $C_1 \ldots C_4$ are defined by the expressions

$$C_1 = [(I_1 - I_3) + (I_1' - I_3')]^2 + [(I_2 - I_4) + (I_2' - I_4')]^2, \tag{17.28}$$

$$C_2 = [(I_1 - I_3) - (I_1' - I_3')]^2 + [(I_2 - I_4) - (I_2' - I_4')]^2, \tag{17.29}$$

$$C_3 = [(I_1 - I_3) + (I_2' - I_4')]^2 + [(I_2 - I_4) + (I_1' - I_3')]^2, \tag{17.30}$$

$$C_4 = [(I_1 - I_3) - (I_2' - I_4')]^2 + [(I_2 - I_4) + (I_1' - I_3')]^2. \tag{17.31}$$

The phase change $\Delta\phi$ can be calculated from these expressions at video rates, over a $512 \times 512$ array of points, using an array processor and look-up tables.

### 17.5.4 Analysis of vibrations

Values of the vibration amplitude can also be obtained from time-average holograms by phase-shifting [Stetson & Brohinsky, 1988].

Three separate sets of four intensity measurements are used. The first set of measurements is made with the object vibrating and additional phase shifts of $0, \pi/2, \pi$ and $3\pi/2$ introduced in the reference beam. The second and third sets of measurements are made with a sinusoidal phase modulation introduced in the reference beam at the vibration frequency. The phase of this modulation is offset from the phase of the vibration by $+\pi/3$ in one case and $-\pi/3$ in the other case.

The intensity at a given point in any one of the twelve frames of data recorded can be written as

$$I_{ji} = I_0[1 + \mathcal{V}\cos(\phi + \delta_i)J_0(\Omega + \beta_j)], \tag{17.32}$$

where $\delta_i$ takes the values $0, \pi/2, \pi$ and $3\pi/2$, $\Omega = \mathbf{K} \cdot \mathbf{L}$, where $\mathbf{K}$ is the sensitivity vector and $\mathbf{L}$ is the amplitude of the vibration, and $\beta_j$ takes the values $-\pi/3, 0$ and $+\pi/3$.

The amplitude of the vibration can then be calculated from the relation

$$\Omega = \arctan\left[\frac{3^{-1/2}(H_1 - H_3)}{2H_2 - H_1 - H_3}\right], \tag{17.33}$$

where

$$H_1 = (I_{11} - I_{13})^2 + (I_{12} - I_{14})^2,$$
$$= 4I_0^2 \mathcal{V}^2 J_0^2 (\Omega - \pi/3), \tag{17.34}$$
$$H_2 = (I_{21} - I_{23})^2 + (I_{22} - I_{24})^2,$$
$$= 4I_0^2 \mathcal{V}^2 J_0^2 (\Omega), \tag{17.35}$$
$$H_3 = (I_{31} - I_{33})^2 + (I_{32} - I_{34})^2,$$
$$= 4I_0^2 \mathcal{V}^2 J_0^2 (\Omega + \pi/3). \tag{17.36}$$

The values of the vibration amplitude obtained by this procedure have to be corrected for the fact that (17.33) assumes that the time-average fringes have a $\cos^2$ intensity distribution, whereas they actually have a $J_0^2$ intensity distribution. The error due to this cause is dependent on the fringe order, which can be determined from the value of $H_1$. The difference between the values of $J_0^2(\Omega)$ and $\cos^2(\Omega)$ can then be read off from a look-up table and used to apply the necessary corrections [Pryputniewicz & Stetson, 1989].

## 17.6 Current trends

Advances in recording materials, such as photorefractive crystals, as well as new techniques, such as electronic holography, now make observations in real time possible. Phase-shifting methods are being used widely to measure three-dimensional displacements [Shellabear & Tyrer, 1991; Watt, Gross & Henning, 1991], as well as to study less stable objects by shearing interferometry [Owner-Petersen, 1991]. A promising area that is being explored is the use of displacement data generated by phase-shifting methods in finite-element analysis programs to allow automatic quantitative analysis of stressed structures [Brown & Zhang, 1991; Hariharan, 1991].

# Appendix 1
## Interference and coherence

### A1.1 Interference

The time-varying electric field $E$ at any point due to a linearly polarized monochromatic light wave propagating in a vacuum in the $z$ direction can be represented by the relation

$$E = a \cos[2\pi \nu(t - z/c)], \qquad (A1.1)$$

where $a$ is the amplitude, $\nu$ the frequency, and $c$ the speed of propagation of light. If $T$ is the period of the vibration,

$$T = 1/\nu = 2\pi/\omega, \qquad (A1.2)$$

where $\omega$ is the circular frequency. The wavelength $\lambda$ is given by the relation

$$\lambda = cT = c/\nu, \qquad (A1.3)$$

and the propagation constant of the wave is

$$k = 2\pi/\lambda. \qquad (A1.4)$$

In a medium of refractive index $n$, the wave propagates with a speed

$$v = c/n \qquad (A1.5)$$

and, since its frequency remains unchanged, its wavelength is

$$\lambda_n = \lambda/n. \qquad (A1.6)$$

Equation (A1.1) can also be written in the form

$$E = \mathrm{Re}\{a \exp[\mathrm{i}2\pi \nu(t - z/c)]\},$$
$$= \mathrm{Re}\{a \exp(-\mathrm{i}\phi) \exp(\mathrm{i}2\pi \nu t)\}, \qquad (A1.7)$$

where $\mathrm{Re}\{...\}$ represents the real part of the expression within the braces, $\mathrm{i} = (-1)^{1/2}$, and $\phi = 2\pi \nu z/c = 2\pi z/\lambda$.

If we assume that all operations on $E$ are linear, we can use the complex representation

$$E = a \exp(-i\phi) \exp(i2\pi vt),$$
$$= A \exp(i2\pi vt), \qquad (A1.8)$$

where $A = a \exp(-i\phi)$ is known as the complex amplitude.

The optical intensity $I$ at a point is given by the time average of the amount of energy which crosses, in unit time, a unit area normal to the energy flow, and is obtained by multiplying the complex amplitude at this point by its complex conjugate, so that

$$I = AA^* = |A|^2. \qquad (A1.9)$$

The complex amplitude at any point due to a number of waves of the same frequency is obtained by summing the complex amplitudes of the individual waves, so that

$$A = A_1 + A_2 + A_3 + \dots. \qquad (A1.10)$$

Accordingly, the intensity at any point due to the interference of two waves is

$$I = |A_1 + A_2|^2,$$
$$= |A_1|^2 + |A_2|^2 + A_1 A_2^* + A_1^* A_2,$$
$$= I_1 + I_2 + 2(I_1 I_2)^{1/2} \cos(\phi_1 - \phi_2). \qquad (A1.11)$$

The visibility of the interference fringes is defined by the relation

$$\mathcal{V} = \frac{I_{max} - I_{min}}{I_{max} + I_{min}},$$
$$= \frac{2(I_1 I_2)^{1/2}}{I_1 + I_2}. \qquad (A1.12)$$

Equations (A1.11) and (A1.12) assume that the electric vectors of the two waves are parallel. If the electric vectors make an angle $\psi$ with each other, the visibility of the interference fringes is

$$\mathcal{V}_\psi = \mathcal{V} \cos \psi, \qquad (A1.13)$$

which drops to zero when $\psi = \pi/2$.


## A1.2  Coherence

We have assumed in the previous section that the light waves are derived from a single point source emitting an infinitely long, monochromatic wave train,

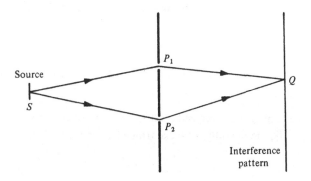

Fig. A1.1. Evaluation of the degree of coherence.

so that the fields due to two light waves are perfectly coherent. However, in reality, the wave fields are only partially coherent.

We can represent the time-varying electric field at any point due to a linearly polarized, quasi-monochromatic, light wave from a source of finite size by the analytic signal [Born & Wolf, 1980]

$$V(t) = \int_0^\infty a(v) \exp[-i\phi(v)] \exp(i2\pi v t) dv, \qquad (A1.14)$$

where $a(v)$ is the amplitude and $\phi(v)$ is the phase of a component with frequency $v$. To evaluate the degree of coherence of the fields at two points illuminated by such a source, we consider the arrangement shown in fig. A1.1. If $V_1(t)$ and $V_2(t)$ are the analytic signals corresponding to the electric fields at $P_1$ and $P_2$, the complex degree of coherence $\gamma_{12}(\tau)$ of the fields for a time delay $\tau$ is defined as the normalized correlation of $V_1(t)$ and $V_2(t)$, which can be written (see Appendix 2) as

$$\gamma_{12}(\tau) = \frac{\langle V_1(t+\tau)V_2^*(t)\rangle}{[\langle V_1(t)V_1^*(t)\rangle\langle V_2(t)V_2^*(t)\rangle]^{1/2}}. \qquad (A1.15)$$

The physical significance of (A1.15) can be understood if the light waves are allowed to emerge through pinholes at $P_1$ and $P_2$, so that they form an interference pattern on a screen. $P_1$ and $P_2$ can then be considered as two secondary sources, so that, from (A1.11), the intensity at $Q$ is

$$\begin{aligned} I &= I_1 + I_2 + \langle V_1(t+\tau)V_2^*(t) + V_1^*(t+\tau)V_2(t)\rangle, \\ &= I_1 + I_2 + 2\text{Re}[\langle V_1(t+\tau)V_2^*(t)\rangle], \end{aligned} \qquad (A1.16)$$

where $I_1$ and $I_2$ are the intensities at $Q$ when $P_1$ and $P_2$ act separately, and $\tau$ is the difference in the transit times for the paths $P_1Q$ and $P_2Q$. From (A1.15), (A1.16) can be rewritten as

$$
\begin{aligned}
I &= I_1 + I_2 + 2(I_1 I_2)^{1/2} \mathrm{Re}[\gamma_{12}(\tau)], \\
&= I_1 + I_2 + 2(I_1 I_2)^{1/2} |\gamma_{12}(\tau)| \cos\phi_{12}(\tau),
\end{aligned}
\qquad \text{(A1.17)}
$$

where $\phi_{12}(\tau)$ is the phase of $\gamma_{12}(\tau)$.

Interference fringes are produced by the variations in $\cos\phi_{12}(\tau)$ across the screen. If $I_1 = I_2$, the visibility of these fringes is, from (A1.12),

$$
\mathcal{V} = |\gamma_{12}(\tau)|. \qquad \text{(A1.18)}
$$

The coherence of the field produced by any light source can then be studied from two aspects.

### A1.3 Spatial coherence

When the difference in the optical paths from the source to $P_1$ and $P_2$ is small, so that $\tau \approx 0$, effects due to the spectral bandwidth of the light can be neglected, and we are concerned essentially with the spatial coherence of the field. A special case of interest is when the dimensions of the source and the separation of $P_1$ and $P_2$ are extremely small compared to the distance from $P_1$ and $P_2$ to the source. If we assume, for convenience, that $P_1$ is located on the $z$ axis in the plane of observation at $(0, 0, z)$, while $P_2$ is at $(x, y, z)$, and define new coordinates in the source plane such that $\xi = x_s/z$ and $\eta = y_s/z$, the complex degree of coherence of the field can be written as

$$
\mu_{12} = \frac{\exp(i\phi_{12}) \iint_s I(\xi, \eta) \exp[ik(x\xi + y\eta)]\mathrm{d}\xi\,\mathrm{d}\eta}{\iint_s I(\xi, \eta)\mathrm{d}\xi\,\mathrm{d}\eta}, \qquad \text{(A1.19)}
$$

where $\phi_{12} = -k(x^2 + y^2)/2z$. The complex degree of coherence is given, in this case, by the normalized Fourier transform (see Appendix 2) of the intensity distribution over the source.

### A1.4 Temporal coherence

If the source is very small (effectively a point source) but radiates over a range of wavelengths, we are concerned with the temporal coherence of the field. In this case, the complex degree of coherence depends only on $\tau$, the difference in the transit times from the source to $P_1$ and $P_2$, and is given by the expression

$$
\gamma(\tau) = \frac{\langle V(t)V^*(t + \tau)\rangle}{\langle V(t)V^*(t)\rangle}, \qquad \text{(A1.20)}
$$

which can be transformed and rewritten as

$$\gamma(\tau) = \mathcal{F}\{S(\nu)\} / \int_{-\infty}^{\infty} S(\nu) d\nu, \qquad (A1.21)$$

where $S(\nu)$ is the frequency spectrum of the radiation.

From (A1.17) and (A1.18), it follows that the degree of temporal coherence can be obtained from the visibility of the interference fringes as the difference in the optical paths from the source is varied.

This argument leads us to the concepts of the coherence time and the coherence length of the radiation. It can be shown that, with radiation having a mean frequency $\nu_0$ and a bandwidth $\Delta\nu$, the visibility of the interference fringes drops to zero for a difference $\Delta\tau$ in the transit times given by the relation

$$\Delta\tau\,\Delta\nu \approx 1. \qquad (A1.22)$$

The time $\Delta\tau$ is called the coherence time of the radiation; the corresponding value of the coherence length is

$$\Delta l \approx c\Delta\tau \approx c/\Delta\nu \approx \lambda_0^2/\Delta\lambda, \qquad (A1.23)$$

where $\lambda_0$ is the mean wavelength and $\Delta\lambda$ the bandwidth of the radiation. For interference fringes with good visibility to be obtained, the optical path difference must be small compared to the coherence length of the radiation.

# Appendix 2

# The Fourier transform, convolution, and correlation

### A2.1 The Fourier transform

The one-dimensional Fourier transform is used widely to study time-varying functions in the frequency domain [Papoulis, 1962; Bracewell, 1978]. Similarly, a function of two orthogonal spatial coordinates can be expressed, by means of a two-dimensional Fourier transform, as a function of two orthogonal spatial frequencies [Goodman, 1968].

The two-dimensional Fourier transform of $g(x, y)$ is defined as

$$\mathcal{F}\{g(x, y)\} = \int_{-\infty}^{\infty} \int_{-\infty}^{\infty} g(x, y) \exp[-i2\pi(\xi x + \eta y)] \mathrm{d}x \mathrm{d}y,$$

$$= G(\xi, \eta). \tag{A2.1}$$

Similarly, the inverse Fourier transform of $G(\xi, \eta)$ is defined as

$$\mathcal{F}^{-1}\{G(\xi, \eta)\} = \int_{-\infty}^{\infty} \int_{-\infty}^{\infty} G(\xi, \eta) \exp[-i2\pi(\xi x + \eta y)] \mathrm{d}\xi \mathrm{d}\eta,$$

$$= g(x, y). \tag{A2.2}$$

These relationships can be written symbolically as

$$g(x, y) \leftrightarrow G(\xi, \eta). \tag{A2.3}$$

The Fourier transform allows us to study the structure of an image in the spatial frequency domain and provides a useful insight into the working of an optical system. This is because the transform effectively decomposes a complex wave front into component plane waves whose propagation can then be analysed as discussed in Appendix 3.

Some of the more important properties of the Fourier transform are summarized below.

317

### A2.1.1  The linearity theorem

$$\mathcal{F}\{ag(x, y) + bh(x, y)\} = aG(\xi, \eta) + bH(\xi, \eta), \qquad \text{(A2.4)}$$

where $a$ and $b$ are constants and $g(x, y) \leftrightarrow G(\xi, \eta), h(x, y) \leftrightarrow H(\xi, \eta)$.

### A2.1.2  The shift theorem

$$\mathcal{F}\{g(x - a, y - b)\} = G(\xi, \eta) \exp[-i2\pi(\xi a + \eta b)]. \qquad \text{(A2.5)}$$

### A2.1.3  The similarity theorem

$$\mathcal{F}\{g(ax, by)\} = (1/|ab|)G(\xi/a, \eta/b). \qquad \text{(A2.6)}$$

### A2.1.4  Rayleigh's theorem

$$\int_{-\infty}^{\infty} \int_{-\infty}^{\infty} |g(x, y)|^2 dx dy = \int_{-\infty}^{\infty} \int_{-\infty}^{\infty} |G(\xi, \eta)|^2 d\xi d\eta. \qquad \text{(A2.7)}$$

This theorem is also known as Plancherel's theorem.

## A2.2  Convolution and correlation

The convolution of two functions $g(x)$ and $h(x)$ is defined as

$$f(x) = \int_{-\infty}^{\infty} g(u)h(x - u)du. \qquad \text{(A2.8)}$$

This can be written as

$$f(x) = g(x) * h(x), \qquad \text{(A2.9)}$$

where the symbol $*$ denotes the convolution operation. The physical interpretation of the convolution operation is simple. The laterally inverted and shifted function $h(x - u)$ is multiplied by $g(u)$; the area under the curve $g(u)h(x - u)$ then gives the value of $f(x)$.

The convolution operation in two dimensions is defined as

$$f(x, y) = \int_{-\infty}^{\infty} \int_{-\infty}^{\infty} g(u, v)h(x - u, y - v)du dv, \qquad \text{(A2.10)}$$

which can also be written as

$$f(x, y) = g(x, y) * h(x, y). \qquad \text{(A2.11)}$$

A function frequently used in conjunction with the convolution operation is the Dirac delta function $\delta(x, y)$. By definition, convolution of a function with the delta function yields the original function, so that

$$\int_{-\infty}^{\infty} \int_{-\infty}^{\infty} f(u, v)\delta(x - u, y - v)dudv = f(x, y). \qquad (A2.12)$$

The delta function can be shown to take the values

$$\delta(x, y) = \infty \quad (x = 0, \text{ and } y = 0)$$
$$\delta(x, y) = 0 \quad (x \neq 0, \text{ or } y \neq 0)$$

and its integral is unity.

The cross-correlation of two functions, $g(x, y)$ and $h(x, y)$, is

$$c(x, y) = \int_{-\infty}^{\infty} \int_{-\infty}^{\infty} g^*(u, v)h(x + u, y + v)dudv, \qquad (A2.13)$$

where $g^*(u, v)$ is the complex conjugate of $g(u, v)$. This equation can be written as

$$c(x, y) = g(x, y) \star h(x, y), \qquad (A2.14)$$

where the symbol $\star$ denotes the correlation operation. A comparison with (A2.10) shows that the cross-correlation can also be expressed as a convolution

$$c(x, y) = g^*(x, y) * h(-x, -y). \qquad (A2.15)$$

It follows that the autocorrelation of a function $g(x, y)$ is

$$a(x, y) = \int_{-\infty}^{\infty} \int_{-\infty}^{\infty} g^*(u; v)g(x + u, y + v)dudv$$
$$= g(x, y) \star g(x, y). \qquad (A2.16)$$

Some useful results that follow are listed below.

### A2.2.1 The convolution theorem

If $g(x, y) \leftrightarrow G(\xi, \eta)$ and $h(x, y) \leftrightarrow H(\xi, \eta)$,

$$\mathcal{F}\{g(x, y) * h(x, y)\} = G(\xi, \eta)H(\xi, \eta). \qquad (A2.17)$$

### A2.2.2 The autocorrelation (Wiener–Khinchin) theorem

$$\mathcal{F}\{g(x, y) \star g(x, y)\} = |G(\xi, \eta)|^2. \qquad (A2.18)$$

## A2.3 Random functions

The correlation function is very useful in the study of randomly varying quantities [Papoulis, 1965]. Since, in this case, an integral such as (A2.13) would be infinite, the cross-correlation of two stationary random functions, typically functions of time, $g(t)$ and $h(t)$, is written as

$$R_{gh}(\tau) = \lim_{\tau \to \infty} \frac{1}{2T} \int_{-T}^{T} g^*(t)h(t + \tau)dt, = \langle g^*(t)h(t + \tau) \rangle. \quad (A2.19)$$

The autocorrelation function of one of these functions, say $g(t)$, is

$$R_{gg}(\tau) = \langle g^*(t)g(t + \tau) \rangle. \quad (A2.20)$$

In physical terms, if $g(t)$ is, for example, the time-varying electric field at a point due to an electromagnetic wave, $R_{gg}(0)$ is the average power at this point.

The power spectrum $S(\omega)$ of a random function $g(t)$ is the Fourier transform of its autocorrelation

$$R_{gg}(\tau) \leftrightarrow S(\omega). \quad (A2.21)$$

It follows that if $R_{gg}(\tau)$ is sharply peaked (in the limit, a delta function), $S(\omega)$ must extend to very high frequencies (and, in the limit, is a constant).

## A2.4 Sampled functions and the discrete Fourier transform

The production of computer-generated holograms involves computation of the Fourier transform $G_s(\xi, \eta)$ of a sampled function $g_s(x, y)$, which is obtained by sampling the object wave $g(x, y)$ (in this case, the wave to be reconstructed) at intervals $(\Delta x, \Delta y)$. If, for simplicity, we consider the one-dimensional case, as shown in fig. A2.1, we can write

$$g_s(x) = g(x) \sum_{m=-\infty}^{\infty} \delta(x - m\Delta x), \quad (A2.22)$$

and

$$G_s(\xi) = (1/\Delta x) \sum_{m=-\infty}^{\infty} G[\xi - (m/\Delta x)], \quad (A2.23)$$

where $G(\xi) \leftrightarrow g(x)$.

As shown in fig. A2.2, the Fourier transform of the sampled function $g_s(x)$ is a regular series of repetitions of the Fourier transform of the original function $g(x)$, shifted in frequency by successive intervals $\Delta\xi = (1/\Delta x)$. Overlap of the shifted Fourier transforms would normally result in aliasing and make it impossible to recover $g(x)$ exactly. However, if the interval at which $g(x)$ is

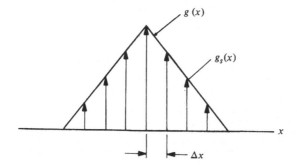

Fig. A2.1. The function $g_s(x)$ consists of an array of delta functions obtained by sampling the function $g(x)$ at intervals of $\Delta x$.

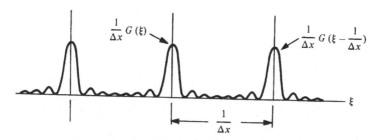

Fig. A2.2. The Fourier transform $G_s(\xi)$ of the sampled function $g_s(x)$ consists of a regular series of repetitions of $G(\xi)$, the Fourier transform of $g(x)$, shifted in frequency by successive intervals $\Delta\xi = 1/\Delta x$.

sampled is chosen so that

$$\Delta x \leq 1/\xi_{\max}, \qquad (A2.24)$$

it is possible to recover $G(\xi)$ from $G_s(\xi)$.

We can use the same argument to show how it is possible to recover the object wave from the image reconstructed by a computed hologram, provided the sampling interval for the hologram is properly chosen. To recover $g(x)$, the values of its sampled Fourier transform $G_s(\xi)$ are required at a series of points in the hologram plane taken at intervals of $\xi$ given by the relation

$$\Delta\xi \leq 1/x_{\max}, \qquad (A2.25)$$

which is analogous to (A2.24). The values of $G_s(\xi)$ at these points can be obtained from the discrete Fourier transform

$$G_s(n/x_{\max}) = (1/\xi_{\max}) \sum_{m=-M/2}^{M/2} g(m/\xi_{\max}) \exp(i2\pi mn/\xi_{\max}x_{\max}). \quad (A2.26)$$

# Appendix 3
# Wave propagation and diffraction

## A3.1 Computation of phase across a spherical wavefront

Consider a point $S(x_1, y_1, 0)$ in the coordinate system shown in fig. A3.1, emitting light waves of wavelength $\lambda$. Their phase at a point $P(x, y, z)$, relative to that at the point $P_0(0, 0, z)$, can be calculated from the difference in the optical paths, and is given by the relation

$$
\begin{aligned}
\phi &= (2\pi/\lambda)(SP - SP_0), \\
&= (2\pi/\lambda)\{[(x - x_1)^2 + (y - y_1)^2 + z^2]^{1/2} - [x_1^2 + y_1^2 + z^2]^{1/2}\} \\
&= (2\pi/\lambda)z\left\{\left[1 + \frac{(x - x_1)^2 + (y - y_1)^2}{z^2}\right]^{1/2} - \left[1 + \frac{x_1^2 + y_1^2}{z^2}\right]^{1/2}\right\}.
\end{aligned}
$$

$$(A3.1)$$

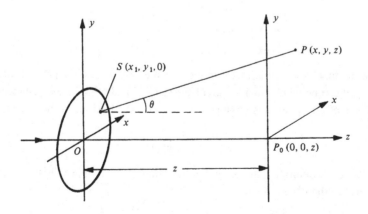

Fig. A3.1. Coordinate system used to evaluate the Fresnel–Kirchhoff integral.

323

If $z$ is large compared to $x_1$, $y_1$, $x$, and $y$, (A3.1) can be written as

$$\phi \approx (2\pi/\lambda)(1/2z)(x^2 + y^2 - 2xx_1 - 2yy_1). \qquad \text{(A3.2)}$$

## A3.2 The Fresnel–Kirchhoff integral

The Fresnel–Kirchhoff integral [O'Neill, 1963; Born & Wolf, 1980] states that if, as shown in fig. A3.1, a plane wave with an amplitude $a$ is incident normally on an object having an amplitude transmittance $\mathbf{t}(x_1, y_1)$ and located in the plane $z = 0$, the complex amplitude at a point $P(x, y, z)$ is

$$a(x, y, z) = (ia/\lambda) \int_{-\infty}^{\infty} \int_{-\infty}^{\infty} \mathbf{t}(x_1, y_1)$$

$$\times \frac{\exp\{(-i2\pi/\lambda)[(x - x_1)^2 + (y - y_1)^2 + z^2]^{1/2}\}}{[(x - x_1)^2 + (y - y_1)^2 + z^2]^{1/2}}$$

$$\times \cos\theta \, dx_1 dy_1. \qquad \text{(A3.3)}$$

If the point $P$ is at a distance $z$ much larger than $(x - x_1)$ and $(y - y_1)$, $\cos\theta \approx 1$, and (A3.3) can be written as

$$a(x, y, z) = (ia/\lambda z) \int_{-\infty}^{\infty} \int_{-\infty}^{\infty} \mathbf{t}(x_1, y_1)$$

$$\times \exp\{(-i\pi/\lambda z)[(x - x_1)^2 + (y - y_1)^2]\} dx_1 dy_1, \quad \text{(A3.4)}$$

if we omit a factor $\exp(-i2\pi/\lambda z)$, which only affects the overall phase. We can then expand (A3.4) to give the result

$$a(x, y, z) = (ia/\lambda z) \int_{-\infty}^{\infty} \int_{-\infty}^{\infty} \mathbf{t}(x_1, y_1) \exp[(-i\pi/\lambda z)(x^2 + y^2)]$$

$$\times \exp[(-i\pi/\lambda z)(x_1^2 + y_1^2)]$$

$$\times \exp\{i2\pi[x_1(x/\lambda z) + y_1(y/\lambda z)]\} dx_1 dy_1. \qquad \text{(A3.5)}$$

Since the first exponential factor in the above expression is independent of $x_1$ and $y_1$, it can be taken outside the integral sign. In addition, if the distance to the plane of observation is large compared to the dimensions of the object, so that

$$z \gg (x_1^2 + y_1^2)/\lambda \qquad \text{(A3.6)}$$

(the far-field condition), the second exponential factor is approximately equal to unity. If, then, we set

$$\xi = x/\lambda z,$$

$$\eta = y/\lambda z, \qquad \text{(A3.7)}$$

where $\xi$ and $\eta$ denote spatial frequencies, (A3.5) becomes

$$a(x, y, z) = (ia/\lambda z) \exp[(-i\pi/\lambda z)(x^2 + y^2)]$$

$$\times \int_{-\infty}^{\infty} \int_{-\infty}^{\infty} \mathbf{t}(x_1, y_1) \exp[i2\pi(\xi x_1 + \eta y_1)] dx_1 dy_1,$$

$$= (ia/\lambda z) \exp[(-i\pi/\lambda z)(x^2 + y^2)] \mathbf{T}(\xi, \eta), \qquad (A3.8)$$

where

$$\mathbf{t}(x_1, y_1) \leftrightarrow \mathbf{T}(\xi, \eta). \qquad (A3.9)$$

It follows that the complex amplitude in the plane of observation is given by the Fourier transform of the amplitude transmittance of the object, multiplied by a spherical phase factor.

### A3.3 Production of a spherical wavefront by a lens

A thin convex lens brings a collimated beam to a focus at a distance $f$ from the lens equal to its focal length. If we neglect absorption, the effect of the lens is merely to introduce a phase delay $\Delta\phi(x, y)$ which varies over the pupil and converts the incident plane wavefront into a spherical wavefront with its centre at the principal focus. From (A3.2) this phase delay can be written as

$$\Delta\phi(x, y) = (\pi/\lambda f)(x^2 + y^2), \qquad (A3.10)$$

so that a thin lens can be considered equivalent to a transparency with a complex amplitude transmittance

$$g(x, y) = \exp[(i\pi/\lambda f)(x^2 + y^2)]. \qquad (A3.11)$$

If, now, a transparency with an amplitude transmittance $\mathbf{t}(x, y)$ is placed in front of the lens, the net transmitted amplitude is $\mathbf{t}(x, y) g(x, y)$. The resultant complex amplitude at a point $(x_f, y_f)$ in the back focal plane of the lens can then be calculated from (A3.5), and is given by the relation

$$a_f(x_f, y_f) = (i/\lambda f) \exp[(-i\pi/\lambda f)(x_f^2 + y_f^2)] \int_{-\infty}^{\infty} \int_{-\infty}^{\infty} \mathbf{t}(x, y)$$

$$\times \exp\{i2\pi[x(x_f/\lambda f) + y(y_f/\lambda f)]\} dx dy. \qquad (A3.12)$$

### A3.4 Fourier transformation by a lens

Consider the optical system shown in fig. A3.2 in which a plane wave with an amplitude $a$ is incident normally on an object with an amplitude transmittance

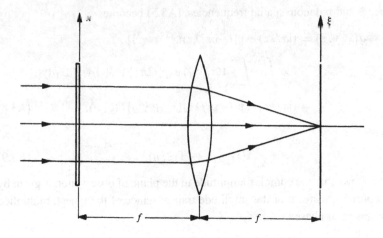

Fig. A3.2. Fourier transformation by a lens

$t(x_1, y_1)$ located in the front focal plane of a lens. From (A3.4) the complex amplitude in the pupil of the lens is

$$a_l(x, y) = (ia/\lambda f) \int_{-\infty}^{\infty} \int_{-\infty}^{\infty} t(x_1 y_1)$$

$$\times \exp\{(-i\pi/\lambda f)[(x - x_1)^2 + (y - y_1)^2]\}dx_1 dy_1.$$

$$(A3.13)$$

From (A3.12), the complex amplitude in the back focal plane of the lens can be written as

$$a_f(x_f, y_f) = (1/i\lambda f) \exp[(-i\pi/\lambda f)(x_f^2 + y_f^2)] \int_{-\infty}^{\infty} \int_{-\infty}^{\infty} a_l(x, y)$$

$$\times \exp\{i2\pi [x(x_f/\lambda f) + y(y_f/\lambda f)]\}dxdy, \qquad (A3.14)$$

or as

$$a_f(\xi, \eta) = (1/i\lambda f) \exp[i\pi\lambda f(\xi^2 + \eta^2)] A_l(\xi, \eta), \qquad (A3.15)$$

where $\xi = x_f/\lambda f, \eta = y_f/\lambda f$ and

$$a_l(x, y) \leftrightarrow A_l(\xi, \eta). \qquad (A3.16)$$

Now, (A3.13) can also be regarded as a convolution

$$a_l(x, y) = (ia/\lambda f)t(x_1, y_1) * \exp[(-i\pi/\lambda f)(x_1^2 + y_1^2)], \qquad (A3.17)$$

so that, if we take the Fourier transforms of both sides, we have

$$A_l(\xi, \eta) = (ia/\lambda f)\mathbf{T}(\xi, \eta)\exp[-i\pi\lambda f(\xi^2 + \eta^2)], \qquad (A3.18)$$

and, when (A3.18) is substituted in (A3.15), we obtain the result

$$a_f(\xi, \eta) = (a/\lambda^2 f^2)\mathbf{T}(\xi, \eta), \qquad (A3.19)$$

which shows that the complex amplitude in the focal plane is, apart from a constant of proportionality, the Fourier transform of the amplitude transmittance of the object.

# Appendix 4
# Speckle

Any diffusely scattering object illuminated by a source of highly coherent light, such as a laser, appears covered with speckle. Speckle is due to the fact that most surfaces are extremely rough on a scale of light wavelengths. All the microscopic elements making up the surface give rise to coherent diffracted waves, but the optical paths from any point of observation to neighbouring elements on the surface exhibit random differences which may amount to several wavelengths. Consequently, when the diffracted waves from these elements interfere with each other, a stationary granular pattern results (see fig. A4.1). The general appearance of such speckle patterns is almost independent of the character of the surface, but the scale of the granularity increases with the viewing distance and the $f$-number of the viewing system.

The statistics of such speckle patterns have been analysed in detail by Goodman [1975]; some of the most important results are summarized in the next few sections.

## A4.1 First-order statistics of speckle patterns

The complex amplitude at any point in the far field of a diffuser illuminated by a coherent source, as shown in fig. A4.2, is the sum of the complex amplitudes of the diffracted waves from all the individual elements on the object and is given by the relation

$$a \exp(-i\phi) = \sum_{n=1}^{n} a_n \exp(-i\phi_n). \qquad (A4.1)$$

If we assume that the moduli of all the individual complex amplitudes are equal, while the phase shifts are large enough that the remainders, after subtracting integral multiples of $2\pi$, are uniformly distributed over the range from 0 to $2\pi$, this reduces to the well-known random-walk problem. The joint

Fig. A4.1. Speckle pattern observed when a diffusing surface is illuminated with coherent light.

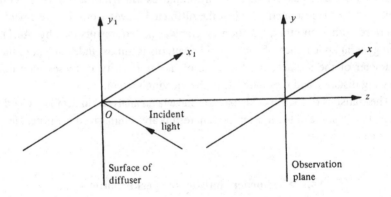

Fig. A4.2. Coordinate system used to study the statistics of speckle patterns.

probability density function of the real and imaginary parts of the complex amplitude is

$$p_{r,i}(a_{r,i}) = (1/2\pi\sigma^2)\exp[-(a_r^2 + a_i^2)/2\sigma^2], \qquad (A4.2)$$

where $\sigma^2$ is a constant. The most common value of the modulus is zero, while the phase has a uniform circular distribution. It can then be shown that the probability density function of the intensity is the negative exponential distribution

$$p(I) = (1/2\sigma^2)\exp(-I/2\sigma^2). \tag{A4.3}$$

From (A4.3) it follows that the mean value of the intensity is

$$\langle I \rangle = 2\sigma^2, \tag{A4.4}$$

while its second moment is

$$\langle I^2 \rangle = 2\langle I \rangle^2. \tag{A4.5}$$

Accordingly, the variance of the intensity is

$$\sigma_I^2 = \langle I^2 \rangle - \langle I \rangle^2,$$
$$= \langle I \rangle^2. \tag{A4.6}$$

If the contrast of the speckle pattern is defined as

$$c = \sigma_I/\langle I \rangle, \tag{A4.7}$$

it is apparent that the contrast of a speckle pattern formed by coherent light is unity.

## A4.2 Second-order statistics of speckle patterns

The two quantities of interest are the autocorrelation function and the power spectral density of the intensity distribution. To evaluate these, we make use of the fact that the complex amplitude $A(x_1, y_1)$ at the scattering surface and the complex amplitude $a(x, y)$ at the observation plane are related by the Fresnel–Kirchhoff integral. The autocorrelation function of the intensity in the speckle pattern is then given by the relation

$$R_I(\Delta x, \Delta y) =$$

$$\langle I \rangle^2 \left\{ 1 + \left| \frac{\int\!\!\int_{-\infty}^{\infty} |A(x_1, y_1)|^2 \exp[(i2\pi/\lambda z)(x_1 \Delta x + y_1 \Delta y)]\mathrm{d}x_1\mathrm{d}y_1}{\int\!\!\int_{-\infty}^{\infty} |A(x_1, y_1)|^2 \mathrm{d}x_1\mathrm{d}y_1} \right|^2 \right\}. \tag{A4.8}$$

For a square scattering surface whose edges have a length $L$, the average dimensions of a speckle calculated from (A4.8) are

$$\delta x = \delta y = \lambda z/L. \tag{A4.9}$$

The power spectral density of the intensity distribution is given by the Fourier transform of the autocorrelation function $R_I(\Delta x, \Delta y)$, which is

$$S_I(s_x, s_y) = \langle I \rangle^2 \left\{ \delta(s_x, s_y) \right.$$
$$\left. + \frac{\displaystyle\iint_{-\infty}^{\infty} |A(x_1, y_1)|^2 |A(x_1 - \lambda z s_x, y_1 - \lambda z s_y)|^2 dx_1 dy_1}{\left[ \displaystyle\iint_{-\infty}^{\infty} |A(x_1, y_1)|^2 dx_1 dy_1 \right]^2} \right\}, \quad (A4.10)$$

where $s_x, s_y$ are spatial frequencies along the $x$ and $y$ axes, respectively. Apart from a delta function at a spatial frequency of zero which contains half the total power, this is the normalized autocorrelation function of the intensity distribution over the scattering surface.

### A4.3  Image speckle

As distinct from the direct scattered field, the statistics of the intensity fluctuations in the image of the scattering surface formed by a lens (or by a hologram) are dependent on the size of the imaging aperture, which acts as a low-pass filter. The entrance pupil of the optical system can be considered as being illuminated by the primary speckle pattern. This random field then appears in the exit pupil, so that the intensity fluctuations in the image can be obtained by treating the exit pupil as a rough object. In this case, for a circular pupil of radius $\rho$, the average size of the speckles in the image is, from (A4.8),

$$\delta x = \delta y = 0.61 \lambda f / \rho, \quad (A4.11)$$

where $f$ is the focal length of the optical system.

### A4.4  Addition of speckle patterns

Two limiting cases are possible when two or more speckle patterns are superimposed. The first is where the light fields are coherent and the amplitudes add. In this case, the first-order statistics of the pattern remain unchanged.

The other case is where the intensities add. If two speckle patterns with equal average intensity $\langle I/2 \rangle$ are superimposed in this manner, it can be shown that the probability density function of the intensity in the resulting pattern is

$$p(I) = (4I/\langle I \rangle^2) \exp(-2I/\langle I \rangle). \quad (A4.12)$$

As can be seen from fig. A4.3, this function differs from the probability density function for a single speckle pattern with the same average intensity,

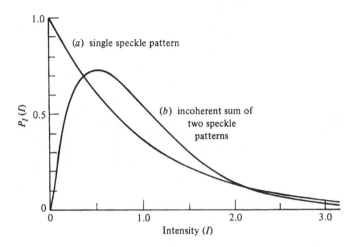

Fig. A4.3. Probability density functions of the irradiance in (*a*) a single speckle pattern with $\langle I \rangle = 1$, and (*b*) the sum of the intensities of two speckle patterns with $\langle I_1 \rangle = \langle I_2 \rangle = 0.5$.

which is also plotted for comparison, mainly in the elimination of most of the dark areas. As a result its contrast is only $2^{-1/2}$. If $N$ speckle patterns with the same average intensity are superimposed, the contrast of the resultant speckle pattern drops to $N^{-1/2}$.

# Appendix 5

## The H & D curve

The response of photographic materials to exposure to light is normally represented by a curve (known as the Hurter and Driffield, or H & D, curve) in which the optical density of the material, after it has been developed and fixed, is plotted against the logarithm of the exposure given to it. The optical density $D$ is defined by the relation

$$D = \log 1/\mathcal{T}, \tag{A5.1}$$

where $\mathcal{T}$ is the transmittance of the material for intensity, while the exposure is defined as the product of $I$, the intensity of the light to which the material has been exposed, and $T$, the exposure time.

Typical H & D curves for two photographic materials are shown in fig. A5.1. As can be seen, the upper portion of the curves is a straight line. One of the parameters commonly used to characterize a photographic material is the value of $\gamma$, the slope of this section. It can be shown that, over this region

$$\mathcal{T} \propto I^{-\gamma}. \tag{A5.2}$$

Although the H & D curve is used almost universally for photography, it is not a convenient way to specify the response of materials for holography, for which a curve showing $\mathbf{t}$, the amplitude transmittance of the material, as a function of the exposure is preferable. Since $\mathbf{t} = \mathcal{T}^{1/2}$, (A5.2) can be rewritten as

$$|\mathbf{t}| \propto I^{-\gamma/2}, \tag{A5.3}$$

over the linear portion of the H & D curve.

If a positive transparency is made from a photographic negative, we have, overall,

$$\gamma = -\gamma_n \gamma_p, \tag{A5.4}$$

where $\gamma_n$ and $\gamma_p$ are the slopes of the H & D curves for the negative and positive

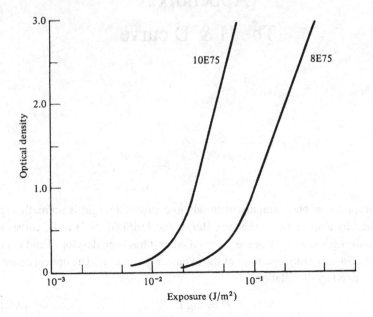

Fig. A5.1. Typical H & D curves for Holotest 10E75 and 8E75 plates ($\lambda = 633$ nm).

materials, respectively. We then have, for the final transparency,

$$|t| \propto I^{\gamma/2}. \tag{A5.5}$$

It should be noted that the value of $\gamma$ determines the response of a photographic material on a macroscopic scale only; its response on a microscopic scale also involves its modulation transfer function (see section 6.2).

# Bibliography

Abramson, N. (1981). *The Making & Evaluation of Holograms*. London: Academic Press.

Beiser, L. (1988). *Holographic Scanning*. New York: Wiley.

Bjelkhagen, H. I. (1993). *Silver Halide Materials for Holography & Their Processing*. Berlin: Springer-Verlag.

Butters, J. N. (1971). *Holography & Its Technology*. London: Peter Peregrinus Ltd.

Cathey, W. T. (1974). *Optical Information Processing & Holography*. New York: Wiley.

Caulfield, H. J., ed. (1979). *Handbook of Optical Holography*. New York: Academic Press.

Caulfield, H. J. & Lu. S. (1970). *The Applications of Holography*. New York: Wiley-Interscience.

Collier, R. J., Burckhardt, C. B. & Lin, L. H. (1971). *Optical Holography*. New York: Academic Press.

Collings, N. (1988). *Optical Pattern Recognition: Using Holographic Techniques*. Reading: Addison-Wesley.

De Velis, J. B. & Reynolds, G. O. (1967). *Theory & Applications of Holography*. Reading: Addison-Wesley.

Erf, R. K., ed. (1974). *Holographic Non-destructive Testing*. New York: Academic Press.

Françon, M. (1969). *Holographie*. Paris: Masson et Cie.

Jones, R. & Wykes, C. (1989). *Holographic & Speckle Interferometry*. Cambridge: Cambridge University Press.

Lehmann, M. (1970). *Holography: Theory & Practice*. London: Focal Press.

Menzel, E., Mirande, W. & Weingartner, I. (1973). *Fourier-Optik und Holographie*. Wien: Springer-Verlag.

Okoshi, T. (1976). *Three-dimensional Imaging Techniques*. New York: Academic Press.

Ostrovsky, Yu, I. (1977). *Holography & Its Applications*. (Transl. G. Leib). Moscow: Mir Publishers.

Ostrovsky, Yu. I., Butusov, M. M. & Ostrovaskaya, G. V. (1980). *Interferometry by Holography*. Berlin: Springer-Verlag.

Ostrovsky, Yu. I. (1991). *Holographic Interferometry in Experimental Mechanics*. Berlin: Springer-Verlag.

338 *Bibliography*

Petit, R., ed (1980), *Electromagnetic Theory of Gratings*. Berlin: Springer-Verlag.

Rastogi, P. K., ed. (1994). *Holographic Interferometry*. Berlin: Springer-Verlag.

Robillard, J. & Caulfield, H. J., eds. (1990). *Industrial Applications of Holography*. New York: Oxford University Press.

Robinson, D. W. & Reid, G. T., eds. (1993). *Interferogram Analysis: Digital Processing Techniques for Fringe Pattern Measurement*. London: IOP.

Saxby, G. (1980). *Holograms*. London: Focal Press.

Saxby, G. (1988). *Practical Holography*. London: Focal Press.

Saxby, G. (1991). *Manual of Practical Holography*. London: Focal Press.

Schumann, W. & Dubas, M. (1979). *Holographic Interferometry*. Berlin: Springer-Verlag.

Schumann, W., Zurcher, J. P. & Cuche, D. (1985). *Holography & Deformation Analysis*. Berlin: Springer-Verlag.

Smith, H. M. (1969). *Principles of Holography*. New York: Wiley-Interscience (2nd edn. 1975).

Smith, H. M. ed. (1977). *Holographic Recording Materials*. Berlin: Springer-Verlag.

Solymar, L. & Cooke, D. J. (1981). *Volume Holography & Volume Gratings*. New York: Academic Press.

Soroko, L. M. (1980). *Holography & Coherent Optics*. (Transl. A. Tybulewicz). New York: Plenum Press.

Stroke, G. W. (1966). *An Introduction to Coherent Optics & Holography*. New York: Academic Press. (2nd edition, 1969).

Syms, R. R. A. (1990). *Practical Volume Holography*. Oxford: Oxford University Press.

Takemoto, S. ed. (1989). *Laser Holography in Geophysics*. Chichester: E. Horwood.

Vasilenko, G. I. & Tsibul'kin, L. M. (1994). *Image Recognition by Holography*. (Transl. A. Tybulewicz). New York: Plenum Press.

Vest, C. M. (1979). *Holographic Interferometry*. New York: John Wiley.

Vikram, C. (1992). *Particle Field Holography*. Cambridge: Cambridge University Press.

von Bally, G. ed. (1979). *Holography in Medicine & Biology*. Berlin: Springer-Verlag.

Wenyon, M. (1978). *Understanding Holography*. Newton Abbot: David & Charles.

Yaroslavskii, L. P. & Merzlyakov, N. S. (1980). *Methods of Digital Holography*. New York: Consultants Bureau, Plenum Publishing Co.

Yu, F. T. S. (1973). *Introduction to Diffraction, Information Processing & Holography*. Cambridge: The MIT Press.

# References

Abramson, N. (1969). The holo-diagram: a practical device for making and evaluating holograms. *Applied Optics*, **8**, 1235–40.

Abramson, N. (1970*a*). The holo-diagram II. A practical device for information retrieval in hologram interferometry. *Applied Optics*, **9**, 97–101.

Abramson, N. (1970*b*). The holo-diagram III. A practical device for predicting fringe patterns in hologram interferometry. *Applied Optics*, **9**, 2311–20.

Abramson, N. (1971). The holo-diagram IV. A practical device for simulating fringe patterns in hologram interferometry. *Applied Optics*, **10**, 2155–61.

Abramson, N. (1972). The holo-diagram V. A device for practical interpreting of hologram fringes. *Applied Optics*, **11**, 1143–7.

Abramson, N. (1974). Sandwich hologram interferometry: a new dimension in holographic comparison. *Applied Optics*, **13**, 2019–25.

Abramson, N. (1975). Sandwich hologram interferometry. 2. Some practical calculations. *Applied Optics*, **14**, 981–4.

Abramson, N. (1976*a*). Sandwich hologram interferometry. 3: Contouring. *Applied Optics*, **15**, 200–5.

Abramson, N. (1976*b*). Holographic contouring by translation. *Applied Optics*, **15**, 1018–22.

Abramson, N. (1977). Sandwich hologram interferometry. 4: Holographic studies of two milling machines. *Applied Optics*, **16**, 2521–31.

Abramson, N. (1978). Light-in-flight recording by holography. *Optics Letters*, **3**, 121–3.

Abramson, N. H. (1983). Light-in-flight recording: high-speed holographic motion pictures of ultrafast phenomena. *Applied Optics*, **22**, 215–32.

Abramson, N. H. & Bjelkhagen, H. (1973). Industrial holographic measurements. *Applied Optics*, **11**, 2792–6.

Abramson, N. H. & Bjelkhagen, H. (1978). Pulsed sandwich holography. 2. Practical application. *Applied Optics*, **17**, 187–91.

Abramson, N. & Bjelkhagen, H. (1979). Sandwich hologram interferometry. 5: Measurement of in-plane displacement and compensation for rigid body motion. *Applied Optics*, **18**, 2870–80.

Abramson, N., Bjelkhagen, H. & Skande, P. (1979). Sandwich holography for storing information interferometrically with a high degree of accuracy. *Applied Optics*, **18**, 2017–21.

Abramson, N. H. & Spears, K. G. (1989). Single pulse light-in-flight recording by holography. *Applied Optics*, **28**, 1834–41.

Agren, C.-H. & Stetson, K. A. (1972). Measuring the resonances of treble viol plates by hologram interferometry and designing an improved instrument. *Journal of the Acoustical Society of America*, **51**, 1971–83.

Aharoni, A., Goodman, J. W. & Amitai, Y. (1993). Beam-correcting holographic doublet for focusing multimode laser diodes. *Optics Letters*, **18**, 179–81.

Aleksandrov, E. G. & Bonch-Bruevich, A. M. (1967). Investigation of surface strains by the hologram technique. *Soviet Physics: Technical Physics*, **12**, 258–65.

Aleksoff, C. C. (1971). Temporally modulated holography. *Applied Optics*, **10**, 1329–41.

Alferness, R. (1975a). Analysis of optical propagation in thick holographic gratings. *Applied Physics*, **7**, 29–33.

Alferness, R. (1975b). Equivalence of the thin-grating decomposition and coupled wave analysis of thick holographic gratings. *Optics Communications*, **15**, 209–12.

Alferness, R. (1976). Analysis of propagation at the second-order Bragg angle of a thick holographic grating. *Journal of the Optical Society of America*, **66**, 353–62.

Alferness, R. & Case, S. K. (1975). Coupling in doubly-exposed, thick holographic gratings. *Journal of the Optical Society of America*, **65**, 730–9.

Altman, J. H. (1966). Pure relief images in type 649 F plates. *Applied Optics*, **5**, 1689–90.

Amadesi, S., Gori, F., Grella, R. & Guattari, G. (1974). Holographic methods for painting diagnostics. *Applied Optics*, **13**, 2009–13.

Amako, J. & Sonehara, T. (1991). Kinoform using an electrically controlled birefringent liquid-crystal spatial light modulator. *Applied Optics*, **30**, 4622–8.

Amitai, Y. & Friesem, A. A. (1988). Combining low aberrations and high diffraction efficiency in holographic optical elements. *Optics Letters*, **13**, 883–5.

Amitai, Y. & Friesem, A. A. (1989). Holographic elements with high efficiency and low aberrations for helmet displays. *Applied Optics*, **28**, 3405–16.

Amitai, Y., Friesem, A. A. & Weiss, V. (1990). Designing holographic lenses with different recording and readout wavelengths. *Journal of the Optical Society of America. A*, **7**, 80–6.

Amitai, Y. & Goodman, J. W. (1991a). Design of substrate-mode holographic interconnects with different recording and readout wavelengths. *Applied Optics*, **30**, 2376–81.

Amitai, Y. & Goodman, J. W. (1991b). Analytic design of an achromatic double grating coupler. *Applied Optics*, **30**, 2970–4.

Amitai, Y., Shariv, I., Kroch, M., Friesem. A. A. & Reinhorn, S. (1993). White-light holographic display based on planar optics. *Optics Letters*, **18**, 1265–7.

Anderson, L. K. (1968). Holographic optical memory for bulk data storage. *Bell Laboratories Record*, **46**, 318–25.

Andreev, R. B., Vorzobova, N. D., Kalintsev, A. G. & Staselko, D. I. (1980). Image pulsed holography recording in the green region of the spectrum. *Optics & Spectroscopy*, **49**, 514–15.

Angell, D. K. (1987). Improved diffraction efficiency of silver halide (sensitized) gelatin. *Applied Optics*, **26**, 4692–702.

ANSI (1986). *American National Standard for the Safe Use of Lasers*, **Z 136.**1–1986. New York: American Standards Institute.

Ansley, D. A. (1970). Techniques for pulsed laser holography of people. *Applied Optics*, **9**, 815–21.

Aoyagi, Y. & Namba, S. (1976). Blazing of holographic grating by ion etching technique. *Japanese Journal of Applied Physics*, **15**, 721–2.

Aoyagi, A., Sano, K. & Namba, S. (1979). High spectroscopic qualities in blazed ion-etched holographic gratings. *Optics Communications*, **29**, 253–5.

Archbold, E. & Ennos, A. E. (1968). Observation of surface vibration modes by stroboscopic hologram interferometry. *Nature*, **217**, 942–3.

Aristov, V. V., Shekhtman, V. Sh. & Timofeev, V. B. (1969). The Borrmann effect and extinction in holography. *Physics Letters*, **28A**, 700–1.

Armstrong, W. T. & Forman, P. R. (1977). Double-pulsed time differential holographic interferometry. *Applied Optics*, **16**, 229–32.

Arnold, S. M. (1985). Electron-beam fabrication of computer-generated holograms. *Optical Engineering*, **24**, 803–7.

Arnold, S. M. (1988). Desktop computer encoding of electron-beam written holograms. In *Computer-Generated Holography II*, Proceedings of the SPIE, vol. 884, ed. S. H. Lee, pp. 23–6. Bellingham: SPIE.

Arnold, S. M. (1989). How to test an asphere with a computer generated hologram. In *Holographic Optics: Optically & Computer Generated*, Proceedings of the SPIE, vol. 1052, eds. I. Cindrich & S. H. Lee, pp. 191–7. Bellingham: SPIE.

Arons, E., Dilworth, D., Shih, M. & Sun, P. C. (1993). Use of Fourier synthesis holography to image through inhomogeneities. *Optics Letters*, **21**, 1852–4.

Assa, A. & Betser, A. A. (1974). The application of holographic multiplexing to record separate isopachic- and isochromatic-fringe patterns. *Experimental Mechanics*, **14**, 502–4.

Bahuguna, R. D., Beaulieu, J. & Arteaga, H. (1992). Reflection display holograms on dichromated gelatin. *Applied Optics*, **31**, 6181–2.

Balasubramanian, N. (1975). Holographic applications in photogrammetry. *Optical Engineering*, **14**, 448–52.

Ballard, G. S. (1968). Double exposure holographic interferometry with separate reference beams. *Journal of Applied Physics*, **39**, 4846–8.

Banyasz, I., Kiss, G. & Varga, P. (1988). Holographic image of a point source in the presence of misalignment. *Applied Optics*, **27**, 1293–7.

Bar-Joseph, I., Hardy, A., Katzir, Y. & Silberberg, Y. (1981). Low-power phase-conjugate interferometry. *Optics Letters*, **6**, 414–16.

Barnard, E. (1988). Optimal error diffusion for computer-generated holograms. *Journal of the Optical Society of America. A*, **5**, 1803–17.

Bartelt, H. & Case, S. K. (1982). High-efficiency hybrid computer-generated holograms. *Applied Optics*, **21**, 2886–90.

Bartolini, R. A. (1972. Improved development for holograms recorded in photoresist. *Applied Optics*, **11**, 1275–6.

Bartolini, R. A. (1974). Characteristics of relief phase holograms recorded in photoresists. *Applied Optics*, **13**, 129–39.

Bartolini, R. A. (1977*a*). Photoresists. In *Holographic Recording Materials*, ed. H. M. Smith, pp. 209–27. Berlin: Springer-Verlag.

Bartolini, R. A. (1977*b*). Optical recording media review. In *Optical Storage Materials*

& *Methods*, Proceedings of the SPIE, vol. 123, ed. L. Beiser & D. Chen, pp. 2–9. Bellingham: SPIE.

Bartolini, R., Hannan, W., Karlsons, D. & Lurie, M. (1970). Embossed hologram motion pictures for television playback. *Applied Optics*, 9, 2283–90.

Bates, H. E. (1973). Burst-mode frequency-doubled YAG : $Nd^{3+}$ laser for time-sequenced high speed photography and holography. *Applied Optics*, 12, 1172–8.

Becker, F. & Yu, Y. H. (1985). Digital fringe reduction techniques applied to the measurement of three-dimensional transonic flow fields. *Optical Engineering*, 24, 429–34.

Beesley, M. J., Castledine, J. G. & Cooper, D. P. (1969). Sensitivity of resist coated silicon slices to argon laser wavelengths. *Electronics Letters*, 5, 257–8.

Beesley, M. J., Foster, H. & Hambleton, K. G. (1968). Holographic projection of microcircuit patterns. *Electronics Letters*, 4, 49–50.

Belvaux, Y. (1967). Hologram duplication. *Annales de Radioelectricité*, 22, 105–8.

Belvaux, Y. (1975). Influence de divers paramètres d'enregistrement lors de la restitution d'un hologramme. *Nouvelle Revue d'Optique*, 6, 137–47.

Benlarbi, B., Cooke, D. J. & Solymar, L. (1980). Higher order modes in thick phase gratings. *Optica Acta*, 27, 885–95.

Bennett, S. J. (1976). Achromatic combinations of hologram optical elements. *Applied Optics*, 15, 542–5.

Benton, S. A. (1969). Hologram reconstructions with extended incoherent sources. *Journal of the Optical Society of America*, 59, 1545–6.

Benton, S. A. (1971). Granularity effects in phase holograms. *Journal of the Optical Society of America*, 61, 649.

Benton, S. A. (1975). Holographic displays – a review. *Optical Engineering*, 14, 402–7.

Benton, S. A. (1977). White light transmission/reflection holographic imaging. In *Applications of Holography & Optical Data Processing*, ed. E. Marom, A. A. Friesem & E. Wiener-Avnear, pp. 401–9. Oxford: The Pergamon Press.

Benton, S. A. (1978). Achromatic images from white-light transmission holograms. *Journal of the Optical Society of America*, 68, 1441.

Benton, S. A. (1980). Holographic displays: 1975–1980. *Optical Engineering*, 19, 686–90.

Benton, S. A. (1981). Achromatic holographic stereograms. *Journal of the Optical Society of America*, 71, 1568A.

Benton, S. A. (1982). Survey of holographic stereograms. In *Processing and Display of Three-Dimensional Data*, Proceedings of the SPIE, vol. 367, ed. J. Pearson, pp. 15–19. Bellingham: SPIE.

Benton, S. A. (1983). Photographic holography. In *Optics in Entertainment*, Proceedings of the SPIE, vol. 391, ed. C. Outwater, pp. 2–9. Bellingham: SPIE.

Benton, S. A. (1988). The principles of reflection holographic stereograms. In *Proceedings of the Third International Symposium on Display Holography*, ed. T. H. Jeong, Lake Forest College, 1988, pp. 593–608.

Benton, S. A., Birner, S. M. & Shirakura, A. (1990). Edge-lit rainbow holograms. In *Practical Holography IV*, Proceedings of the SPIE, vol. 1212, ed. S. A. Benton, pp. 149–57. Bellingham: SPIE.

Benton, S. A., Mingace, Jr., H. S. & Walter, W. R. (1979). In *Optics and Photonics*

*Applied to 3-Dimensional Imagery*, Proceedings of the SPIE, vol. 212, eds. M. Grosmann & P. Mayrueis, pp. 2–7. Bellingham: SPIE.

Bergman, L. A., Wu, W. H., Johnston, A. R., Nixon. R., Esener, S. C., Guest, C. C., Yu, P., Brabik, T. J., Feldman, M. & Lee, S. H. (1986). Holographic optical interconnects for VLSI. *Optical Engineering*, **25**, 1109–18.

Biedermann, K. (1969). A function characterizing photographic film that directly relates to brightness of holographic images. *Optik*, **28**, 160–76.

Biedermann, K. (1970). The scattered flux spectrum of photographic materials for holography. *Optik*, **31**, 367–89.

Biedermann, K. & Holmgren, O. (1977), Large-size distortion-free computer-generated holograms in photoresist. *Applied Optics*, **16**, 2014–16.

Biedermann, K. & Johansson, S. (1972). Evaluation of the modulation transfer function of photographic emulsions by means of a multiple-sine-slit microdensitometer. *Optik*, **35**, 391–403.

Biedermann, K. & Johansson, S. (1975). A universal instrument for the evaluation of the MTF and other recording parameters of photographic materials. *Journal of Physics E: Scientific Instruments*, **8**, 751–7.

Biedermann, K. & Molin, N.-E. (1970). Combining hypersensitization and *in situ* processing for time-average observation in real-time hologram interferometry. *Journal of Physics E: Scientific Instruments*, **3**, 669–80.

Binfield, P., Galloway, R. & Watson, J. (1993). Reciprocity failure in continuous-wave holography. *Applied Optics*, **32**, 4337–43.

Birch, K. G. & Green, F. J. (1972). The application of computer-generated holograms to testing optical elements. *Journal of Physics D: Applied Physics*, **5**, 1982–92.

Bjelkhagen, H. (1974). Holographic time-average vibration study of a structure dynamic model of an airplane fin. *Optics & Laser Technology*, **6**, 117–23.

Bjelkhagen, H. I. (1977*a*). Experiences with large-scale reflection and transmission holograms. In *Three-Dimensional Imaging*, Proceedings of the SPIE, vol. 120, ed. S. A. Benton, pp. 122–6. Redondo Beach: SPIE.

Bjelkhagen, H. (1977*b*). Pulsed sandwich holography. *Applied Optics*, **16**, 1727–31.

Bjelkhagen, H. I. (1992). Holographic portraits made by pulse lasers. *Leonardo*, **25**, 443–8.

Bloom, A. L. (1968). *Gas Lasers*. New York: John Wiley.

Blyth, J. (1991). Methylene blue sensitized dichromated gelatin holograms: a new electron donor for their increased photosensitivity. *Applied Optics*, **30**, 1598–602.

Boj, P. G., Crespo, J. & Quintana, J. A. (1992). Broadband reflection holograms in dichromated gelatin. *Applied Optics*, **31**, 3302–5.

Bone, D. J. (1991). Fourier fringe analysis: the two-dimensional phase unwrapping problem. *Applied Optics*, **30**, 3627–32.

Bone, D. J., Bachor, H.-A. & Sandeman, R. J. (1986). Fringe-pattern analysis using a 2-D Fourier transform. *Applied Optics*, **25**, 1653–60.

Boone, P. M. (1975). Use of reflection holograms in holographic interferometry and speckle correlation for measurement of surface displacement. *Optica Acta*, **22**, 579–89.

Boone, P. & Verbiest, R. (1969). Application of hologram interferometry to plate deformation and translation measurements. *Optica Acta*, **16**, 555–67.

Booth, B. L. (1975). Photopolymer material for holography. *Applied Optics*, **14**, 593–601.

Booth, B. L. (1977). Photopolymer laser recording materials. *Journal of Applied Photographic Engineering*, **3**, 24–30.

Born, M. & Wolf, E. (1980). *Principles of Optics*. Oxford: Pergamon Press.

Bracewell, R. (1978). *The Fourier Transform & Its Applications*. Tokyo: McGraw-Hill Kogakusha.

Brady, D. J. & Psaltis, D. (1991). Holographic interconnections in photorefractive waveguides. *Applied Optics*, **30**, 2324–33.

Bragg, W. L. (1939). A new type of 'X-ray microscope'. *Nature*, **143**, 678.

Bragg, W. L. (1942). The x-ray microscope. *Nature*, **149**, 470–1.

Brandes, R. G., Francois, E. E. & Shankoff, T. A. (1969). Preparation of dichromated gelatin films for holography. *Applied Optics*, **8**, 2346–48.

Breidne, M., Johansson, S., Nilsson, L.-E. & Åhlen, H. (1979). Blazed holographic gratings. *Optica Acta*, **26**, 1427–41.

Brenner, K.-H. & Sauer, F. (1988). Diffractive-reflective optical interconnects. *Applied Optics*, **27**, 4251–4.

Briers, J. D. (1976). The interpretation of holographic interferograms. *Optics & Quantum Electronics*, **8**, 469–501.

Broja, M., Wyrowski, F. & Bryngdahl, O. (1989). Digital halftoning by iterative procedure. *Optics Communications*, **69**, 205–10.

Brooks, R. E., Heflinger, L. O. & Wuerker, R. F. (1965). Interferometry with a holographically reconstructed comparison beam. *Applied Physics Letters*, **7**, 248–9.

Brown, B. R. & Lohmann, A. W. (1966). Complex spatial filtering with binary masks. *Applied Optics*, **5**, 967–9.

Brown, B. R. & Lohmann, A. W. (1969). Computer generated binary holograms. *IBM Journal of Research & Development*, **13**, 160–7.

Brown, G. M. & Zhang, J. (1991). FEM/CAH study of wrist pin insertion into a connecting rod. In *Proceedings of the 10th Invitational UFEM Symposium*, ed. R. J. Pryputniewicz, pp. 494–504. Worcester: Worcester Polytechnic Institute.

Brucker, E. B. (1991). Search for short-lived particles using holography. *Practical Holography V*, Proceedings of the SPIE, vol. 1461, ed. S. A. Benton, pp. 206–14. Bellingham: SPIE.

Brumm, D. G. (1967). Double images in copy holograms. *Applied Optics*, **6**, 588–9.

Bryngdahl, O. (1967). Polarizing holography, *Journal of the Optical Society of America*, **57**, 545–7.

Bryngdahl, O. (1969a). Longitudinally reversed shearing interferometry. *Journal of the Optical Society of America*, **59**, 142–6.

Bryngdahl, O. (1969b). Holography with evanescent waves. *Journal of the Optical Society of America*, **59**, 1645–50.

Bryngdahl, O. (1973). Evanescent waves in optical imaging. In *Progress in Optics*, vol. 11, ed. E. Wolf, pp. 169–221. Amsterdam: North-Holland.

Bryngdahl, O. (1974a). Optical map transformations. *Optics Communications*, **10**, 164–8.

Bryngdahl, O. (1974b). Geometrical transformations in optics. *Journal of the Optical Society of America*, **64**, 1092–9.

Bryngdahl, O. (1975). Computer generated holograms as generalized optical components. *Optical Engineering*, **14**, 426–35.

Bryngdahl, O. & Lohmann, A. W. (1968a). Interferograms are image holograms. *Journal of the Optical Society of America*, **58**, 141–2.

Bryngdahl, O. & Lohmann, A. (1968b). One dimensional holography with spatially incoherent light. *Journal of the Optical Society of America*, **58**, 625–8.

Bryngdahl, O. & Lohmann, A. (1968c). Non-linear effects in holography. *Journal of the Optical Society of America*, **58**, 1325–34.

Bryngdahl, O. & Lohmann, A. (1970a). Variable magnification in incoherent holography. *Applied Optics*, **9**, 231–2.

Bryngdahl, O. & Lohmann, A. (1970b). Holography in white light. *Journal of the Optical Society of America*, **60**, 281–3.

Bryngdahl, O. & Wyrowski, F. (1990). Digital holography – computer-generated holograms. In *Progress in Optics*, vol. 28, ed. E. Wolf, pp. 1–86. Amsterdam: North-Holland.

BSI (1983). *British Standard Guide on Protection of Personnel Against Hazards from Laser Radiation*, BS4803: 1983. London: British Standards Institution.

Budzinski, J. (1992). SNOP: a method for skeletonization of a fringe pattern along the fringe direction. *Applied Optics*, **31**, 3109–13.

Burch, J. J. (1967). A computer algorithm for the synthesis of spatial frequency filters. *Proceedings of the IEEE*, **55**, 599–601.

Burch, J. M. (1965). The application of lasers in production engineering. *The Production Engineer*, **44**, 431–42.

Burch, J. M. & Palmer, D. A. (1961). Interferometric methods for the photographic production of large gratings. *Optica Acta*, **8**, 73–80.

Burckhardt, C. B. (1966a). Diffraction of a plane wave at a sinusoidally stratified dielectric grating. *Journal of the Optical Society of America*, **56**, 1502–9.

Burckhardt, C. B. (1966b). Display of holograms in white light. *Bell Systems Technical Journal*, **45**, 1841–4.

Burckhardt, C. B. (1967). Efficiency of a dielectric grating. *Journal of the Optical Society of America*, **57**, 601–3.

Burckhardt, C. B. (1970). A simplification of Lee's method of generating holograms by computer. *Applied Optics*, **9**, 1949.

Burke, W. J., Staebler, D. L., Phillips, W. & Alphonse, G. A. (1978). Volume phase holographic storage in ferroelectric crystals. *Optical Engineering*, **17**, 308–16.

Burns, J. R. (1985). Large-format embossed holograms. In *Applications of Holography*, Proceedings of the SPIE, vol. 523, ed. L. Huff, pp. 7–14. Bellingham: SPIE.

Buschmann, H. T. (1971). The production of low noise, bright, phase holograms by bleaching. *Optik*, **34**, 240–53.

Buschmann, H. T. (1972). The wavelength dependence of the transfer properties of photographic materials for holography. In *Optical & Acoustical Holography*, ed. E. Camatini, pp. 151–72. New York: The Plenum Press.

Buschmann, H. T. & Metz, H. J. (1971). Die wellenlängenabhängigkeit der übertragungseigenschaften photographischer materialen für die holographie. *Optics Communications*, **2**, 373–6.

Butters, J. N. & Leendertz, J. A. (1971). A double exposure technique for speckle pattern interferometry. *Journal of Physics E: Scientific Instruments*, **4**, 277–9.

Carraba, M. M., Spencer, K. M., Rich, C. & Rault, D. (1990). The utilization of a holographic Bragg diffraction filter for Rayleigh line rejection in Raman Spectroscopy. *Applied Spectroscopy*, **44**, 1558–61.

Cartwright, S. L., Dunn, P. & Thompson, B. J. (1980). Particle sizing using far-field holography: new developments. *Optical Engineering*, **19**, 727–33.

Casasent, D. ed. (1978). *Optical Data Processing*, Topics in Applied Physics, vol. 23. Berlin: Springer-Verlag.

Casasent, D. (1985). Computer generated holograms in pattern recognition: a review. *Optical Engineering*, **24**, 724–30.

Casasent, D. & Caimi, F. (1977). Photodichroic crystals for coherent optical data processing. *Optics & Laser Technology*, **9**, 63–8.

Case, S. K. (1975). Coupled wave theory for multiply exposed thick holographic gratings. *Journal of the Optical Society of America*, **65**, 724–9.

Case, S. K. & Alferness, R. (1976). Index modulation and spatial harmonic generation in dichromated gelatin films. *Applied Physics*, **10**, 41–51.

Case, S. K., Haugen, P. R. & Løkberg, O. J. (1981). Multifacet holographic optical elements for wavefront transformations. *Applied Optics*, **20**, 2670–5.

Cathey Jr., W. T. (1965). Three dimensional wavefront reconstruction using a phase hologram. *Journal of the Optical Society of America*, **55**, 457.

Caulfield, H. J. (1972). Multiplexing double-exposure holographic interferograms. *Applied Optics*, **11**, 2711–12.

Caulfield, H. J. & Beyen, W. J. (1967). Birefringent beam splitting for holography. *Review of Scientific Instruments*, **38**, 977–8.

Cederquist, J. & Tai, A. M. (1984). Computer-generated holograms for geometric transformations. *Applied Optics*, **23**, 3099–104.

Cescato, L., Gluch, E. & Streibl, N. (1990). Holographic quarterwave plates. *Applied Optics*, **29**, 3286–90.

Cha, S. & Vest, C. M. (1979). Interferometry and reconstruction of strongly refracting asymmetric refractive-index fields. *Optics Letters*, **4**, 311–13.

Cha, S. & Vest, C. M. (1981). Tomographic reconstruction of strongly refracting fields and its application to interferometric measurement of boundary layers. *Applied Optics*, **20**, 2787–94.

Champagne, E. B. (1967). Non-paraxial imaging, magnification and aberration properties in holography. *Journal of the Optical Society of America*, **57**, 51–5.

Champagne, E. B. & Massey, N. G. (1969). Resolution in holography. *Applied Optics*, **8**, 1879–85.

Chang, B. J. (1973). Holography with non-coherent light. *Optics Communications*, **9**, 357–9.

Chang, B. J. (1976). Post-processing of developed dichromated gelatin holograms. *Optics Communications*, **17**, 270–1.

Chang, B. J. (1979). Dichromated gelatin as a holographic storage medium. In *Optical Information Storage*, Proceedings of the SPIE, vol. 177, ed. K. G. Leib, pp. 71–81. Bellingham: SPIE.

Chang, B. J. & Leith, E. N. (1979). Space-invariant multiple-grating interferometers in holography. *Journal of the Optical Society of America*, **69**, 689–96.

Chang, B. J. & Leonard, C. D. (1979). Dichromated gelatin for the fabrication of holographic optical elements. *Applied Optics*, **18**, 2407–17.

Chang, B. J. & Winick, K. (1980). Silver-halide gelatin holograms. In *Recent Advances in Holography*, Proceedings of the SPIE, vol. 215, eds. T. C. Lee & P. N. Tamura, pp. 172–7. Bellingham: SPIE.

Chang, M. & George, N. (1970). Holographic dielectric grating: theory and practice. *Applied Optics*, **9**, 713–19.

Chang, M. P. & Ersoy, O. K. (1993). Iterative interlacing error diffusion for synthesis of computer-generated holograms. *Applied Optics*, **32**, 3122–9.

Chatelain, B. (1973). Holographic photo-elasticity: independent observation of the isochromatic and isopachic fronges for a single model subjected to only one process. *Optics & Laser Technology*, **5**, 201–4.

Chau, H. H. M. (1968). Holographic interferometer for isopachic stress analysis. *Review of Scientific Instruments*, **39**, 1789–92.

Chau, H. M. (1970). A full view holographic system. *Applied Optics*, **9**, 1479–80.

Chen, H. (1979). Astigmatic one-step rainbow hologram process. *Applied Optics*, **18**, 3728–30.

Chen, H., Tai. A. & Yu, F. T. S. (1978). Generation of color images with one-step rainbow holograms. *Applied Optics*, **17**, 1490–1.

Chen, H. & Yu, F. T. S. (1978). One-step rainbow hologram. *Optics Letters*, **2**, 85–7.

Chen, H., Hershey, R. R. & Leith, E. (1987). Design of a holographic lens for the infrared. *Applied Optics*, **26**, 1983–8.

Chen, Y., Chen, H., Dilworth, D., Leith, E., Lopez, J., Shih, M., Sun. P. C. & Vossler, G. (1993). Evaluation of holographic methods for imaging through biological tissue. *Applied Optics*, **32**, 4330–6.

Chomat, M. & Miler, M. (1973). Application of holography to the analysis of mechanical vibration in electronic components. *TESLA Electronics*, **3**, 83–93.

Chu, D. C., Fienup, J. R. & Goodman, J. W. (1973). Multi-emulsion, on-axis, computer generated holograms. *Applied Optics*, **12**, 1386–8.

Chu, R. S. & Tamir, T. (1970). Guided-wave theory of light diffraction by acoustic microwaves. *IEEE Transactions on Microwave Theory & Techniques*, **MTT-18**, 486–503.

Cindrich, I. (1967). Image scanning by rotation of a hologram. *Applied Optics*, **6**, 1531–4.

Close, D. H. (1975). Holographic optical elements. *Optical Engineering*, **14**, 408–19.

Cochran, G. (1966). New method of making Fresnel transforms with incoherent light. *Journal of the Optical Society of America*, **56**, 1513–17.

Cochran, W. T., Cooley, J. W., Favin, D. L., Helms, H. D., Kaenel, R. A., Lang, W. W., Maling Jr., G. C., Nelson, D. E., Rader, C. M. & Welch, P. D. (1967). What is the fast Fourier transform? *Proc. IEEE*, **55**, 1664–74.

Colburn, W. S. & Dubow, J. B. (1973). Photoplastic recording materials. Technical Report AFAL-TR-73-255. Ann Arbor: Harris Electro-Optics Center of Radiation.

Coleman, D. J. & Magariños, J. (1981). Controlled shifting of the spectral response of reflection holograms. *Applied Optics*, **20**, 2600–1.

Collier, R. J., Burckhardt, C. B. & Lin, L. H. (1971). *Optical Holography*. New York: Academic Press.

Collier, R. J., Doherty, E. T. & Pennington, K. S. (1965). Application of moiré techniques to holography. *Applied Physics Letters*, **7**, 223–5.

Collier, R. J. & Pennington, K. S. (1966). Ghost imaging in holograms formed in the near field. *Applied Physics Letters*, **8**, 44–6.

Collier, R. J. & Pennington, K. S. (1967). Multicolor imaging from holograms formed on two-dimensional media *Applied Optics*, **6**, 1091–5.

Collins, L. F. (1968). Difference holography. *Applied Optics*, **7**, 203–5.

Cooke, D. J. & Ward, A. A. (1984). Reflection hologram processing for high efficiency in silver halide emulsions. *Applied Optics*, **23**, 934–41.

Coupland, J. M. & Halliwell, N. A. (1988). Particle image velocimetry: rapid transparency analysis using optical correlation. *Applied Optics*, **27**, 1919–21.

Coupland, J. M. & Halliwell, N. A. (1992). Particle-image velocimetry: three-dimensional velocity measurements using holographic recording and optical correlation. *Applied Optics*, **31**, 1005–7.

Couture, J. J. A. & Lessard, R. A. (1984). Diffraction efficiency changes induced by coupling effects between gratings of transmission holograms. *Optik*, **68**, 69–80.

Cox, A. J. & Dibble, D. C. (1991). Holographic reproduction of a diffraction-free beam. *Applied Optics*, **30**, 1330–2.

Creath, K. (1985). Phase-shifting speckle interferometry. *Applied Optics*, **24**, 3053–8.

Creath, K. (1988). Phase-measurement interferometry techniques. In *Progress in Optics*, vol. 26, ed. E. Wolf, pp. 349–93. Amsterdam: Elsevier.

Credelle, T. L. & Spong, F. W. (1972). Thermoplastic media for holographic recording. *RCA Review*, **33**, 206–26.

Cunha, A. & Leith, E. (1988). One-way phase conjugation with partially coherent light and super-resolution. *Optics Letters*, **13**, 1105–7.

Curran, R. K. & Shankoff, T. A. (1970). The mechanism of hologram formation in dichromated gelatin. *Applied Optics*, **9**, 1651–7.

Cutrona, L. J. (1960). Optical data processing and filtering systems. *IEEE Transactions on Information Theory*, **IT-6**, 386–400.

Dainty, J. C. & Welford, W. T. (1971). Reduction of speckle in image plane hologram reconstruction by moving pupils. *Optics Communications*, **3**, 289–94.

Dallas, W. J. (1971a). Phase quantization – a compact derivation. *Applied Optics*, **10**, 673–4.

Dallas, W. J. (1971b). Phase quantization in holograms – a few illustrations. *Applied Optics*, **10**, 674–6.

Dallas, W. J. (1980). Computer-generated holograms. In *The Computer in Optical Research*, Topics in Applied Physics, vol. 41, ed. B. R. Frieden, pp. 291–366. Berlin: Springer-Verlag.

Dallas, W. J. & Lohmann, A. W. (1975). Deciphering vibration holograms. *Optics Communications*, **13**, 134–7.

Damman, H. (1970). Blazed synthetic phase-only holograms. *Optik*, **31**, 95–104.

Damman, H. & Gortler, K. (1971). High-efficiency multiple-imaging by means of multiple phase gratings. *Optics Communications*, **3**, 312–15.

Dändliker, R. (1980). Heterodyne holographic interferometry. In *Progress in Optics*, vol. 17, ed. E. Wolf, pp. 1–84. Amsterdam: North-Holland.

Dändliker, R., Ineichen, B. & Mottier, F. M. (1973). High resolution hologram interferometry by electronic phase measurement. *Optics Communications*, **9**, 412–16.

Dändliker, R., Thalmann, R. & Willemin, J.-F. (1982). Fringe interpolation by two-reference-beam holographic interferometry: reducing sensitivity to hologram misalignment. *Optics Communications*, **42**, 301–6.

D'Auria, L., Huignard, J. P., Slezak, C. & Spitz, E. (1974). Experimental holographic read-write memory using 3-D storage. *Applied Optics*, **13**, 808–18.

De Bitteto, D. J. (1966). White light viewing of surface holograms by simple dispersion compensation. *Applied Physics Letters*, **9**, 417–18.

De Bitteto, D. J. (1969). Holographic panoramic stereograms synthesized from white light recordings. *Applied Optics*, **8**, 1740–1.

De Bitteto, D. J. (1970). A front lighted 3-D holographic movie. *Applied Optics*, **9**, 498–9.

Deen, L. M., Walkup, J. F. & Hagler, M. O. (1975). Representations of space-variant optical systems using volume holograms. *Applied Optics*, **14**, 2438–46.

De Mattia, P. & Fossati-Bellani, V. (1978). Holographic contouring by displacing the object and the illumination beam. *Optics Communications*, **26**, 17–21.

Denisyuk, Yu. N. (1962). Photographic reconstruction of the optical properties of an object in its own scattered radiation field. *Soviet Physics – Doklady*, **7**, 543–5.

Denisyuk, Yu. N. (1963). On the reproduction of the optical properties of an object by the wave field of its scattered radiation. *Optics & Spectroscopy*, **15**, 279–84.

Denisyuk, Yu. N. (1965). On the reproduction of the optical properties of an object by the wave field of its scattered radiation. II. *Optics & Spectroscopy*, **18**, 152–7.

Der, V. K., Holloway, D. C. & Fourney, W. L. (1973). Four-exposure holographic moiré technique. *Applied Optics*, **12**, 2552–4.

Dickson, L. D. & Sincerbox, G. T. (1991). Holographic scanners for bar code readers. In *Optical Scanning*, ed. G. E. Marshall, pp. 159–211. New York: Marcel Dekker.

Dietrich, H. F., Raine, R. J. & O'Brien, R. N. (1976). A 5-minute monobath for Kodak 649 F plates used in holography and holographic interferometry. *Journal of Photographic Science*, **24**, 120–3.

Dörband, B. & Tiziani, H. J. (1985). Testing aspheric surfaces with computer generated holograms: analysis of adjustment and shape errors. *Applied Optics*, **24**, 2604–11.

Dubas, M. & Schumann, W. (1974). Sur la détermination holographique de l'état de deformation à la surface d'un corps non-transparent. *Optica Acta*, **21**, 547–62.

Dubas, M. & Schumann, W. (1975). On direct measurement of strain and rotation in holographic interferometry using the line of complete localization. *Optica Acta*, **22**, 807–19.

Dubas, M. & Schumann, W. (1977). Contribution a l'étude théorique des images et des franges produites par deux hologrammes en sandwich. *Optica Acta*, **24**, 1193–209.

Duncan, S. S., McQuoid, J. A. & McCartney, D. J. (1985). Holographic filters in dichromated gelatin position tuned over the near-infrared region. *Optical Engineering*, **24**, 781–5.

Duncan Jr., R. C. & Staebler, D. L. (1977). Inorganic photochromic materials. In *Holographic Recording Materials*, Topics in Applied Physics, vol. 20, ed. H. M. Smith, pp. 133–60. Berlin: Springer-Verlag.

Dunn, P. & Thompson, B. J. (1982). Object shape, fringe visibility, and resolution in far-field holography. *Optical Engineering*, **21**, 327–32.

Ebbeni, J., Coenen, J. & Hermanne, A. (1976). New analysis of holophotoelastic patterns and their application. *Journal of Strain Analysis*, **11**, 11–17.

Eismann, M. T., Tai, A. M. & Cederquist, J. N. (1989). Iterative design of a holographic beamformer. *Applied Optics*, **28**, 2641–50.

Ekberg, M., Larsson, M., Hård, S. & Nilsson, B. (1990). Multilevel phase holograms manufactured by electron-beam lithography. *Optics Letters*, **15**, 568–9.

Elias, P., Grey, D. S. & Robinson, D. Z. (1952). Fourier treatment of optical processes. *Journal of the Optical Society of America*, **42**, 127–34.

El-Sum, H. M. A. & Kirkpatrick, P. (1952). Microscopy by reconstructed wavefronts. *Physical Review*, **85**, 763.

Enger, R. C. & Case, S. K. (1983). High-frequency holographic transmission gratings in photoresist. *Journal of the Optical Society of America*, **73**, 1113–18.

Ennos, A. E. (1968). Measurement of in-plane surface strain by hologram interferometry. *Journal of Physics E: Scientific Instruments*, **1**, 731–4.

Ennos, A. E., Robinson, D. W. & Williams, D. C. (1985). Automatic fringe analysis in holographic interferometry. *Optica Acta*, **32**, 135–45.

Erf, R. K. (1974). *Holographic Non-destructive Testing*. New York: Academic Press.

Ersoy, O. K., Zhuang, J. Y. & Brede, J. (1992). Iterative interlacing approach for synthesis of computer-generated holograms. *Applied Optics*, **31**, 6894–901.

Eschbach, R. (1991). Comparison of error diffusion methods for computer-generated holograms. *Applied Optics*, **30**, 3702–10.

Eschler, H. (1975). Multifrequency acousto-optic page composers for holographic data storage. *Optics Communications*, **13**, 148–53.

Ewan, B. C. R. (1979). Particle velocity distribution measurement by holography. *Applied Optics*, **18**, 3156–60.

Fagan, W. F., ed. (1990). *Optical Security & Anticounterfeiting Systems*, Proceedings of the SPIE, vol. 1210. Bellingham: SPIE.

Fainman, Y., Lenz, E. & Shamir, J. (1981). Contouring by phase conjugation. *Applied Optics*, **20**, 158–63.

Fairchild, R. C. & Fienup, J. R. (1982). Computer-originated aspheric holographic optical elements. *Optical Engineering*, **21**, 133–40.

Faklis, D. & Morris, G. M. (1989). Broadband imaging with holographic lenses. *Optical Engineering*, **28**, 592–8.

Farmer, W. J., Benton, S. A. & Klug, M. A. (1991). The application of the edge-lit format to holographic stereograms. In *Practical Holography V*, Proceedings of the SPIE, vol. 1461, ed. S. A. Benton, pp. 215–16. Bellingham: SPIE.

Faulde, M., Fercher, A. F., Torge, R. & Wilson, R. N. (1973). Optical testing by means of synthetic holograms and partial lens compensation. *Optics Communications*, **7**, 363–5.

Favre, H. (1929). Sur une nouvelle methode optique de détermination de tensions intérieures. *Revue d'Optique*, **8**, 193–213, 241–61, 289–307.

Feldman, M. R. & Guest, C. C. (1987). Computer generated holographic optical elements for optical interconnection of very large scale integrated circuits. *Applied Optics*, **26**, 4377–84.

Fercher, A. F. (1976). Computer generated holograms for testing optical elements: error analysis and error compensation. *Optica Acta*, **23**, 347–65.

Flanders, D. C. (1983). Submicrometer periodicity gratings as artificial anisotropic dielectrics. *Applied Physics Letters*, **42**, 492–4.

Fimia, A., Beléndez, A. & Pascual, I. (1991). Silver halide (sensitized) gelatin in

Agfa-Gevaert plates: the optimized procedure. *Journal of Modern Optics*, **38**, 2043–51.

Fimia, A., Pascual, I. & Beléndez, A. (1992). Optimized spatial frequency response in silver halide sensitized gelatin. *Applied Optics*, **31**, 4625–7.

Fisher, R. L. (1989). Design methods for a holographic head-up display curved combiner. *Optical Engineering*, **28**, 616–21.

Forshaw, M. R. B. (1973). The imaging properties and aberrations of thick transmission holograms. *Optica Acta*, **20**, 669–86.

Forshaw, M. R. B. (1974). Diffraction of a narrow laser beam by a thick hologram: experimental results. *Optics Communications*, **12**, 279–81.

Forshaw, M. R. B. (1975). Explanation of the diffraction fine structure in overexposed thick holograms. *Optics Communications*, **15**, 218–21.

Fourney, M. E. (1968). Application of holography to photoelasticity. *Experimental Mechanics*, **8**, 33–8.

Fourney, M. E. & Mate, K. V. (1970). Further applications of holography to photoelasticity. *Experimental Mechanics*, **10**, 177–86.

Fournier, J.-M., Tribillon, G. & Viénot, J.-C. (1977). Recording of large size holograms in photographic emulsion: image reconstruction. In *Three-Dimensional Imaging*, Proceedings of the SPIE, vol. 120, ed. S. A. Benton, pp. 116–21. Redondo Beach: SPIE.

Frejlich, J., Kamshilin, A. A., Kulikov, V. V. & Mokrushina, E. V. (1989). Adaptive holographic vibrometry using photorefractive crystals. *Optics Communications*, **70**, 82–6.

Frère, C., Leseberg, D. & Bryngdahl, O. (1986). Computer-generated holograms of three-dimensional objects composed of line segments. *Journal of the Optical Society of America. A*, **3**, 726–30.

Friesem, A. A. & Federowicz, R. J. (1966). Recent advances in multicolor wavefront reconstruction. *Applied Optics*, **5**, 1085–6.

Friesem, A. A. & Federowicz, R. J. (1967). Multicolor wavefront reconstruction. *Applied Optics*, **6**, 529–36.

Friesem, A. A. & Levy, U. (1976). Fringe formation in two-wavelength contour holography. *Applied Optics*, **15**, 3009–20.

Friesem, A. A. & Walker, J. L. (1970). Thick absorption recording media in holography. *Applied Optics*, **9**, 201–14.

Friesem, A. A. & Zelenka, J. S. (1967). Effects of film non-linearities in holography. *Applied Optics*, **6**, 1755–9.

Funnell, W. R. J. (1981). Image processing applied to the interactive analysis of interferometric fringes. *Applied Optics*, **20**, 3245–50.

Füzessy, Z. & Gyimesi, F. (1984). Difference holographic interferometry: displacement measurement. *Optical Engineering*, **23**, 780–3.

Gabor, D. (1948). A new microscopic principle. *Nature*, **161**, 777–8.

Gabor, D. (1949). Microscopy by reconstructed wavefronts. *Proceedings of the Royal Society A*, **197**, 454–87.

Gabor, D. (1951). Microscopy by reconstructed wavefronts. II. *Proceedings of the Physical Society (Lond.) B*, **64**, 449–69.

Gabor, D. (1969). Associative holographic memories. *IBM Journal of Research & Development*, **13**, 156–9.

Gabor, D. (1970). Laser speckle and its elimination. *IBM Journal of Research & Development*, **14**, 509–14.

Gale, M. T. & Knop, K. (1976). Colour-encoded focused image holograms. *Applied Optics*, **15**, 2189–98.

Gale, M. T., Knop, K. & Russell, J. P. (1975). A colour micro-storage and display system using focused image holograms. *Optics & Laser Technology*, **7**, 234–6.

Gara, A. D., Majkowski, R. F. & Stapleton, T. T. (1973). Holographic system for automatic surface mapping. *Applied Optics*, **12**, 2172–9.

Gåsvik, K. (1975). Holographic reconstruction of the state of polarization. *Optica Acta*, **22**, 189–206.

Gates, J. W. C. (1968). Holographic phase recording by interference between reconstructed wavefronts from separate holograms. *Nature*, **220**, 473–4.

Gates, J. W. C., Hall, R. G. N. & Ross, I. N. (1970). Holographic recording using frequency-doubled radiation at 530 nm. *Journal of Physics E: Scientific Instruments*, **3**, 89–94.

Gates, J. W. C., Hall, R. G. N. & Ross, I. N. (1972). Holographic interferometry of impact-loaded objects using a double-pulse laser. *Optics & Laser Technology*, **4**, 72–5.

Gaylord, T. K. & Moharam, M. G. (1982). Planar dielectric grating diffraction theories. *Applied Physics. B*, **28**, 1–14.

Gaylord, T. K. & Moharam, M. G. (1985). Analysis and applications of optical diffraction by gratings. *Proceedings of the IEEE*, **73**, 894–937.

George, N. (1970). Full view holograms. *Optics Communications*, **1**, 457–9.

George, N. & Jain, A. (1973). Speckle reduction using multiple tones of illumination. *Applied Optics*, **12**, 1202–12.

Georgekutty, T. G. & Liu, H. K. (1987). Simplified dichromated gelatin hologram recording process. *Applied Optics*, **26**, 372–6.

Gerasimova, S. A. & Zakharchenko, V. M. (1981). Holographic processor for associative information retrieval. *Soviet Journal of Optical Technology*, **48**, 404–6.

Gerritsen, H. J., Hannan, W. J. & Ramberg, E. G. (1968). Elimination of speckle noise in holograms with redundancy. *Applied Optics*, **7**, 2301–11.

Ghandeharian, H. & Boerner, W. M. (1977). Autocorrelation of transmittance of holograms made of diffuse objects. *Optica Acta*, **24**, 1087–97.

Ghandeharian, H. & Boerner, W. M. (1978). Degradation of holographic images due to depolarization of reflected light. *Journal of the Optical Society of America*, **68**, 931–4.

Ghiglia, D. G., Mastin, G. A. & Romero, L. A. (1987). Cellular-automata method for phase unwrapping. *Journal of the Optical Society of America. A.*, **4**, 267–80.

Glaser, I. (1973). Anamorphic imagery in holographic stereograms. *Optics Communications*, **7**, 323–6.

Glaser, I. & Friesem, A. A. (1977). Imaging properties of holographic stereograms. In *Three-Dimensional Imaging*, Proceedings of the SPIE, vol. 120, ed. S. A. Benton, pp. 150–62. Redondo Beach: SPIE.

Glass, A. M. (1978). The photorefractive effect. *Optical Engineering*, **17**, 470–9.

Golbach, H. (1973). Reduction of speckle in holographic reflected-light microscopy. *Optik*, **37**, 45–9.

Goldberg, J. L. (1975). A holographic interferometer for the measurement of the vector displacement of a slowly deforming rough surface. *Japanese Journal of Applied Physics*, **14** (Supplement 14-1), 253–8.

Goodman, J. W. (1967). Film grain noise in wavefront reconstruction imaging. *Journal of the Optical Society of America*, **57**, 493–502.

Goodman, J. W. (1968). *Introduction to Fourier Optics*. New York: McGraw-Hill.

Goodman, J. W. (1975). Statistical properties of laser speckle patterns. In *Laser Speckle & Related Phenomena*, Topics in Applied Physics, vol. 9, ed. J. C. Dainty, pp. 9–75. Berlin: Springer-Verlag.

Goodman, J. W. (1981). Linear space-variant optical data processing. In *Optical Information Processing: Fundamentals*, Topics in Applied Physics, vol. 48, ed. S. H. Lee, pp. 235–60. Berlin: Springer-Verlag.

Goodman, J. W., Huntley, W. H., Jackson, D. W. & Lehmann, M. (1966). Wavefront reconstruction imaging through random media. *Applied Physics Letters*, **8**, 311–13.

Goodman, J. W. & Knight, G. R. (1968). Effects of film nonlinearities on wavefront-reconstruction images of diffuse objects. *Journal of the Optical Society of America*, **58**, 1276–83.

Goodman, J. W., Leonberger, F. J., Kung, S. Y. & Athale, R. (1984). Optical interconnections for VLSI systems. *Proceedings of the IEEE*, **72**, 850–66.

Goodman, J. W. & Silvestri, A. M. (1970). Some effects of Fourier domain phase quantization. *IBM Journal of Research & Development*, **14**, 478–84.

Graube, A. (1973). Holograms recorded with red light in dye sensitized dichromated gelatin. *Optics Communications*, **8**, 251–3.

Graube, A. (1974). Advances in bleaching methods for photographically recorded holograms. *Applied Optics*, **13**, 2942–6.

Graver, W. R., Gladden, J. W. & Estes, J. W. (1980). Phase holograms formed by silver halide (sensitized) gelatin processing. *Applied Optics*, **19**, 1529–36.

Greenaway, D. L. (1980). Cards and card-readers for voucher and access control systems. *Landis & Gyr Review*, **27**, 20–5.

Greguss, P. (1975). *Holography in Medicine*. London: IPC Press.

Greguss, P. (1976). Holographic interferometry in biomedical sciences. *Optics & Laser Technology*, **8**, 153–9.

Groh, G. (1968). Multiple imaging by means of point holograms. *Applied Optics*, **7**, 1643–4.

Groh, G. & Kock, M. (1970). 3-D display of X-ray images by means of holography. *Applied Physics*, **9**, 775–7.

Gupta, P. C. & Singh, K. (1975a). Characteristic fringe function for time-average holography of periodic nonsinusoidal vibrations. *Applied Optics*, **14**, 129–33.

Gupta, P. C. & Singh, K. (1975b). Time-average hologram interferometry of periodic, non-cosinusoidal vibrations. *Applied Physics*, **6**, 233–40.

Gupta, P. C. & Singh, K. (1976). Hologram interferometry of vibrations represented by the square of a Jacobian elliptic function. *Nouvelle Revue d'Optique*, **7**, 95–100.

Guther, R. & Kusch, S. (1974). Ein beitrag zum Intermodulationsrauschen in der Volumenholographie. *Experimentelle Technik der Physik*, **22**, 119–41.

Gyimesi, F. & Füzessy, Z. (1988). Difference holographic interferometry: theory. *Journal of Modern Optics*, **35**, 1699–716.

Haig, N. D. (1973). Three dimensional holograms by rotational multiplexing of two-dimensional films. *Applied Optics*, **12**, 419–20.

Haines, K. A. & Hildebrand, B. P. (1965). Contour generation by wavefront reconstruction. *Physics Letters*, **19**, 10–11.

Haines, K. A. & Hildebrand, B. P. (1966). Surface-deformation measurement using the wavefront reconstruction technique. *Applied Optics*, **5**, 595–602.

Halle, M. W., Benton, S. A., Klug, M. A. & Underkoffler, J. S. (1991). The Ultragram: a generalized holographic stereogram. In *Practical Holography V*, Proceedings of the SPIE, vol. 1461, ed. S. A. Benton, pp. 142–55. Bellingham: SPIE.

Hamasaki, J. (1968). Signal-to-noise ratios for hologram images of subjects in strong incoherent light. *Applied Optics*, **7**, 1613–20.

Han, C.-Y., Ishii, Y. & Murata, K. (1983). Reshaping collimated laser beams with Gaussian profile to uniform profiles. *Applied Optics*, **22**, 3644–7.

Hannan, W. J., Flory, R. E., Lurie, M. & Ryan, R. J. (1973). Holotape: a low-cost prerecorded television system using holographic storage. *Journal of the Society of Motion Picture and Television Engineers*, **82**, 905–15.

Haridas, P., Hafen, E., Pless, I., Harton, J., Dixit, S., Goloskie, D. & Benton, S. (1985). Detection of short-lived particles using holography. *Optical Engineering*, **24**, 741–5.

Hariharan, P. (1971). Reversal processing technique for phase holograms. *Optics Communications*, **3**, 119–21.

Hariharan, P. (1972). Bleached reflection holograms. *Optics Communications*, **6**, 377–9.

Hariharan, P. (1973). Application of holographic subtraction to time-average hologram interferometry of vibrating objects. *Applied Optics*, **12**, 143–6.

Hariharan, P. (1976a). Longitudinal distortion in images reconstructed by reflection holograms. *Optics Communications*, **17**, 52–4.

Hariharan, P. (1976b). Comment on: Sensitivity improvement by step-biasing in holographic interferometry. *Optical Engineering*, **15**, 279.

Hariharan, P. (1977a). Hologram interferometry: identification of the sign of surface displacements. *Optica Acta*, **24**, 989–90.

Hariharan, P. (1977b). Simple full-view rainbow holograms. *Optical Engineering*, **16**, 520–2.

Hariharan, P. (1978). Hologram recording geometry: its influence on image luminance. *Optica Acta*, **25**, 527–30.

Hariharan, P. (1979a). Intermodulation noise in amplitude holograms: the effect of hologram thickness. *Optica Acta*, **26**, 211–15.

Hariharan, P. (1979b). Volume phase reflection holograms: the effect of hologram thickness on image luminance. *Optica Acta*, **26**, 1443–7.

Hariharan, P. (1980a). Improved techniques for multicolour reflection holograms. *Journal of Optics (Paris)*, **11**, 53–5.

Hariharan, P. (1980b). Holographic recording materials: recent developments. *Optical Engineering*, **19**, 636–41.

Hariharan, P. (1980c). Pseudocolour images with volume reflection holograms. *Optics Communications*, **35**, 42–4.

Hariharan, P. (1982). Concentric etalon for single-frequency operation of high-power ion lasers. *Optics Letters*, **7**, 274–5.

Hariharan, P. (1983). Colour holography. In *Progress in Optics*, vol. 20, ed. E. Wolf, pp. 265–324. Amsterdam: North-Holland.

Hariharan, P. (1986*a*). Bleached photographic phase holograms: the influence of drying procedures on diffraction efficiency. *Optics Communications*, **56**, 318–20.

Hariharan, P. (1986*b*). Silver-halide sensitized gelatin holograms: mechanism of hologram formation. *Applied Optics*, **25**, 2040–2.

Hariharan, P. (1990*a*). Basic processes involved in the production of bleached holograms. *Journal of Photographic Science*, **38**, 76–81.

Hariharan, P. (1990*b*). Rehalogenating bleaches for photographic phase holograms. 3: Mechanism of material transfer. *Applied Optics*, **29**, 2983–5.

Hariharan, P. (1991). Strain analysis: towards a synthesis of computational and experimental methods. In *Proceedings of the 10th Invitational UFEM Symposium*, ed. R. J. Pryputniewicz, pp. 471–90. Worcester: Worcester Polytechnic Institute.

Hariharan, P. & Chidley, C. M. (1987*a*). Photographic phase holograms: the influence of developer composition on scattering and diffraction efficiency. *Applied Optics*, **26**, 1230–4.

Hariharan, P. & Chidley, C. M. (1987*b*). Rehalogenating bleaches for photographic phase holograms: the influence of halide type and concentration on diffraction efficiency and scattering. *Applied Optics*, **26**, 3895–8.

Hariharan, P. & Chidley, C. M. (1988*a*). Photographic phase holograms: spatial frequency effects with conventional and reversal bleaches. *Applied Optics*, **27**, 3065–7.

Hariharan, P. & Chidley, C. M. (1988*b*). Rehalogenating bleaches for photographic phase holograms. 2: Spatial frequency effects. *Applied Optics*, **27**, 3852–4.

Hariharan, P. & Chidley, C. M. (1989). Bleached reflection holograms: a study of color shifts due to processing. *Applied Optics*, **28**, 422–4.

Hariharan, P. & Hegedus, Z. S. (1973). Simple multiplexing technique for double-exposure hologram interferometry. *Optics Communications*, **9**, 152–5.

Hariharan, P. & Hegedus, Z. S. (1974*a*). Reduction of speckle in coherent imaging by spatial frequency sampling. *Optica Acta*, **21**, 345–56.

Hariharan, P. & Hegedus, Z. S. (1974*b*). Reduction of speckle in coherent imaging by spatial frequency sampling. II. Random spatial frequency sampling. *Optica Acta*, **21**, 683–95.

Hariharan, P. & Hegedus, Z. S. (1975*a*). Four-exposure hologram moiré interferometry and speckle-pattern interferometry: a comparison. *Applied Optics*, **14**, 22–3.

Hariharan, P. & Hegedus, Z. S. (1975*b*). Relative phase shift of images reconstructed by phase and amplitude holograms. *Applied Optics*, **14**, 273–4.

Hariharan, P. & Hegedus, Z. S. (1976). Two-hologram interferometry: a simplified sandwich technique. *Applied Optics*, **15**, 848–9.

Hariharan, P., Hegedus, Z. S. & Steel, W. H. (1979). One-step multicolour rainbow holograms with wide angle of view. *Optica Acta*, **26**, 289–91.

Hariharan, P., Kaushik, G. S. & Ramanathan, C. S. (1972). Reduction of scattering in photographic phase holograms. *Optics Communications*, **5**, 59–61.

Hariharan, P., Oreb, B. F. & Brown, N. (1982). A digital phase-measurement system for real-time holographic interferometry. *Optics Communications*, **41**, 393–6.

Hariharan, P., Oreb, B. F. & Brown, N. (1983). Real-time holographic interferometry:

a microcomputer system for the measurement of vector displacements. *Applied Optics*, **22**, 876–80.

Hariharan, P. & Oreb, B. F. (1984). Two-index holographic contouring: application of digital techniques. *Optics Communications*, **51**, 142–4.

Hariharan, P. & Oreb, B. F. (1986). Stroboscopic holographic interferometry: application of digital techniques. *Optics Communications*, **59**, 83–6.

Hariharan, P., Oreb, B. F. & Eiju, T. (1987). Digital phase-shifting interferometry: a simple error-compensating phase calculation algorithm. *Applied Optics*, **26**, 2504–6.

Hariharan, P., Oreb, B. F. & Freund, C. H. (1987). Stroboscopic holographic interferometry: measurements of vector components of a vibration. *Applied Optics*, **26**, 3899–903.

Hariharan, P. & Ramanathan, C. S. (1971). Suppression of printout effect in photographic phase holograms. *Applied Optics*, **10**, 2197–9.

Hariharan, P., Ramanathan, C. S. & Kaushik, G. S. (1971). Simplified processing technique for photographic phase holograms. *Optics Communications*, **3**, 246–7.

Hariharan, P., Ramanathan, C. S. & Kaushik, G. S. (1973). Monobath processing for holography. *Applied Optics*, **12**, 611–12.

Hariharan, P. & Ramprasad, B. S. (1972). Simplified optical system for holographic subtraction. *Journal of Physics E: Scientific Instruments*, **5**, 967–8.

Hariharan, P. & Ramprasad, B. S. (1973a). Wavefront tilter for double-exposure holographic interferometry. *Journal of Physics E: Scientific Instruments*, **6**, 173–5.

Hariharan, P. & Ramprasad, B. S. (1973b). Rapid *in-situ* processing for real-time holographic interferometry. *Journal of Physics E: Scientific Instruments*, **6**, 699–701.

Hariharan, P. & Sen, D. (1961). Radial shearing interferometer. *Journal of Scientific Instruments*, **38**, 428–32.

Hariharan, P., Steel, W. H. & Hegedus, Z. S. (1977). Multicolor holographic imaging with a white light source. *Optics Letters*, **1**, 8–9.

Harris Jr., F. S., Sherman, G. C. & Billings, B. H. (1966). Copying holograms. *Applied Optics*, **5**, 665–6.

Hart, S., Mendes, G., Bazargan, K. & Xu, S. (1988). Deep-red holography using a junction laser and silver-halide holographic emulsion. *Optics Letters*, **13**, 955–7.

Hasegawa, S., Yamagishi, F., Ikeda, H. & Inagaki, T. (1989). Straight-line scanning analysis of an all holographic scanner. *Applied Optics*, **28**, 5317–25.

Haskell, R. E. (1973). Computer-generated binary holograms with minimum quantization errors. *Journal of the Optical Society of America*, **63**, 504.

Haskell, R. E. & Culver, B. C. (1972). New coding technique for computer-generated holograms. *Applied Optics*, **11**, 2712–14.

Hatakoshi, G. & Goto, K. (1985). Grating lenses for the semiconductor laser wavelength. *Applied Optics*, **24**, 4307–11.

Hauck, R. & Bryngdahl, O. (1984). Computer-generated holograms with pulse-density modulation. *Journal of the Optical Society of America. A*, **1**, 5–10.

Heaton, J. M., Mills, P. A., Paige, E. G. S., Solymar, L. & Wilson, T. (1984). Diffraction efficiency and angular selectivity of volume phase holograms recorded in photorefractive materials. *Optica Acta*, **31**, 885–901.

Heaton, J. M. & Solymar, L. (1985a). Wavelength and angular selectivity of high diffraction efficiency reflection holograms in silver halide photographic emulsion. *Applied Optics*, **24**, 2931–6.

Heaton, J. M. & Solymar, L. (1985*b*). Transient energy transfer during hologram formation in photorefractive crystals. *Optica Acta*, **32**, 397–408.

Heflinger. L. O., Stewart, G. L. & Booth, C. R. (1978). Holographic motion pictures of microscopic plankton. *Applied Optics*, **17**, 951–4.

Henshaw, P. D. & Ezekiel, S. (1974). High resolution holographic contour generation with white light reconstruction. *Optics Communications*, **12**, 39–42.

Hercher, M. (1969). Tunable single mode operation of gas lasers using intracavity tilted etalons. *Applied Optics*, **8**, 1103–6.

Hermann, J. P., Herriau, J. P. & Huignard, J. P. (1981). Nanosecond four-wave mixing and holography in BSO crystals. *Applied Optics*, **20**, 2173–5.

Herriau, J. P. & Huignard, J. P. (1986). Hologram fixing process at room temperature in photorefractive $Bi_{12}SiO_{20}$ crystals. *Applied Physics Letters*, **49**, 1140–2.

Herriau, J. P., Huignard, J. P. & Aubourg, P. (1978). Some polarization properties of volume holograms in $Bi_{12}SiO_{20}$ crystals and applications. *Applied Optics*, **17**, 1851–2.

Herzig, H. P. & Dändliker, R. (1988). Holographic optical scanning elements with minimum aberrations. *Applied Optics*, **27**, 4739–46.

Herzig, H. P., Ehbets, P., Prongué, D. & Dändliker, R. (1992). Fan-out elements recorded as volume holograms: optimized recording conditions. *Applied Optics*, **31**, 5716–23.

Hesselink, L. & Bashaw, M. C. (1993). Optical memories implemented with photorefractive media. *Optical & Quantum Electronics*, **25**, S611–61.

Hildebrand, B. P. & Haines, K. A. (1966). The generation of 3-dimensional contour maps by wavefront reconstruction. *Physics Letters*, **21**, 422–3.

Hildebrand, B. P. & Haines, K. A. (1967). Multiple-wavelength and multiple-source holography applied to contour generation. *Journal of the Optical Society of America*, **57**, 155–62.

Hioki, R. & Suzuki, T. (1965). Reconstruction of wavefronts in all directions. *Japanese Journal of Applied Physics*, **4**, 816.

Holloway, D. C. & Johnson, R. H. (1971). Advancements in holographic photoelasticity. *Experimental Mechanics*, **11**, 57–63.

Honda, T., Okada, K. & Tsujiuchi, J. (1981). 3-D distortion of observed images reconstructed from a cylindrical holographic stereogram: (1) laser light reconstruction type. *Optics Communications*, **36**, 11–16.

Hopf, F. A. (1980). Interferometry using conjugate-wave generation. *Journal of the Optical Society of America*, **70**, 1320–3.

Hopkins, H. H. (1950). *Wave Theory of Aberrations*. Oxford: The Clarendon Press.

Horner, J. L., ed. (1984). Special Issue on Optical Pattern Recognition. *Optical Engineering*, **23**, 687–747.

Horwitz, C. M. (1974). A new solar selective surface. *Optics Communications*, **11**, 210–12.

Hovanesian, J. D. (1974). Variable isochromatic/isopachic fringe visibility. *Experimental Mechanics*, **14**, 233–6.

Hovanesian, J. D., Brcic, V. & Powell, R. L. (1968). A new stress-optic method: stress-holo interferometry. *Experimental Mechanics*, **8**, 362–8.

Hubel, P. & King, M. A. (1992). Color holography using multiple layers of Du Pont photopolymer. In *Practical Holography V*, Proceedings of the SPIE, vol. 1667, ed. S. A. Benton, pp. 215–24. Bellingham: SPIE.

Hubel, P. M. & Solymar, L. (1991). Color-reflection holography: theory and experiment. *Applied Optics*, **30**, 4190–203.

Huff, L. & Fusek, R. L. (1980). Color holographic stereograms. *Optical Engineering*, **19**, 691–5.

Huff, L. & Fusek, R. L. (1981). Optical techniques for increasing image width in cylindrical holographic stereograms. *Optical Engineering*, **20**, 241–5.

Huignard, J. P. (1981). Phase conjugation, real time holography and degenerate four-wave mixing in photoreactive BSO crystals. In *Current Trends in Optics*, ed. F. T. Arecchi and F. R. Aussenegg, pp. 150–60. London: Taylor & Francis.

Huignard, J. P., Herriau, J. P., Aubourg, P. & Spitz, E. (1979). Phase-conjugate wavefront generation via real-time holography in $Bi_{12}SiO_{20}$ crystals. *Optics Letters*, **4**, 21–3.

Huignard, J. P., Herriau, J. P., Pichon, L. & Marrakchi, A. (1980). Speckle-free imaging in four-wave mixing experiments with $Bi_{12}SiO_{20}$ crystals. *Optics Letters*, **5**, 436–7.

Huignard, J. P., Herriau, J. P., Rivet, G. & Günter, P. (1980). Phase conjugation and spatial frequency dependence of wavefront reflectivity in $Bi_{12}SiO_{20}$ crystals. *Optics Letters*, **5**, 102–4.

Huignard, J. P., Herriau, J. P. & Valentin, T. (1977). Time average holographic interferometry with photoconductive electroptic $Bi_{12}SiO_{20}$ crystals. *Applied Optics*, **16**, 2796–8.

Huignard, J. P. & Marrakchi, A. (1981). Two-wave mixing and energy transfer in $Bi_{12}SiO_{20}$ crystals: application to image amplification and vibration analysis. *Optics Letters*, **6**, 622–4.

Huignard, J. P. & Micheron, F. (1976). High sensitivity read-write volume holographic storage in $Bi_{12}SiO_{20}$ and $Bi_{12}GeO_{20}$ crystals. *Applied Physics Letters*, **29**, 591–3.

Hung, Y. Y., Hu, C. P., Henley, D. R. & Taylor, C. E. (1973). Two improved methods of surface displacement measurements by holographic interferometry. *Optics Communications*, **8**, 48–51.

Hutley, M. C. (1975). Blazed interference diffraction gratings for the ultraviolet. *Optica Acta*, **22**, 1–13.

Hutley, M. C. (1976). Interference (holographic) diffraction gratings. *Journal of Physics E: Scientific Instruments*, **9**, 513–20.

Hutley, M. C. (1982). *Diffraction Gratings*. London: Academic Press.

Ichioka, Y. & Lohmann, A. W. (1972). Interference testing of large optical components with circular computer holograms. *Applied Optics*, **11**, 2597–602.

Ih, C. S. (1975). Multicolor imagery from holograms by spatial filtering. *Applied Optics*, **14**, 438–44.

Ih, C. S. (1977). Holographic laser beam scanners utilizing an auxiliary reflector. *Applied Optics*, **16**, 2137–46.

Ih, C. S. & Baxter, L. A. (1978). Improved random spatial phase modulation for speckle elimination. *Applied Optics*, **17**, 1447–54.

Ineichen, B., Liegeois, C. & Meyrueis, P. (1982). Thermoplastic film camera for holographic recording of extended objects in industrial applications. *Applied Optics*, **21**, 2209–14.

Ingwall, R. T. & Fielding, H. L. (1985). Hologram recording with a new photopolymer system. *Optical Engineering*, **24**, 808–12.

Ingwall, R. T. & Troll, M. (1989). Mechanism of hologram formation in DMP-128 photopolymer. *Optical Engineering*, **28**, 586–91.

Ingwall, R. T., Troll, M. & Vetterling, W. T. (1987). Properties of reflection holograms recorded in Polaroid's DMP-128 photopolymer. In *Practical Holography II*, Proceedings of the SPIE, vol. 747, ed. T. H. Jeong, pp. 67–73. Bellingham: SPIE.

Iwata, F. & Ohnuma, K. (1985). Brightness and contrast of a surface relief rainbow hologram for an embossing master. In *Applications of Holography*, Proceedings of the SPIE, vol. 523, ed. L. Huff, pp. 15–17. Bellingham: SPIE.

Iwata, F. & Tsujiuchi, J. (1974). Characteristics of a photoresist hologram and its replica. *Applied Optics*, **13**, 1327–36.

Ja, Y. H. (1982). Observation of interference between a signal and its conjugate in a four-wave mixing experiment using $Bi_{12}GeO_{20}$ crystals. *Optical & Quantum Electronics*, **14**, 367–9.

Jacobson, A. D. & McClung, F. J. (1965). Holograms produced with pulsed laser illumination. *Applied Optics*, **4**, 1509–10.

Jahoda, F. C., Jeffries, R. A. & Sawyer, G. A. (1967). Fractional fringe holographic plasma interferometry. *Applied Optics*, **6**, 1407–10.

Jannson, T. (1974). Impulse response and Shannon number of holographic optical systems. *Optics Communications*, **10**, 232–7.

Jannson, T., Savant, G. & Qiao, Y. (1989). Bragg holographic structures for XUV applications: a new approach. *Optics Letters*, **14**, 344–6.

Janta, J. & Miler, M. (1972). Time-average holographic interferometry of damped oscillations. *Optik*, **36**, 185–95.

Javidi, B. & Horner, J. L. (1994). Optical pattern recognition system for validation and security verification. *Optical Engineering*, **33**, 1752–6.

Jenkins, B. K., Sawchuk, A. A., Strand, T. C., Forchheimer, R. & Soffer, B. H. (1984). Sequential optical logic implementation. *Applied Optics*, **23**, 3455–64.

Jeong, M. H., Song, J. B. & Lee, I. W. (1991). Simplified processing method of dichromated gelatin holographic recording material. *Applied Optics*, **30**, 4172–3.

Jeong, T. H. (1967). Cylindrical holography and some proposed applications. *Journal of the Optical Society of America*, **57**, 1396–8.

Jeong, T. H., Rudolf, P. & Luckett, A. (1966). 360° holography. *Journal of the Optical Society of America*, **56**, 1263–4.

Jo, J. C. & Lee, S. S. (1982). Holographic image restoration by using an unconstrained single deblurring filter. *Optica Acta*, **29**, 1231–6.

Johnson, K. M., Hesselink, L. & Goodman, J. W. (1984). Holographic reciprocity law failure. *Applied Optics*, **23**, 218–27.

Johnson, R. V. & Tanguay, A. R. (1988). Stratified volume holographic optical elements. *Optics Letters*, **13**, 189–91.

Joly, L. (1983). Grain growth during rehalogenating bleaching. *Journal of Photographic Science*, **31**, 143–7.

Joly, L. & Vanhorebeek, R. (1980). Development effects in white-light reflection holography. *Photographic Science & Engineering*, **24**, 108–13.

Jones, M. I., Walkup, I. F. & Hagler, M. O. (1982). Multiplex hologram representations of space-variant optical systems using ground-glass encoded reference beams. *Applied Optics*, **21**, 1291–7.

Jordan, M. P. & Solymar, L. (1978). A note on volume holograms. *Electronics Letters*, **14**, 271–2.

Judge, Th. R., Quan, Ch. & Bryanston-Cross, P. J. (1992). Holographic deformation measurements by Fourier transform techniques with automatic phase unwrapping. *Optical Engineering*, **31**, 533–43.

Kakichashvili, Sh. D. (1972). Principles of recording polarization holograms. *Optics & Spectroscopy*, **33**, 324–7.

Kalestynski, A. (1973). Holographic multiplication in one exposure by the use of a multibeam reference field. *Applied Optics*, **12**, 1946–50.

Kalestynski, A. (1976). Multiplying lensless Fourier holograms recorded using a multibeam reference light field. *Applied Optics*, **15**, 853–5.

Kamshilin, A. A., Frejlich, J. & Cescato, L. (1986). Photorefractive crystals for stabilization of holographic setup. *Applied Optics*, **25**, 2376–81.

Kamshilin, A. A. & Miteva, M. G. (1981). Effect of infra-red irradiation on holographic recording in bismuth silicon oxide. *Optics Communications*, **36**, 429–33.

Kamshilin, A. A. & Mokrushina, E. V. (1986). Possible use of photorefractive crystals in holographic vibrometry. *Soviet Technical Physics Letters*, **12**, 149–51.

Kamshilin, A. A. & Mokrushina, E. V. (1989). Polarization discrimination of a signal after self-diffraction in an external field in an $Bi_{12}TiO_{20}$ crystal. *Soviet Physics Technical Physics*, **34**, 493–4.

Kamshilin, A. A., Mokrushina, E. V. & Petrov, M. P. (1989). Adaptive holographic interferometers operating through self-diffraction of recording beams in photorefractive crystals. *Optical Engineering*, **28**, 580–5.

Kamshilin, A. A. & Petrov, M. P. (1985). Continuous reconstruction of holographic interferograms through anisotropic diffraction in photorefractive crystals. *Optics Communications*, **53**, 23–6.

Kasahara, T., Kimura, Y., Hioki, R. & Tanaka, S. (1969). Stereoradiography using holographic techniques. *Japanese Journal of Applied Physics*, **8**, 124–5.

Kaspar, F. G. (1973). Diffraction by thick periodically stratified gratings with complex dielectric constant. *Journal of the Optical Society of America*, **63**, 37–45.

Kaspar, F. G. (1974). Computation of light transmitted by a thick grating for application to contact printing. *Journal of the Optical Society of America*, **64**, 1623–30.

Kato, M. & Okino, Y. (1973). Speckle reduction by double recorded holograms. *Applied Optics*, **12**, 1199–201.

Kermisch, D. (1969). Nonuniform sinusoidally modulated dielectric gratings. *Journal of the Optical Society of America*, **59**, 1409–14.

Kermisch, D. (1970). Image reconstruction from phase information only. *Journal of the Optical Society of America*, **60**, 15–17.

Kermisch, D. (1971). Efficiency of photochromic gratings. *Journal of the Optical Society of America*, **61**, 1202–6.

Kerr, D., Mendoza-Santoyo, F. & Tyrer, J. R. (1990). Extraction of phase data from electronic speckle pattern interferometric fringes using a single-phase-step method: a novel approach. *Journal of the Optical Society of America. A*, **7**, 820–6.

Kessler, S. & Kowarschik, R. (1975). Diffraction efficiency of volume holograms. Part 1. Transmission holograms. *Optical & Quantum Electronics*, **7**, 1–14.

Kiemle, U. (1974). Considerations on holographic memories in the gigabyte region. *Applied Optics*, **13**, 803–7.

Killat, U. (1977*a*). Coupled wave theory of hologram gratings with arbitrary attenuation. *Optics Communications*, **21**, 110–11.

Killat, U. (1977*b*). Holographic microfiche of picture-like information. *Optica Acta*, **24**, 453–62.

King, M. C. (1968). Multiple exposure hologram recording of a 3-D image with a 360° view. *Applied Optics*, **7**, 1641–2.

King, M. C., Noll, A. M. & Berry, D. H. (1970). A new approach to computer generated holography. *Applied Optics*, **9**, 471–5.

Klein, W. R. & Cook, B. D. (1967). Unified approach to ultrasonic light diffraction. *IEEE Transactions on Sonics & Ultrasonics*, **SU-14**, 123–34.

Klimenko, I. S., Matinyan, E. G. & Dubitskii, L. G. (1975). Use of focused image holography for the nondestructive testing of electronic parts. *Soviet Journal of Nondestructive Testing*, **10**, 696–9.

Klug, M. A., Halle, M. W. & Hubel, P. M. (1992). Full color Ultragrams. In *Practical Holography VI*, Proceedings of the SPIE, vol. 1667, ed. S. A. Benton, pp. 110–19. Bellingham: SPIE.

Klug, M. A. & Kihara, N. (1994). 'Reseau' full-color one-step holographic stereograms. In *Proceedings of the Fifth International Symposium on Display Holography*, ed. T. H. Jeong (to appear). Lake Forest: Lake Forest College.

Knight, G. (1974). Page-oriented associative holographic memory. *Applied Optics*, **13**, 904–12.

Knight, G. (1975*a*). Holographic associative memory and processor. *Applied Optics*, **14**, 1088–92.

Knight, G. R. (1975*b*). Holographic memories. *Optical Engineering*, **14**, 453–9.

Kock, M. & Tiemens, U. (1973). Tomosynthesis: a holographic method for variable depth display. *Optics Communications*, **7**, 260–5.

Koechner, W. (1979*a*). Solid state lasers. In *Handbook of Optical Holography*, ed. H. J. Caulfield, pp. 257–67. New York: Academic Press.

Koechner, W. (1979*b*). Pulsed holography. In *Laser Handbook*, ed. M. L. Stitch, pp. 578–626. Amsterdam: North-Holland.

Kogelnik, H. (1965). Holographic image projection through inhomogeneous media. *Bell System Technical Journal*, **44**, 2451–5.

Kogelnik, H. (1967). Reconstructing response and efficiency of hologram gratings. In *Proceedings of the Symposium on Modern Optics*, pp. 605–17. Brooklyn: Polytechnic Press.

Kogelnik, H. (1969). Coupled wave theory for thick hologram gratings. *Bell System Technical Journal*, **48**, 2909–47.

Kogelnik, H. (1972). Optics at Bell Laboratories – lasers in technology. *Applied Optics*, **11**, 2426–34.

Kogelnik, H. & Pennington, K. (1968). Holographic image through a random medium. *Journal of the Optical Society of America*, **58**, 273–274.

Kohler, B., Bernet, S., Renn, A. & Wild, U. P. (1993). Storage of 2000 holograms in a photochemical hole-burning system. *Optics Letters*, **18**, 2144–46.

Komar, V. G. (1977). Progress on the holographic movie process in the USSR. In *Three-Dimensional Imaging*, Proceedings of the SPIE, vol. 120, ed. S. A. Benton, pp. 127–44. Redondo Beach: SPIE.

Komar, V. G. & Serov, O. B. (1989). Works on the holographic cinematography in the USSR. In *Holography '89*, Proceedings of the SPIE, vol. 1183, eds. Yu. N. Denisyuk & T. H. Jeong, pp. 170–82. Bellingham: SPIE.

Kostuk, R. (1989). Comparison of models for multiplexed holograms. *Applied Optics*, **28**, 771–7.

Kostuk, R. K. (1991a). Factorial optimization of bleach constituents for silver halide holograms. *Applied Optics*, **30**, 1611–6.

Kostuk, R. K. (1991b). Practical design considerations and performance characteristics of high numerical aperture holographic lenses. In *Practical Holography V*, Proceedings of the SPIE, vol. 1461, ed. S. A. Benton, pp. 24–34. Bellingham: SPIE.

Kostuk, R. K., Goodman, J. W. & Hesselink, L. (1985). Optical imaging applied to microelectronic chip-to-chip interconnections. *Applied Optics*, **24**, 2851–8.

Kostuk, R. K., Goodman, J. W. & Hesselink, L. (1987). Design considerations for holographic optical interconnects. *Applied Optics*, **26**, 3947–53.

Kostuk, R. K., Huang, Y-T., Hetherington, D. & Kato, M. (1989). Reducing alignment and chromatic sensitivity of holographic interconnects with substrate-mode holograms. *Applied Optics*, **28**, 4939–44.

Kowarschik, R. (1976). Diffraction efficiency of attenuated sinusoidally modulated gratings in volume holograms. *Optica Acta*, **23**, 1039–51.

Kowarschik, R. (1978a). Diffraction efficiency of sequentially stored gratings in transmission volume holograms. *Optica Acta*, **25**, 67–81.

Kowarschik, R. (1978b). Diffraction efficiency of sequentially stored gratings in reflection volume holograms. *Optical & Quantum Electronics*, **10**, 171–8.

Kowarschik, R. & Kessler, S. (1975). Zum Beugungswirkungsgrad von Volumenhologrammen. Teil II. Reflexionshologramme. *Optical & Quantum Electronics*, **7**, 399–411.

Kozma, A. (1966). Photographic recording of spatially modulated coherent light. *Journal of the Optical Society of America*, **56**, 428–32.

Kozma, A. (1968a). Effects of film grain noise in holography. *Journal of the Optical Society of America*, **58**, 436–8.

Kozma, A. (1968b). Analysis of the film non-linearities in hologram recording. *Optica Acta*, **15**, 527–51.

Kozma, A., Jull, G. W. & Hill, K. O. (1970). An analytical and experimental study of non-linearities in hologram recording. *Applied Optics*, **9**, 721–31.

Kozma, A. & Massey, N. (1969). Bias level reduction of incoherent holograms. *Applied Optics*, **8**, 393–7.

Kozma, A. & Zelenka, J. S. (1970). Effect of film resolution and size in holography. *Journal of the Optical Society of America*, **60**, 34–43.

Kramer, C. J. (1981). Holographic laser scanners for nonimpact printing. *Laser Focus*, **17**, No. 6, 70–82.

Kramer, C. J. (1991). Holographic deflector for graphic arts systems. In *Optical Scanning*, ed. G. E. Marshall, pp. 213–349. New York: Marcel Dekker.

Kreis, T. M. & Kreitlow, H. (1980). Digital processing of holographic interference patterns. In *Technical Digest on Hologram Interferometry & Speckle Metrology*, pp. TuB2-1–TuB2-4. Washington: The Optical Society of America.

Krile, T. F., Marks II, R. J., Walkup, J. F. & Hagler, M. O. (1977). Holographic representations of space-variant systems using phase-coded reference beams. *Applied Optics*, **16**, 3131–5.

Kubo, H. & Nagata, R. (1976a). Holographic photoelasticity with depolarized object wave. *Japanese Journal of Applied Physics*, **15**, 641–4.

Kubo, H. & Nagata, R. (1976b). Application of polarization holography by the Kurtz's method to photoelasticity. *Japanese Journal of Applied Physics*, **15**, 1095–9.

Kubota, K., Ono, Y., Kondo, M., Sugama, S., Nishida, N. & Sakaguchi, M. (1980). Holographic disk with high data transfer rate: its application to an audio response memory. *Applied Optics*, **19**, 944–51.

Kubota, T. (1978). Characteristics of thick hologram gratings recorded in absorptive medium. *Optica Acta*, **25**, 1035–53.

Kubota, T. (1986). Recording of high quality color holograms. *Applied Optics*, **25**, 4141–5.

Kubota, T. (1989). Control of the reconstruction wavelength of Lippman holograms recorded in dichromated gelatin. *Applied Optics*, **28**, 1845–9.

Kubota, T., Fujioka, K. & Kitagawa, M. (1992). Method for reconstructing a hologram using a compact device. *Applied Optics*, **31**, 4734–7.

Kubota, T. & Ose, T. (1979a). Methods of increasing the sensitivity of methylene blue sensitized dichromated gelatin. *Applied Optics*, **18**, 2538–9.

Kubota, T. & Ose, T. (1979b). New technique for recording a Lippman hologram. *Opt. Commun.*, **28**, 159–62.

Kubota, T. & Ose, T. (1979c). Lippmann color holograms recorded in methylene-blue sensitized dichromated gelatin. *Optics Letters*, **4**, 289–91.

Küchel, F. M. & Tiziani, H. J. (1981). Real-time contour holography using BSO crystals. *Optics Communications*, **38**, 17–20.

Kurtz, C. N. (1968). Copying reflection holograms. *Journal of the Optical Society of America*, **58**, 856–7.

Kurtz, C. N. (1969). Holographic polarization recording with an encoded reference beam. *Applied Physics Letters*, **14**, 59–61.

Kuo, C. P., Aye, T., Pelka, D. G., Jannson, J. & Jannson, T. (1990). Tunable holographic Fabry-Perot étalon fabricated from poor quality glass substrates. *Optics Letters*, **15**, 351–3.

Labeyrie, A. & Flamand, J. (1969). Spectrographic performance of holographically made diffraction gratings. *Optics Communications*, **1**, 5–8.

La Macchia, J. T. & White, D. L. (1968). Coded multiple-exposure holograms. *Applied Optics*, **7**, 91–4.

Lamberts, R. L. & Kurtz, C. N. (1971). Reversal bleaching for low flare light in holograms. *Applied Optics*, **19**, 1342–7.

Landry, M. J. (1967). The effect of two hologram-copying parameters on the quality of copies. *Applied Optics*, **6**, 1947–56.

Lang, M. & Eschler, H. (1974). Gigabytye capacities for holographic memories. *Optics & Laser Technology*, **6**, 219–24.

Langbein, U. & Lederer, F. (1980). Modal theory for thick holographic gratings with sharp boundaries. I. General treatment. *Optica Acta*, **27**, 171–82.

*References*

Langdon, R. M. (1970). A high capacity holographic memory. *The Marconi Review*, **33**, 113–30.

Latta, J. N. (1971a). Computer-based analysis of hologram imagery and aberrations. I. Hologram types and their nonchromatic aberrations. *Applied Optics*, **10**, 599–608.

Latta, J. N. (1971b). Computer-based analysis of hologram imagery and aberrations. II. Aberrations induced by a wavelength shift. *Applied Optics*, **10**, 609–18.

Latta, J. N. (1971c). Computer-based analysis of holography using ray tracing. *Applied Optics*, **10**, 2698–710.

Latta, M. R. & Pole, R. V. (1979). Design techniques for forming 488 nm holographic lenses with reconstruction at 633 nm. *Applied Optics*, **18**, 2418–21.

Lederer, F. & Langbein, U. (1977). Attenuated thick hologram gratings. Part I. Diffraction efficiency. *Optical & Quantum Electronics*, **9**, 473–85.

Lederer, F. & Langbein, U. (1980). Modal theory for thick holographic gratings with sharp boundaries. II. Unslanted transmission and reflection gratings. *Optica Acta*, **27**, 183–200.

Lee, S. H., ed. (1981). *Optical Information Processing: Fundamentals*, Topics in Applied Physics, vol. 48. Berlin: Springer-Verlag.

Lee, T. C., Lin, J. W. & Tufte, O. N. (1977). Thermoplastic photoconductor for optical recording and storage – new developments. In *Optical Storage & Methods*, Proceedings of the SPIE, vol. 123, eds. L. Beiser & D. Chen, pp. 74–7. Bellingham: SPIE.

Lee, W. H. (1970). Sampled Fourier transform hologram generated by computer. *Applied Optics*, **9**, 639–43.

Lee, W. H. (1974). Binary synthetic holograms. *Applied Optics*, **13**, 1677–82.

Lee, W. H. (1978). Computer-generated holograms: techniques and applications. In *Progress in Optics*, vol. 16, ed. E. Wolf, pp. 121–232. Amsterdam: North-Holland.

Lee, W. H. (1979). Binary computer-generated holograms. *Applied Optics*, **18**, 3661–9.

Lee, W. H. (1989). Holographic optical head for compact disk applications. *Optical Engineering*, **28**, 650–3.

Lee, W. H. & Streifer, W. (1978a). Diffraction efficiency of evanescent-wave holograms. I. TE polarization. *Journal of the Optical Society of America*, **68**, 795–801.

Lee, W. H. & Streifer, W. (1978b). Diffraction efficiency of evanescent-wave holograms. II. TM polarization. *Journal of the Optical Society of America*, **68**, 802–9.

Leith, E. N. & Chang, B. J. (1973). Space invariant holography with quasi-coherent light. *Applied Optics*, **12**, 1957–63.

Leith, E. N. & Chen, H. (1978). Deep image rainbow holograms. *Optics Letters*, **2**, 82–4.

Leith, E. N., Chen, H. & Roth, J. (1978). White light hologram technique. *Applied Optics*, **17**, 3187–8.

Leith, E. N., Kozma, A., Upatnieks, J., Marks, J. & Massey, N. (1966). Holographic data storage in three-dimensional media. *Applied Optics*, **5**, 1303–12.

Leith, E. N. & Upatnieks, J. (1962). Reconstructed wavefronts and communication theory. *Journal of the Optical Society of America*, **52**, 1123–30.

Leith, E. N. & Upatnieks, J. (1963). Wavefront reconstruction with continuous-tone objects. *Journal of the Optical Society of America*, **53**, 1377–81.

Leith, E. N. & Upatnieks, J. (1964). Wavefront reconstruction with diffused illumination and three-dimensional objects. *Journal of the Optical Society of America*, **54**, 1295–301.

Leith, E. N. & Upatnieks, J. (1965). Microscopy by wavefront reconstruction. *Journal of the Optical Society of America*, **55**, 569–70.

Leith, E. N. & Upatnieks, J. (1966). Holographic imagery through diffusing media. *Journal of the Optical Society of America*, **56**, 523.

Leith, E. N. & Upatnieks, J. (1967). Holography with achromatic fringe systems. *Journal of the Optical Society of America*, **57**, 975–80.

Leith, E. N., Upatnieks, J. & Haines, K. A. (1965). Microscopy by wavefront reconstruction. *Journal of the Optical Society of America*, **55**, 981–6.

Lengyel, B. A. (1971). *Lasers*. New York: Wiley-Interscience.

Leseberg, D. (1986). Computer-generated holograms: display using one-dimensional transforms. *Journal of the Optical Society of America. A*, **3**, 1846–51.

Leseberg, D. & Bryngdahl, O. (1984). Computer-generated rainbow holograms. *Applied Optics*, **23**, 2441–7.

Leseberg, D. & Frère, C. (1988). Computer-generated holograms of 3-D objects composed of tilted planar segments. *Applied Optics*, **27**, 3020–4.

Lesem, L. B., Hirsch, P. M. & Jordan Jr., J. A. (1969). The kinoform: a new wavefront reconstruction device. *IBM Journal of Research & Development*, **13**, 150–5.

Lessard, R. A., Som, S. C. & Boivin, A. (1973). New technique of color holography. *Applied Optics*, **12**, 2009–11.

Leung, K. M., Lee, T. C., Bernal, E. & Wyant, J. C. (1979). Two-wavelength contouring with the automated thermoplastic holographic camera. In *Interferometry*, Proceedings of the SPIE, vol. 192, ed. G. W. Hopkins, pp. 184–189. Bellingham: SPIE.

Leung, K. M., Lindquist, J. C. & Shepherd, L. T. (1980). E-beam computer generated holograms for aspheric testing. In *Recent Advances in Holography*, Proceedings of the SPIE, vol. 215, ed. T. C. Lee & P. N. Tamura, pp. 70–75. Bellingham: SPIE.

Levitt, J. A. & Stetson, K. A. (1976). Mechanical vibrations: mapping their phase with hologram interferometry. *Applied Optics*, **15**, 195–9.

Lin, F., Chou, H., Strzelecki, E. & Shellan, J. B. (1992). Multiplexed holographic Fabry-Perot étalons. *Applied Optics*, **31**, 2478–84.

Lin, F., Strzelecki, E. M. & Jansson, T. (1990). Optical multiplanar VLSI interconnects based on multiplexed waveguide holograms. *Applied Optics*, **29**, 1126–33.

Lin, L. H. (1969). Hologram formation in hardened dichromated gelatin films. *Applied Optics*, **8**, 963–6.

Lin, L. H. (1970). Edge-illuminated hologram. *Journal of the Optical Society of America*, **60**, 714A.

Lin, L. H. (1971). Method of characterizing hologram-recording materials. *Journal of the Optical Society of America*, **61**, 203–8.

Lin, L. H. & Beauchamp, H. L. (1970a). An automatic shutter for holography. *Review of Scientific Instruments*, **41**, 1438–40.

Lin, L. H. & Beauchamp, H. L. (1970b). Write-read-erase in situ optical memory using thermoplastic holograms. *Applied Optics*, **9**, 2088–92.

Lin, L. H. & Lo Bianco, C. V. (1967). Experimental techniques in making multicolor white light reconstructed holograms. *Applied Optics*, **6**, 1255–8.

Lin, L. H., Pennington, K. S., Stroke, G. W. & Labeyrie, A. E. (1966). Multicolor holographic image reconstruction with white light illumination. *Bell System Technical Journal*, **45**, 659–60.

Livanos, A. C., Katzir, A., Shellan, J. B. & Yariv, A. (1977). Linearity and enhanced sensitivity of the Shipley AZ-1350 B photoresist. *Applied Optics*, **16**, 1633–5.

Loewen, E., Maystre, D., McPhedran, R. & Wilson, I. (1975). Correlation between efficiency of diffraction gratings and theoretical calculations over a wide range. *Japanese Journal of Applied Physics*, **14** (Supplement 14-1), 143–52.

Lohmann, A. (1956). Optische Einseitenbandübertragung angewandt auf das Gabor-Mikroskop. *Optica Acta*, **3**, 97–9.

Lohmann, A. W. (1965a). Reconstruction of vectorial wavefronts. *Applied Optics*, **4**, 1667–8.

Lohmann, A. W. (1965b). Wavefront reconstruction for incoherent objects. *Journal of the Optical Society of America*, **55**, 1555–6.

Lohmann, A. W. (1986). What classical optics can do for the digital optical computer. *Applied Optics*, **25**, 1543–49.

Lohmann, A. W. & Paris, D. P. (1967). Binary Fraunhofer holograms generated by computer. *Applied Optics*, **6**, 1739–48.

Loomis, J. S. (1980). Computer-generated holography and optical testing. *Optical Engineering*, **19**, 679–85.

Lowenthal, S. & Joyeux, D. (1971). Speckle removal by a slowly moving diffuser associated with a motionless diffuser. *Journal of the Optical Society of America*, **61**, 847–51.

Lowenthal, S., Serres, J. & Froehly, C. (1969). Enregistrement d'hologrammes en lumière spatialement incohérente. *Comptes Rendus des Seances de l'Academie des Sciences, Paris, B*, **268**, 841–4.

Lu, S. (1968). Generating multiple images for integrated circuits by Fourier-transform holograms. *Proceedings of the IEEE*, **56**, 116–17.

Lucie-Smith, E. (1992). *Alexander*. London: Art Books International.

Lukin, A. V. & Mustafin, K. S. (1979). Holographic methods of testing aspherical surfaces. *Soviet Journal of Optical Technology*, **46**, 237–44.

Lukosz, W. & Wüthrich, A. (1974). Holography with evanescent waves. I. Theory of the diffraction efficiency for s-polarized light. *Optik*, **41**, 191–211.

Lukosz, W. & Wüthrich, A. (1976). Hologram recording and read-out with the evanescent field of guided waves. *Optics Communications*, **19**, 232–5.

MacDonald, R. P., Chrostowski, J., Boothroyd, S. A. & Syrett, B. A. (1993). Holographic formation of a diode laser nondiffracting beam. *Applied Optics*, **32**, 6470–4.

MacGovern, A. J. & Wyant, J. C. (1971). Computer-generated holograms for testing optical elements. *Applied Optics*, **10**, 619–24.

MacQuigg, D. R. (1977). Hologram fringe stabilization method. *Applied Optics*, **16**, 291–2.

Macovski, A., Ramsey, S. D. & Schaefer, L. F. (1971). Time-lapse interferometry and contouring using television systems. *Applied Optics*, **10**, 2722–7.

Macy Jr., W. W. (1983). Two-dimensional fringe pattern analysis. *Applied Optics*, **22**, 3898–901.

Magariños, J. R. & Coleman, D. J. (1987). Holographic optical configurations for eye protection against lasers. *Applied Optics*, **26**, 2575–81.

Magnusson, R. & Gaylord, T. K. (1977). Analysis of multiwave diffraction of thick gratings. *Journal of the Optical Society of America*, **67**, 1165–70.

Magnusson, R. & Gaylord, T. K. (1978a). Diffraction regimes of transmission gratings. *Journal of the Optical Society of America*, **68**, 809–14.

Magnusson, R. & Gaylord, T. K. (1978b). Equivalence of multiwave coupled-wave theory and modal theory for periodic-media diffraction. *Journal of the Optical Society of America*, **68**, 1777–9.

Mait, J. N. (1990). Design of binary-phase and multiphase Fourier gratings for array generation. *Journal of the Optical Society of America, A.*, **7**, 1514–28.

Mait, J. N. & Himes, G. S. (1989). Computer-generated holograms by means of a magnetooptic spatial light modulator. *Applied Optics*, **28**, 4879–87.

Malin, M. & Morrow, H. E. (1981). Wavelength scaling holographic elements. *Optical Engineering*, **20**, 756–8.

Mallick, S. (1975). Pulse holography of uniformly moving objects. *Applied Optics*, **14**, 602–5.

Mandel, L. (1965). Color imagery by wavefront reconstruction. *Journal of the Optical Society of America*, **55**, 1697–8.

Maréchal, A. & Croce, P. (1953). A filter of spatial frequencies for the improvement of contrast of optical images. *Comptes Rendus des Seances de l'Academie des Sciences*, Serie II, **237**, 607–9.

Marom, E. (1967). Color imagery by wavefront reconstruction. *Journal of the Optical Society of America*, **57**, 101–2.

Marrakchi, A., Huignard, J. P. & Günter, P. (1981). Diffraction efficiency and energy transfer in two-wave mixing experiments with $Bi_{12}SiO_{20}$ crystals. *Applied Physics*, **24**, 131–4.

Marrakchi, A., Huignard, J. P. & Herriau, J. P. (1980). Application of phase conjugation in $Bi_{12}SiO_{20}$ crystals to mode pattern visualisation of diffuse vibrating structures. *Optics Communications*, **34**, 15–18.

Marrakchi, A., Johnson, R. V. & Tanguay, A. R. (1986). Polarization properties of photorefractive diffraction in electrooptic and optically active sillenite crystals (Bragg regime). *Journal of the Optical Society of America, B.*, **3**, 321–36.

Marrone, E. S. & Ribbens, W. B. (1975). Dual-index holographic contour mapping over a large range of contour spacings. *Applied Optics*, **14**, 23–4.

Martienssen, W. & Spiller, S. (1967). Holographic reconstruction without granulation. *Physics Letters*, **24A**, 126–8.

Masajada, J. & Nowak, J. (1991). Third-order aberrations for holograms and holographic lenses recorded on quadrics of revolution. *Applied Optics*, **30**, 1791–5.

Matsuda, K., Freund, C. H. & Hariharan, P. (1981). Phase difference amplification using longitudinally reversed shearing interferometry: an experimental study. *Applied Optics*, **20**, 2763–5.

Matsumoto, T., Iwata, K. & Nagata, R. (1973). Measuring accuracy of three-dimensional displacements in holographic interferometry. *Applied Optics*, **12**, 961–7.

Matsumoto, K. & Takashima, M. (1970). Phase-difference amplification by nonlinear holograms. *Journal of the Optical Society of America*, **60**, 30–3.

Matsumura, M. (1975). Speckle noise reduction by random phase shifters. *Applied Optics*, **14**, 660–5.

Mazakova, M., Pancheva, M., Kandilarov, P. & Sharlandjiev, P. (1982a). Dichromated gelatin for volume holographic recording with high sensitivity. Part I. *Optical & Quantum Electronics*, **14**, 311–15.

Mazakova, M., Pancheva, M., Kandilarov, P. & Sharlandjiev, P. (1982b). Dichromated gelatin for volume holographic recording with high sensitivity. Part II. *Optical & Quantum Electronics*, **14**, 317–20.

McCauley, D. G., Simpson, C. E. & Murbach, W. J. (1973). Holographic optical element for visual display applications. *Applied Optics*, **12**, 232–42.

McClung, F. J., Jacobson, A. D. & Close, D. H. (1970). Some experiments performed with a reflected light pulsed laser holography system. *Applied Optics*, **9**, 103–6.

McCrickerd, J. T. & George, N. (1968). Holographic stereogram from sequential component photographs. *Applied Physics Letters*, **12**, 10–12.

McKechnie, T. S. (1975a). Reduction of speckle in an image by a moving aperture: second order statistics. *Optics Communications*, **13**, 29–34.

McKechnie, T. S. (1975b). Reduction of speckle by a moving aperture: first order statistics. *Optics Communications*, **13**, 35–9.

McKechnie, T. S. (1975c). Speckle reduction. In *Laser Speckle & Related Phenomena*, Topics in Applied Physics, vol. 9, ed. J. C. Dainty, pp. 123–70. Berlin: Springer-Verlag.

McMahon, D. H., Franklin, A. R. & Thaxter, J. B. (1969). Light beam deflection using holographic scanning techniques. *Applied Optics*, **8**, 399–402.

McMahon, D. H. & Maloney, W. T. (1970). Measurements of the stability of bleached photographic phase holograms. *Applied Optics*, **9**, 1363–8.

McPhedran, R. C., Wilson, I. J. & Waterworth, M. D. (1973). Profile formation in holographic diffraction gratings. *Optics & Laser Technology*, **5**, 166–71.

McPhedran, R. C., Derrick, G. H. & Botten, L. C. (1980). Theory of crossed gratings. In *Electromagnetic Theory of Gratings*, Topics in Current Physics, vol. 22, ed. R. Petit, pp. 227–76. Berlin: Springer-Verlag.

Meier, R. W. (1965). Magnification and third-order aberrations in holography. *Journal of the Optical Society of America*, **55**, 987–92.

Menzel, E. (1974). Comment to the methods of contour holography. *Optik*, **40**, 557–9.

Mercier, R. & Lowenthal, S. (1980). Comparison of in-line and carrier frequency holograms in aspheric testing. *Optics Communications*, **33**, 251–6.

Mertz, L. & Young, N. O. (1962). Fresnel transformations of images. In *Proceedings of the Conference on Optical Instruments and Techniques*, London, 1961, ed. K. J. Habell, pp. 305–12. London: Chapman & Hall.

Meyerhofer, D. (1972). Phase holograms in dichromated gelatin. *RCA Review*, **33**, 110–30.

Micheron, F. & Bismuth, G. (1972). Electrical control of fixation and erasure of holographic patterns in ferroelectric materials. *Applied Physics Letters*, **20**, 79–81.

Micheron, F. & Bismuth, G. (1973). Field and time thresholds for the electrical fixation of holograms recorded in $(Sr_{0.75}Ba_{0.25})Nb_2O_6$ crystals. *Applied Physics Letters*, **23**, 71–2.

Miles, J. F. (1972). Imaging and magnification properties in holography. *Optica Acta*, **19**, 165–86.

Miles, J. F. (1973). Evaluation of the wavefront aberration in holography. *Optica Acta*, **20**, 19–31.

Miridonov, S. V., Petrov, M. P. & Stepanov, S. I. (1978). Light diffraction by volume holograms in optically active photorefractive crystals. *Soviet Technical Physics Letters*, **4**, 393–4.

Moharam, M. G. & Gaylord, T. K. (1981). Rigorous coupled-wave analysis of planar-grating diffraction. *Journal of the Optical Society of America*, **71**, 811–18.

Moharam, M. G. & Gaylord, T. K. (1982). Diffraction analysis of dielectric surface-relief gratings. *Journal of the Optical Society of America*, **72**, 1385–92.

Moharam, M. G., Gaylord, T. K., Sincerbox, G. T., Werlich, H. & Yung, B. (1984). Diffraction characteristics of photoresist surface-relief gratings. *Applied Optics*, **23**, 3214–20.

Moharam, M. G., Gaylord, T. K. & Magnusson, R. (1980*a*). Criteria for Bragg regime diffraction by phase gratings. *Optics Communications*, **32**, 14–18.

Moharam, M. G., Gaylord, T. K. & Magnusson, R. (1980*b*). Criteria for Raman-Nath regime diffraction by phase gratings. *Optics Communications*, **32**, 19–23.

Moharam, M. G. & Young, L. (1978). Criterion for Bragg and Raman-Nath diffraction regimes. *Applied Optics*, **17**, 1757–9.

Mok, F. H. (1993). Angle-multiplexed storage of 5000 holograms in lithium niobate. *Optics Letters*, **18**, 915–17.

Mok, F., Diep, J., Liu, H. K. & Psaltis, D. (1986). Real-time computer-generated hologram by means of liquid-crystal television spatial light modulator. *Optics Letters*, **11**, 748–50.

Molin, N. E. & Stetson, K. A. (1969). Measuring combination mode vibration patterns by hologram interferometry. *Journal of Physics E: Scientific Instruments*, **2**, 609–12.

Molin, N. E. & Stetson, K. A. (1970*a*). Measurement of fringe loci and localization in hologram interferometry for pivot motion, in-plane rotation and in-plane translation. Part I. *Optik*, **31**, 157–77.

Molin, N. E. & Stetson, K. A. (1970*b*). Measurement of fringe loci and localization in hologram interferometry for pivot motion, in-plane rotation and in-plane translation. Part II. *Optik*, **31**, 281–91.

Molin, N. E. & Stetson, K. A. (1971). Fringe localization in hologram interferometry of mutually independent and dependent rotations around orthogonal, non-intersecting axes. *Optik*, **33**, 399–422.

Mosyakin, Yu. S. & Skrotskii, G. V. (1972). Holographic optical elements. *Soviet Journal of Quantum Electronics*, **2**, 199–206.

Mottier, F. M. (1969). Holography of randomly moving objects. *Applied Physics Letters*, **15**, 44–5.

Munch, J. & Wuerker, R. (1989). Holographic technique for correcting aberrations in a telescope. *Applied Optics*, **28**, 1313–17.

Nakadate, S. (1986). Vibration measurement using phase-shifting time-average holographic interferometry. *Applied Optics*, **25**, 4155–61.

Nakadate, S., Magome, N., Honda, T. & Tsujiuchi, J. (1981). Hybrid holographic interferometer for measuring three-dimensional deformations. *Optical Engineering*, **20**, 246–52.

Nakadate, S. & Saito, H. (1985). Fringe-scanning speckle-pattern interferometry. *Applied Optics*, **24**, 2172–80.

Nakadate, S., Saito, H. & Nakajima, T. (1986). Vibration measurement using phase-shifting stroboscopic holographic interferometry. *Optica Acta*, **33**, 1295–309.

Namioka, T., Seya, M. & Noda, H. (1976). Design and performance of holographic concave gratings. *Japanese Journal of Applied Physics*, **15**, 1181–97.

Nassenstein, H. (1968a). Copying of holograms. *Optik*, **27**, 327–34.

Nassenstein, H. (1968b). Holographie und Interferenzversuche mit inhomogenen Oberflächenwellen. *Physics Letters*, **28A**, 249–51.

Nassenstein, H. (1969). Rekonstruktion von Hologrammen mit höherem Beugungswirkungsgrad. *Optik*, **30**, 201–5.

Nassenstein, H. (1970). Superresolution by diffraction of subwaves. *Optics Communications*, **2**, 231–4.

Nath, N. S. N. (1938). Diffraction of light by supersonic waves. *Proceedings of the Indian Academy of Sciences*, **8A**, 499–503.

Nelson, R. H., Vander Lugt, A. & Zech, R. G. (1974). Holographic data storage and retrieval. *Optical Engineering*, **13**, 429–30.

Neumann, D. B. (1966). Geometrical relationships between the original object and the two images of a hologram reconstruction. *Journal of the Optical Society of America*, **56**, 858–61.

Neumann, D. B. (1968). Holography of moving scenes. *Journal of the Optical Society of America*, **58**, 447–54.

Neumann, D. B., Jacobson, C. F. & Brown, G. M. (1970). Holographic technique for determining the phase of vibrating objects. *Applied Optics*, **9**, 1357–68.

Neumann, D. B. & Penn, R. C. (1972). Object motion compensation using reflection holography. *Journal of the Optical Society of America*, **62**, 1373.

Neumann, D. B. & Rose, H. W. (1967). Improvement of recorded holographic fringes by feedback control. *Applied Optics*, **6**, 1097–1104.

Newswanger, C. & Outwater, C. (1985). Large format holographic stereograms and their applications. In *Applications of Holography*, Proceedings of the SPIE, vol. 523, ed. L. Huff, pp. 26–32. Bellingham: SPIE.

Nikolova, L. & Todorov, T. (1984). Diffraction efficiency and selectivity of polarization holographic recording. *Optica Acta*, **31**, 579–88.

Nikolova, L., Todorov, T., Tomova, N. & Dragostinova, V. (1988). Polarization preserving wavefront reversal by four-wave mixing in photoanisotropic materials. *Applied Optics*, **27**, 1598–602.

Nishihara, H. & Koyama, J. (1979). New technique to record and playback holographic color video memories. *Optics Communications*, **31**, 16–20.

Nisida, M. & Saito, H. (1964). A new interferometric method of two-dimensional stress analysis. *Experimental Mechanics*, **4**, 366–76.

Nobis, D. & Vest, C. M. (1978). Statistical analysis of errors in holographic interferometry. *Applied Physics*, **17**, 2198–204.

Nordin, G. P. & Tanguay, A. R. (1992). Photopolymer-based stratified volume holographic optical elements. *Optics Letters*, **17**, 1709–11.

Norman, S. L. & Singh, M. P. (1975). Spectral sensitivity and linearity of Shipley AZ-1350J photoresist. *Applied Optics*, **14**, 818–20.

Nowak, J. & Zajac, M. (1983). Investigations of the influence of hologram aberrations on the light intensity distribution in the image plane. *Optica Acta*, **30**, 1749–67.

Nugent, K. A. (1985). Interferogram analysis using an accurate fully automatic algorithm. *Applied Optics*, **24**, 3101–5.

Okada, K., Honda, T. & Tsujiuchi, J. (1981). 3-D distortion of observed images reconstructed from a cylindrical holographic stereogram (2) white light reconstruction type. *Optics Communications*, **36**, 17–21.

Okada, K., Honda, T. & Tsujiuchi, J. (1982). Image blur of multiplex holograms. *Optics Communications*, **41**, 397–402.

Okoshi, T. (1977). Projection-type holography. In *Progress in Optics*, vol. 15, ed. E. Wolf, pp. 141–85. Amsterdam: North-Holland.

Oliva, J., Boj, P. G. & Pardo, M. (1984). Dichromated gelatin holograms derived from Agfa 8E75 HD plates. *Applied Optics*, **23**, 196–7.

Olson, D. W. (1989). The elementary plane-wave model for hologram ray tracing. *American Journal of Physics*, **57**, 445–55.

O'Neill, E. L. (1956). Spatial filtering in optics. *IEEE Transactions on Information Theory*, **IT-2**, 56–65.

O'Neill, E. L. (1963). *Introduction to Statistical Optics*. Reading: Addison-Wesley Publishing Company.

Optical Society of America (1953). *The Science of Color*. New York: Thomas Y. Crowell.

Oreb, B. F. & Hariharan, P. (1981). Improved integrating exposure-control system for color holography. *Optical Engineering*, **20**, 749–52.

O'Regan, R. & Dudderar, T. D. (1971). A new holographic interferometer for stress analysis. *Experimental Mechanics*, **11**, 241–7.

Östlund, L. A. & Biedermann, K. (1977). Laser speckle reduction: equivalence of the moving aperture method and incoherent spatial filtering. *Applied Optics*, **16**, 685–90.

Ostrovskaya, G. V. & Ostrovskii, Yu. I. (1971). Two wave-length hologram method for studying the dispersion properties of phase objects. *Soviet Physics – Technical Physics*, **15**, 1890–2.

Owechko, Y. (1993). Cascaded-grating holography for artificial neural networks. *Applied Optics*, **32**, 1380–98.

Owechko, Y. & Soffer, B. H. (1991). Optical interconnection method for neural networks using self-pumped phase conjugate mirrors. *Optics Letters*, **16**, 675–7.

Owen, M. P. & Solymar, L. (1980). Efficiency of volume phase reflection holograms recorded in an attenuating medium. *Optics Communications*, **34**, 321–6.

Owner-Petersen, M. (1991). Digital speckle pattern shearing interferometry: limitations and prospects. *Applied Optics*, **30**, 2730–8.

Özkul, C., Allano, D. & Trinité, M. (1986). Filtering effects in far-field in-line holography. *Optical Engineering*, **25**, 1142–8.

Palais, J. C. & Wise, J. A. (1971). Improving the efficiency of very low efficiency holograms by copying. *Applied Optics*, **10**, 667–8.

Papoulis, A. (1962). The Fourier Integral & Its Applications. New York: McGraw-Hill.

Papoulis, A. (1965). Probability, Random Variables & Stochastic Processes, New York: McGraw-Hill.

Parker, R. J. (1978). A new method of frozen-fringe holographic interferometry using thermoplastic recording media. Optica Acta, 25, 787–92.

Pastor, J. (1969). Hologram interferometry and optical technology. Applied Optics, 8, 525–31.

Peng, K. & Frankena, H. J. (1986). Nonparaxial theory of curved holograms. Applied Optics, 25, 1319–26.

Pennington, K. S., Harper, J. S. & Laming, F. P. (1971). New phototechnology suitable for recording phase holograms and similar information in hardened gelatine. Applied Physics Letters, 18, 80–4.

Pennington, K. S. & Lin, L. H. (1965). Multicolor wavefront reconstruction. Applied Optics, 7, 56–7.

Petrov, M. P., Miridonov, S. V., Stepanov, S. I. & Kulikov, V. V. (1979). Light diffraction and nonlinear image processing in electro-optic $Bi_{12}SiO_{20}$ crystals. Optics Communications, 31, 301–4.

Petrov, M. P., Stepanov, S. I. & Kamshilin, A. A. (1979a). Light diffraction from volume holograms in electro-optic birefringent crystals. Optics Communications, 29, 44–8.

Petrov, M. P., Stepanov, S. I. & Kamshilin, A. A. (1979b). Holographic storage of information and peculiarities of light diffraction in birefringent electro-optic crystals. Optics & Laser Technology, 11, 149–51.

Phillips, N. J. & van der Werf, R. A. J. (1985). The creation of efficient reflective Lippmann layers in ultra-fine grain silver halide materials using non-laser sources. Journal of Photographic Science, 33, 22–8.

Phillips, N. J., Ward, A. A., Cullen, R. & Porter, D. (1980). Advances in holographic bleaches. Photographic Science & Engineering, 24, 120–4.

Pitlak, R. T. & Page, R. (1985). Pulsed lasers for holography. Optical Engineering, 24, 639–44.

Podbielska, H., ed. (1991). Holography, Interferometry & Optical Pattern Recognition in Biomedicine, Proceedings of the SPIE, vol. 1429, Bellingham: SPIE.

Podbielska, H., ed. (1992). Holography, Interferometry & Optical Pattern Recognition in Biomedicine II, Proceedings of the SPIE, vol. 1647, Bellingham: SPIE.

Pole, R. V., Werlick, H. W. & Krusche, R. J. (1978). Holographic light deflection. Applied Optics, 17, 3294–7.

Pole, R. V. & Wolenmann, H. P. (1975). Holographic laser beam deflector. Applied Optics, 14, 976–80.

Powell, R. L. & Stetson, K. A. (1965). Interferometric vibration analysis by wavefront reconstruction. Journal of the Optical Society of America, 55, 1593–8.

Prikryl, I. & Kvapil, J. (1980). A note on hologram synthesis from 2-D transparencies. Journal of Optics (Paris), 11, 231–3.

Prongué, D., Herzig, H. P., Dändliker, R. & Gale, M. T. (1992). Optimized kinoform structures for highly efficient fan-out elements. Applied Optics, 31, 5706–11.

Pryputniewicz, R. J. (1978). Holographic strain analysis: an experimental implementation of the fringe vector theory. Applied Optics, 17, 3613–18.

Pryputniewicz, R. J. (1980). The properties of fringes in hologram interferometry. In

*Technical Digest on Hologram Interferometry & Speckle Metrology*, pp. MB1-1–MB1-7. Washington: The Optical Society of America.

Pryputniewicz, R. J. (1985). Heterodyne holography applications in studies of small components. *Optical Engineering*, **24**, 849–54.

Pryputniewicz, R. J. (1987). Review of methods for automatic analysis of fringes in hologram interferometry. In *Interferometric Metrology, Critical Reviews*, Proceedings of the SPIE, vol. 816, ed. N. A. Massie, pp. 140–8. Bellingham: SPIE.

Pryputniewicz, R. J. & Bowley, W. W. (1978). Techniques of holographic displacement measurement: an experimental comparison. *Applied Optics*, **17**, 1748–56.

Pryputniewicz, R. J. & Stetson, K. A. (1976). Holographic strain analysis: extension of fringe-vector method to include perspective. *Applied Optics*, **15**, 725–8.

Pryputniewicz, R. J. & Stetson, K. A. (1980). Determination of sensitivity vectors in hologram interferometry from two known rotations of the object. *Applied Optics*, **19**, 2201–5.

Pryputniewicz, R. J. & Stetson, K. A. (1989). Measurement of vibration patterns using electro-optic holography. In *Laser Interferometry: Quantitative Analysis of Interferograms: Third in a Series*, Proceedings of the SPIE, vol. 1162, ed. R. J. Pryputniewicz, pp. 456–67. Bellingham: SPIE.

Psaltis, D., Brady, D. & Wagner, K. (1988). Adaptive optical networks using photorefractive crystals. *Applied Optics*, **27**, 1752–9.

Psaltis, D., Brady, D., Gu, X.-G. & Lin, S. (1990). Holography in artificial neural networks. *Nature (London)*, **343**, 325–30.

Psaltis, D., Paek, E. G. & Venkatesh, S. S. (1984). Optical image correlation with a binary spatial light modulator. *Optical Engineering*, **23**, 698–704.

Qiao, Y., Orlov, S., Psaltis, D. & Neurgaonkar, R. R. (1993). Electrical fixing of photorefractive holograms in $Sr_{0.75}Ba_{0.25}Nb_2O_6$. *Optics Letters*, **18**, 1004–6.

Radley Jr., R. J. (1975). Two-wavelength holography for measuring plasma electron density. *Physics of Fluids*, **18**, 175–9.

Ragnarsson, S. I. (1970). A new holographic method of generating a high efficiency, extended range spatial filter with application to restoration of defocussed images. *Physica Scripta*, **2**, 145–53.

Ragnarsson, S. I. (1978). Scattering phenomena in volume holograms with strong coupling. *Applied Optics*, **17**, 116–27.

Rajbenbach, H., Bann, S. & Huignard, J.-P. (1992). Long-term readout of photorefractive memories by using a storage/amplification two-crystal configuration. *Optics Letters*, **17**, 1712–14.

Rakuljic, G. A. & Leyva, V. (1993). Volume holographic narrow-band optical filter. *Optics Letters*, **18**, 459–61.

Rallison, R. D. & Schicker, S. R. (1992). Polarization properties of gelatin holograms. In *Practical Holography VI*, Proceedings of the SPIE, vol. 1667, ed. S. A. Benton, pp. 266–75. Bellingham: SPIE.

Rastani, K. & Hubbard, W. M. (1992). Large interconnects in photorefractives: grating erasure problem and a proposed solution. *Applied Optics*, **31**, 598–605.

Rastogi, P. K. (1984a). A real-time holographic moiré technique for the measurement of slope change. *Optica Acta*, **31**, 159–67.

Rastogi, P. K. (1984b). Comparative holographic moiré interferometry in real time. *Applied Optics*, **23**, 924–7.

Rastogi, P. K. (1986). Live slope measurement featuring continuously variable sensitivity and slope change direction by holographic moiré. *Optics Communications*, **58**, 1–3.

Rastogi, P. K. (1991). Visualization and measurement of slope and curvature fields using holographic interferometry: an application to flaw detection. *Journal of Modern Optics*, **38**, 1251–63.

Rastogi, P. K. (1992a). Phase-shifting applied to four-wave holographic interferometers. *Applied Optics*, **31**, 1680–1.

Rastogi, P. K. (1992b). Measurement of the difference and sum of phases in holographic moiré using phase shifting technique. *Optics Communications*, **93**, 336–8.

Rastogi, P. K. (1993a). Modification of the Carré phase-stepping method to suit four-wave holographic interferometry. *Optical Engineering*, **32**, 190–1.

Rastogi, P. K. (1993b). Phase-shifting holographic moiré: phase-shifter error-insensitive algorithms for the extraction of the difference. *Applied Optics*, **32**, 3669–75.

Rastogi, P. K., ed. (1994). *Holographic Interferometry*. Berlin: Springer-Verlag.

Rastogi, P. K., Barillot, M. & Kaufmann, G. H. (1991). Comparative phase shifting holographic interferometry. *Applied Optics*, **30**, 722–8.

Rastogi, P. K. & Pflug, L. (1991a). A holographic technique featuring broad range sensitivity to contour diffuse objects. *Journal of Modern Optics*, **38**, 1673–83.

Rastogi, P. K. & Pflug, L. (1991b). Real-time holographic phase organization technique to obtain customized contouring of diffuse surfaces. *Applied Optics*, **30**, 1603–10.

Rau, J. E. (1966). Detection of differences in real distributions. *Journal of the Optical Society of America*, **56**, 1490–4.

Rebane, A. & Aaviksoo, J. (1988). Holographic interferometry of ultrafast transients by photochemical hole burning. *Optics Letters*, **13**, 993–5.

Rebane, A., Kaarli, R., Saari, P., Anijalg, A. & Timpmann, K. (1983). Photochemical time-domain holography of weak picosecond pulses. *Optics Communications*, **47**, 173–6.

Redman, J. D., Wolton, W. P. & Shuttleworth, E. (1968). Use of holography to make truly three-dimensional x-ray images. *Nature*, **220**, 58–60.

Reich, S., Rav-Noy, Z. & Friesem, A. A. (1977). Frost supression in photoconductor-thermoplastic holographic recording devices. *Applied Physics Letters*, **31**, 654–6.

Renn, A. & Wild, U. P. (1987). Spectral hole burning and hologram storage. *Applied Optics*, **26**, 4040–2.

Rhodes, W. T. (1981). Space-variant optical systems and processing. In *Applications of the Optical Fourier Transform*, ed. F. Stark, pp. 333–69. New York: Academic Press.

Rich, C. & Cook, D. (1991). Lippmann volume holographic filters for Rayleigh line rejection in Raman spectroscopy. In *Practical Holography V*, Proceedings of the SPIE, vol. 1461, ed. S. A. Benton, pp. 2–7. Bellingham: SPIE.

Rich, C. & Petersen, J. (1992). Broadband IR Lippmann holograms for solar control applications. In *Practical Holography VI*, Proceedings of the SPIE, vol. 1667, ed. S. A. Benton, pp. 165–71. Bellingham: SPIE.

Roberts, H. N., Watkins, J. W. & Johnson, R. H. (1974). High-speed holographic digital recorder. *Applied Optics*, **13**, 841–56.

Roberts, N. C. (1989). Beam shaping by holographic filters. *Applied Optics*, **28**, 31–2.

Robertson, B., Restall, E. J., Taghizadeh, M. R. & Walker, A. C. (1991*a*). Space-variant holographic optical elements in dichromated gelatin. *Applied Optics*, **30**, 2368–75.

Robertson, B., Turunen, J., Ichikawa, H., Miller, J. M., Taghizadeh, M. R. & Vasara, A. (1991*b*). Hybrid kinoform fanout holograms in dichromated gelatin. *Applied Optics*, **30**, 3711–20.

Robinson, D. W. & Reid, G. T., eds. (1993). *Interferogram Analysis: Digital Processing Techniques for Fringe Pattern Measurement*. London: IOP.

Robinson, D. W. & Williams, D. C. (1986). Digital phase stepping speckle interferometry. *Optics Communications*, **57**, 26–30.

Rogers, G. L. (1952). Experiments in diffraction microscopy. *Proceedings of the Royal Society of Edinburgh*, **63A**, 193–221.

Rogers, G. L. (1966). Polarization effects in holography. *Journal of the Optical Society of America*, **56**, 831.

Rosen, L. (1966). Focused-image holography with extended sources. *Applied Physics Letters*, **9**, 337–9.

Rosen, L. (1967). The pseudoscopic inversion of holograms. *Proceedings of the IEEE*, **55**, 118.

Rotz, F. B. & Friesem, A. A. (1966). Holograms with non-pseudoscopic real images. *Applied Physics Letters*, **8**, 146–8.

Rowley, D. M. (1979). A holographic interference camera. *Journal of Physics E: Scientific Instruments*, **12**, 971–5.

Rowley, D. M. (1981). Interferometry with miniature format volume reflection holograms. *Optica Acta*, **28**, 907–15.

Rudolph, D. & Schmahl, G. (1967). Verfahren zur Herstellung von Röntgenlinsen und beugungsgittern. *Umschau in Wissenschaft und Technik*, **7**, 225.

Russell, P. St. J. (1981). Optical volume holography. *Physics Reports*, **71**, 209–312.

Russell, P. St. J. & Solymar, L. (1980). Borrmann-like anomalous effects in volume holography. *Applied Physics*, **22**, 335–53.

Růžek, J. & Fiala, P. (1979). Reflection holographic portraits. *Optica Acta*, **26**, 1257–64.

Saari, P., Kaarli, R. & Rebane, A. (1986). Picosecond time- and space-domain holography by photochemical hole burning. *Journal of the Optical Society of America. B*, **3**, 527–33.

Saito, T., Imamura, T., Honda, T. & Tsujiuchi, J. (1980). Solvent vapour method in thermoplastic photoconductor media. *Journal of Optics (Paris)*, **11**, 285–92.

Saito, T., Imamura, T., Honda, T. & Tsujiuchi, J. (1981). Enhancement of sensitivity by stratifying a photoconductor on thermoplastic-photoconductor media. *Journal of Optics (Paris)*, **12**, 49–58.

Salminen, O. & Keinonen, T. (1982). On absorption and refractive index modulations of dichromated gelatin gratings. *Optica Acta*, **29**, 531–40.

Sanford, R. J. & Durelli, A. J. (1971). Interpretation of fringes in stress-holo-interferometry. *Experimental Mechanics*, **11**, 161–6.

Sato, T., Ogawa, H. & Ueda, M. (1974). Contour generation of vibrating object by weighted subtraction of holograms. *Applied Optics*, **13**, 1280–2.

Satoh, I., Kato, M., Fujito, K. & Tateishi, F. (1989). Holographic memory system for Kanji character generation. *Applied Optics*, **28**, 2635–40.

Sauer, F. (1989). Fabrication of diffractive-reflective optical interconnects for infrared operation based on total internal reflection. *Applied Optics*, **28**, 386–8.

Savander, P. & Sheridan, J. T. (1993). Diffraction by crossed grating in pre-exposed photoresist. *Optik*, **94**, 101–13.

Schlüter, M. (1980). Analysis of holographic interferograms with a TV picture system. *Optics & Laser Technology*, **12**, 93–5.

Schmahl, G. (1975). Holographically made diffraction gratings for the visible, UV and soft X-ray region. *Journal of the Spectroscopical Society of Japan*, **23**, Supplement No. 1, 3–11.

Schmahl, G. & Rudolph, D. (1976). Holographic diffraction gratings. In *Progress in Optics*, vol. 14, ed. E. Wolf, pp. 196–244. Amsterdam: North-Holland.

Schwar, M. R. J., Pandya, T. P. & Weinberg, F. J. (1967). Point holograms as optical elements. *Nature*, **215**, 239–41.

Seldowitz, M. A., Allebach, J. P. & Sweeney, D. W. (1987). Synthesis of digital holograms by direct binary search. *Applied Optics*, **26**, 2788–98.

Shajenko, P. & Johnson, C. D. (1968). Stroboscopic holographic interferometry. *Applied Physics Letters*, **13**, 44–6.

Shamir, J., Caulfield, H. J. & Johnson, R. B. (1989). Massive holographic interconnection networks and their limitations. *Applied Optics*, **28**, 311–24.

Shankoff, T. A. (1968). Phase holograms in dichromated gelatin. *Applied Optics*, **7**, 2101–5.

Shariv, I., Amitai, Y. & Friesem, A. A. (1993). Compact holographic beam expander. *Optics Letters*, **18**, 1268–70.

Shellabear, M. C. & Tyrer, J. R. (1991). Application of ESPI to three-dimensional vibration measurements. *Optics & Lasers in Engineering*, **15**, 43–56.

Sheridon, N. K. (1968). Production of blazed holograms. *Applied Physics Letters*, **12**, 316–18.

Sherman, G. C. (1967). Hologram copying by Gabor holography of transparencies. *Applied Optics*, **6**, 1749–53.

Shibayama, K. & Uchiyama, H. (1971). Measurement of three-dimensional displacements by hologram interferometry. *Applied Optics*, **10**, 2150–4.

Shvartsman, F. P. (1991). Holographic optical elements by dry photopolymer embossing. In *Practical Holography V*, Proceedings of the SPIE, vol. 1461, ed. S. A. Benton, pp. 313–20. Bellingham: SPIE.

Siebert, L. D. (1967). Front lighted pulse laser holography. *Applied Physics Letters*, **11**, 326–8.

Siebert, L. D. (1968). Large scene front-lighted hologram of a human subject. *Proceedings of the IEEE*, **56**, 1242–3.

Simova, E. & Sainov, V. (1989). Comparative holographic moiré interferometry for nondestructive testing: comparison with conventional holographic interferometry. *Optical Engineering*, **28**, 261–6.

Simova, E. S. & Stoev, K. N. (1992*a*). Phase-stepping holographic moiré: simultaneous in-plane and out-of-plane displacement measurement. *Applied Optics*, **31**, 2405–8.

Simova, E. S. & Stoev, K. N. (1992*b*). Phase-stepping automatic fringe analysis in holographic moiré. *Applied Optics*, **31**, 5965–74.

Sirohi, R. S., Blume, H. & Rosenbruch, K. J. (1976). Optical testing using synthetic holograms. *Optica Acta*, **23**, 229–36.

Sjölinder, S. (1981). Dichromated gelatin and the mechanism of hologram formation. *Photographic Science & Engineering*, **25**, 112–18.

Slinger, C. W., Syms, R. R. A. & Solymar, L. (1987). Multiple holographic transmission gratings in silver halide emulsion. *Applied Physics. B*, **42**, 121–8.

Smigielski, P., Fagot, H. & Albe, F. (1985). Progress in holographic cinematography. In *Progress in Holographic Applications*, Proceedings of the SPIE, vol. 600, ed. J. P. Ebbeni, pp. 186–93. Bellingham: SPIE.

Smith, H. M. (1968). Photographic relief images. *Journal of the Optical Society of America*, **58**, 533–9.

Smith, H. M., ed. (1977). *Holographic Recording Materials*, Topics in Applied Physics, vol. 20. Berlin: Springer-Verlag.

Smith, R. W. (1977). Astigmatism-free holographic lens elements. *Optics Communications*, **21**, 102–5.

Smothers, W. K., Monroe, B. M., Weber, A. M. & Keys, D. E. (1990). Photopolymers for holography. In *Practical Holography IV*, Proceedings of the SPIE, vol. 1212, ed. S. A. Benton, pp. 20–9. Bellingham: SPIE.

Snow, K. & Vandewarker, R. (1968). An application of holography to interference microscopy. *Applied Optics*, **7**, 549–54.

Snow, K. & Vandewarker, R. (1970). On using holograms for test glasses. *Applied Optics*, **9**, 822–7.

Sollid, J. E. (1969). Holographic interferometry applied to measurements of small static displacements of diffusely reflecting surfaces. *Applied Optics*, **8**, 1587–95.

Solymar, L. (1977*a*). A general two-dimensional theory for volume holograms. *Applied Physics Letters*, **31**, 820–2.

Solymar, L. (1977*b*). Two-dimensional $N$-coupled wave theory for volume holograms. *Optics Communications*, **23**, 199–202.

Solymar, L. (1978). A two-dimensional volume hologram theory including the effect of varying average dielectric constant. *Optics Communications*, **26**, 158–60.

Solymar, L. & Cooke, D. J. (1981). *Volume Holography & Volume Gratings*. New York: Academic Press.

Som, S. C. & Budhiraja, C. J. (1975). Noise reduction by continuous addition of subchannel holograms. *Applied Optics*, **14**, 1702–5.

Song, Q. W., Lee, M. C., Talbot, P. J. & Tam, E. (1991). Optical switching with photorefractive polarization holograms. *Optics Letters*, **16**, 1228–30.

Song, S. H., Carey, C. D., Selviah, D. R., Midwinter, J. E. & Lee, E. H. (1993). Optical perfect shuffle interconnection using a computer-generated hologram. *Applied Optics*, **32**, 5022–5.

Sopori, B. L. & Chang, W. S. C. (1971). 3-D hologram synthesis from 2-D pictures. *Applied Optics*, **10**, 2789–90.

Spitz, E. (1967). Holographic reconstruction of objects through a diffusing medium in motion. *Comptes Rendus de l'Academie des Sciences B*, **264**, 1449–51.

St.-Hilaire, P., Benton, S. A., Lucente, M., Jepsen, M. L., Kollin, J., Yoshikawa, H. & Underkoffler, J. (1990). Electronic display system for computational holography. In *Practical Holography IV*, Proceedings of the SPIE, vol.1212, ed. S. A. Benton, pp. 174–82. Bellingham: SPIE.

St.-Hilaire, P., Benton, S. A. & Lucente, M. (1992). Synthetic aperture holography: a novel approach to three-dimensional displays. *Journal of the Optical Society of America. A*, **9**, 1969–77.

St.-Hilaire, P., Benton, S. A., Lucente, M. & Hubel, P. M. (1992). Color images with the MIT holographic display. In *Practical Holography VI*, Proceedings of the SPIE, vol.1667, ed. S. A. Benton, pp. 73–84. Bellingham: SPIE.

Staebler, D. L. (1977). Ferroelectric crystals. In *Holographic Recording Materials*, Topics in Applied Physics, vol. 20, ed. H. M. Smith, pp. 101–32. Berlin: Springer-Verlag.

Steel, W. H. (1970). Fringe localization and visibility in classical and hologram interferometers. *Optica Acta*, **17**, 873–81.

Stepanov, S. I., Kulikov, V. V. & Petrov, M. P. (1982). Intensification of traveling holograms in $Bi_{12}SiO_{20}$ crystals. *Soviet Technical Physics Letters*, **8**, 229–30.

Stepanov, S. I. & Petrov, M. P. (1984). Photorefractive crystals of the $Bi_{12}SiO_{20}$ type for interferometry, wavefront conjugation and processing of non-stationary images. *Optica Acta*, **31**, 1335–43.

Stepanov, S. I. & Petrov, M. P. (1985). Efficient unstationary holographic recording in photorefractive crystals under an external alternating electric field. *Optics Communications*, **53**, 292–5.

Stetson, K. A. (1967a). Holography with total internally reflected light. *Applied Physics Letters*, **11**, 225–6.

Stetson, K. A. (1967b). Holographic fog penetration. *Journal of the Optical Society of America*, **57**, 1060–1.

Stetson, K. A. (1968a). Improved resolution and signal-to-noise ratios in total internal reflection holograms. *Applied Physics Letters*, **12**, 362–4.

Stetson, K. A. (1968b). Holographic surface contouring by limited depth of focus. *Applied Optics*, **7**, 987–9.

Stetson, K. A. (1969). A rigorous theory of the fringes of hologram interferometry. *Optik*, **29**, 386–400.

Stetson, K. A. (1970a). Moiré method for determining bending moments from hologram interferometry. *Optics & Laser Technology*, **2**, 80–4.

Stetson, K. A. (1970b). The argument of the fringe function in hologram interferometry of general deformations. *Optik*, **31**, 576–91.

Stetson, K. A. (1970c). Effects of beam modulation on fringe loci and localization in time-average hologram interferometry. *Journal of the Optical Society of America*, **60**, 1378–88.

Stetson, K. A. (1971). Hologram interferometry of nonsinusoidal vibrations analysed by density functions. *Journal of the Optical Society of America*, **61**, 1359–62.

Stetson, K. A. (1972a). Fringes of hologram interferometry for simple nonlinear oscillations. *Journal of the Optical Society of America*, **62**, 297–8.

Stetson, K. A. (1972b). Method of stationary phase for analysis of fringe functions in hologram interferometry. *Applied Optics*, **11**, 1725–31.

Stetson, K. A. (1974). Fringe interpretation for hologram interferometry of rigid body motions and homogeneous deformations. *Journal of the Optical Society of America*, **64**, 1–10.

Stetson, K. A. (1975a). Fringe vectors and observed-fringe vectors in hologram interferometry. *Applied Optics*, **14**, 272–3.

Stetson, K. A. (1975b). Homogeneous deformations: determination by fringe vectors in hologram interferometry. *Applied Optics*, **14**, 2256–9.

Stetson, K. A. (1976). Holographic strain analysis by fringe localization planes. *Journal of the Optical Society of America*, **66**, 627.

Stetson, K. A. (1978). The use of an image derotator in hologram interferometry and speckle photography of rotating objects. *Experimental Mechanics*, **18**, 67–73.

Stetson, K. A. (1979). Use of projection matrices in hologram interferometry. *Journal of the Optical Society of America*, **69**, 1705–10.

Stetson, K. A. (1982). Method of vibration measurements in heterodyne interferometry. *Optics Letters*, **7**, 233–4.

Stetson, K. A. (1992) Phase-step interferometry of irregular shapes by using an edge-following algorithm. *Applied Optics*, **31**, 5320–5.

Stetson, K. A. & Brohinsky, W. R. (1986). Measurement of phase change in heterodyne interferometry: a novel scheme. *Applied Optics*, **25**, 2643–4.

Stetson, K. A. & Brohinsky, W. R. (1987). Electro-optic holography system for vibration analysis and nondestructive testing. *Optical Engineering*, **26**, 1234–9.

Stetson, K. A. & Brohinsky, W. R. (1988). Fringe-shifting technique for numerical analysis of time-average holograms of vibrating objects. *Journal of the Optical Society of America. A*, **5**, 1472–6.

Stetson, K. A. & Powell, R. L. (1965). Interferometric hologram evaluation and real-time vibration analysis of diffuse objects. *Journal of the Optical Society of America*, **55**, 1694–5.

Stetson, K. A. & Taylor, P. A. (1971). The use of normal mode theory in holographic vibration analysis with application to an asymmetrical circular disk. *Journal of Physics E: Scientific Instruments*, **4**, 1009–15.

Stewart, W. C., Mezrich, R. S., Cosentino, L. S., Nagle, E. M., Wendt, F. S. & Lohman, R. D. (1973). An experimental read-write holographic memory. *RCA Review*, **34**, 3–44.

Stirn, B. A. (1975). Recording 360° holograms in the undergraduate laboratory. *American Journal of Physics*, **43**, 297–300.

Streibl, N. (1989). Beamshaping with optical array generators. *Journal of Modern Optics*, **36**, 1559–73.

Streibl, N., Völkel, R., Schwider, J., Habel, P. & Lindlein, N. (1993). Parallel optoelectronic interconnections with high packing density through a light-guiding plate using grating couplers and field lenses. *Optics Communications*, **99**, 167–71.

Stroke, G. W. (1965). Lensless Fourier-transform method for optical holography. *Applied Physics Letters*, **6**, 201–3.

Stroke, G. W. (1966). White-light reconstruction of holographic images using transmission holograms recorded with conventionally focused images and in-line background. *Physics Letters*, **23**, 325–7.

Stroke, G. W., Brumm, D. & Funkhouser, A. (1965). Three-dimensional holography with 'lensless' Fourier-transform holograms and coarse P/N Polaroid film *Journal of the Optical Society of America*, **55**, 1327–8.

Stroke, G. W., & Restrick, R. C. (1965). Holography with spatially non-coherent light. *Applied Physics Letters*, **7**, 229–30.

Stroke, G. W. Restrick, R., Funkhouser, A. & Brumm, D. (1965). Resolution-retrieving compensation of source effects by correlation reconstruction in high resolution holography. *Physics Letters*, **18**, 274–5.

Stroke, G. W. & Zech, R. G. (1966). White light reconstruction of colour images from black and white volume holograms recorded on sheet film. *Applied Physics Letters*, **9**, 215–18.

Stroke, G. & Zech, R. G. (1967). A posteriori image-correcting 'deconvolution' by holographic Fourier-transform division. *Physics Letters*, **25A**, 89–90.

Su, F. & Gaylord, T. K. (1972). Calculation of arbitrary-order diffraction efficiencies of thick gratings with arbitrary grating shape. *Journal of the Optical Society of America*, **62**, 802–6.

Sugaya, T., Ishikawa, M., Hoshino, I. & Iwamoto, A. (1981). Holographic system for filing and retrieving patents. *Applied Optics*, **20**, 3104–8.

Suhara, T., Nishihara, H. & Koyama, J. (1975). The modulation transfer function in the hologram copying process. *Optics Communications*, **14**, 35–8.

Sutherlin, K. K., Lauer, J. P. & Olenick, R. W. (1974). Holoscan: a commercial holographic ROM. *Applied Optics*, **13**, 1345–54.

Suzuki, M., Saito, T. & Matsuoka, T. (1978). Multicolor rainbow hologram. *Kogaku*, **7**, 29–31.

Sweatt, W. C. (1977). Achromatic triplet using holographic optical elements. *Applied Optics*, **16**, 1390–1.

Sweeney, D. W. & Vest, C. M. (1973). Reconstruction of three-dimensional refractive index fields from multidirectional interferometric data. *Applied Optics*, **12**, 2649–64.

Syms, R. R. A. & Solymar, L. (1982). Noise gratings in photographic emulsions. *Optics Communications*, **43**, 107–10.

Syms, R. R. A. & Solymar, L. (1983). Planar volume phase holograms formed in bleached photographic emulsions. *Applied Optics*, **22**, 1479–96.

Tai, A., Yu, F. T. S. & Chen, H. (1979). Multislit one-step rainbow holographic interferometry. *Applied Optics*, **18**, 6–7.

Takai, N., Yamada, M. & Idogawa, T. (1976). Holographic interferometry using a reference wave with a sinusoidally modulated amplitude. *Optics & Laser Technology*, **8**, 21–3.

Takeda, M., Ina, H. & Kobayashi, S. (1982). Fourier-transform method of fringe-pattern analysis for computer-based topography and interferometry. *Journal of the Optical Society of America*, **72**, 156–60.

Tamura, P. N. (1977). Multicolor image from superposition of rainbow holograms. In *Clever Optics: Innovative Applications of Optics*, Proceedings of the SPIE, vol. 126, ed. N. Balasubramanian & J. C. Wyant, pp. 59–66. Redondo Beach: SPIE.

Tamura, P. N. (1978a). Pseudocolor encoding of holographic images using a single wavelength. *Applied Optics*, **17**, 2532–6.

Tamura, P. N. (1978*b*). One step rainbow holography with a field lens. *Applied Optics*, **17**, 3343.

Tanner, L. H. (1966). Some applications of holography in fluid mechanics. *Journal of Scientific Instruments*, **43**, 81–3.

Tatsuno, K. & Arimoto, A. (1980). Hologram recording by visible diode lasers. *Applied Optics*, **19**, 2096–7.

Tedesco, J. M. (1989). Holographic laser-protective filters and eyewear. *Optical Engineering*, **28**, 609–15.

Thalmann, R. & Dändliker, R. (1985). Holographic contouring using electronic phase measurement. *Optical Engineering*, **24**, 930–5.

Thalmann, R. & Dändliker, R. (1987). Strain measurement by heterodyne holographic interferometry. *Applied Optics*, **26**, 1964–71.

Thaxter, J. B. & Kestigian, M. (1974). Unique properties of SBN and their use in a layered optical memory. *Applied Optics*, **13**, 913–24.

Thinh, V. N. & Tanaka, S. (1973). Real time interferometry using thermoplastic hologram. *Japanese Journal of Applied Physics*, **12**, 1954–5.

Thompson, B. J. (1963). Fraunhofer diffraction patterns of opaque objects with coherent background. *Journal of the Optical Society of America*, **53**, 1350.

Thompson, B. J. (1974). Holographic particle sizing techniques. *Journal of Physics E: Scientific Instruments*, **7**, 781–8.

Thompson, B. J., Ward, J. H. & Zinky, W. R. (1967). Application of hologram techniques for particle size analysis. *Applied Optics*, **6**, 519–26.

Tischer, F. J. (1970). Analysis of ghost images in holography by the use of Chebyshev polynomials. *Applied Optics*, **9**, 1369–74.

Todorov, T. & Nikolova, L. (1989). Polarization holography: properties, registration materials and applications. In *Holography '89*, Proceedings of the SPIE, vol. 1183, eds. Yu. N. Denisyuk & T. H. Jeong, pp. 249–52. Bellingham: SPIE.

Todorov, T., Nikolova, L., Stoyanova, K. & Tomova, N. (1985). Polarization holography. 3: Some applications of polarization holographic recording. *Applied Optics*, **24**, 785–8.

Tomlinson, W. J. & Aumiller, G. D. (1975). Technique for measuring refractive index changes in photochromic materials. *Applied Optics*, **14**, 1100–4.

Tonin, R. & Bies, D. A. (1978). General theory of time-averaged holography for the study of three-dimensional vibrations at a single frequency. *Journal of the Optical Society of America*, **68**, 924–31.

Tontchev, D. & Zhivkova, S. (1992). Enhancement of the signal-to-noise ratio during holographic recording in sillenite. *Optics Letters*, **23**, 1715–17.

Tontchev, D., Zhivkova, S. & Miteva, M. (1990). Holographic interferometric microscope on the basis of a $Bi_{12}TiO_{20}$ crystal. *Applied Optics*, **29**, 4753–6.

Toyooka, S., Nishida, H. & Takezaki, J. (1989). Automatic analysis of holographic and shearographic fringes to measure flexural strains in plates. *Optical Engineering*, **28**, 55–60.

Trolinger, J. D. (1975*a*). Particle field holography. *Optical Engineering*, **14**, 383–92.

Trolinger, J. D. (1975*b*). Flow visualization holography. *Optical Engineering*, **14**, 470–81.

Troth, R. C. & Dainty, J. C. (1991). Holographic interferometry using anisotropic self-diffraction in $Bi_{12}SiO_{20}$. *Optics Letters*, **16**, 53–5.

Trukhmanova, T. D. & Denisyuk, O. V. (1977). Investigation of the applicability of locally manufactured fine-grain emulsions for obtaining relief structures. *Zhurnal Nauchnoi i Prikladnoi Fotografii i Kinematografii*, **22**, 178–81.

Tsujiuchi, J. (1963). Correction of optical images by compensation of aberrations and by spatial frequency filtering. In *Progress in Optics*, vol. 2, ed. E. Wolf, pp. 133–82. Amsterdam: North-Holland.

Tsukamoto, K., Ishii, A., Ishida, A., Sumi M. & Uchida, N. (1974). Holographic information retrieval system. *Applied Optics*, **13**, 869–74.

Tsunoda, Y., Tatsumo, K. & Kataoka, K. (1976). Holographic video disk: an alternative approach to optical video disks. *Applied Optics*, **15**, 1398–403.

Tsuruta, T., Shiotake, N. & Itoh, Y. (1968). Hologram interferometry using two reference beams. *Japanese Journal of Applied Physics*, **7**, 1092–100.

Tsuruta, T., Shiotake, N., Tsujiuchi, J. & Matsuda, K. (1967). Holographic generation of contour map of diffusely reflecting surface by using immersion method. *Japanese Journal of Applied Physics*, **6**, 661–2.

Uchida, N. (1973). Calculation of diffraction efficiency in hologram gratings attenuated along the direction perpendicular to the grating vector. *Journal of the Optical Society of America*, **63**, 280–7.

Uozato, H. & Nagata, R. (1977). Holographic photoelasticity by using dual-hologram method. *Japanese Journal of Applied Physics*, **16**, 95–100.

Upatnieks, J. (1967). Improvement of two-dimensional image quality in coherent optical systems. *Applied Optics*, **6**, 1905–10.

Upatnieks, J. (1988). Compact holographic sight. In *Holographic Optics: Design & Applications*, Proceedings of the SPIE, vol. 883, ed. L. Cindrich, pp. 171–6. Bellingham: SPIE.

Upatnieks, J. (1992). Edge-illuminated holograms. *Applied Optics*, **31**, 1048–52.

Upatnieks, J. & Embach, J. T. (1980). 360-degree hologram displays. *Optical Engineering*, **19**, 696–704.

Upatnieks, J. & Leonard, C. (1969). Diffraction efficiency of bleached, photographically recorded interference patterns. *Applied Optics*, **8**, 85–9.

Upatnieks, J. & Leonard, C. (1970). Efficiency and image contrast of dielectric holograms. *Journal of the Optical Society of America*, **60**, 297–305.

Upatnieks, J., Marks, J. & Federowicz, R. (1966). Color holograms for white light reconstruction. *Applied Physics Letters*, **8**, 286–7.

Upatnieks, J., Vander Lugt, A. & Leith, E. (1966). Correction of lens aberrations by means of holograms. *Applied Optics*, **5**, 589–93.

Urbach, J. C. (1977). Thermoplastic hologram recording. In *Holographic Recording Materials*, Topics in Applied Physics, vol. 20, ed. H. M. Smith, pp. 161–207. Berlin: Springer-Verlag.

Urbach, J. C. & Meier, R. W. (1966). Thermoplastic xerographic holography. *Applied Optics*, **5**, 666–7.

Urquhart, K. S., Lee, S. H., Guest, C. C., Feldman, M. R. & Farhoosh, H. (1989). Computer aided design of computer generated holograms for electron beam fabrication. *Applied Optics*, **28**, 3387–96.

Urquhart, K. S., Stein, R. & Lee, S. H. (1993). Computer-generated holograms fabricated by direct write of positive electron-beam resist. *Optics Letters*, **18**, 308–10.

Vagin, L. N., Nazarova, L. G., Arseneva, T. M. & Vanin, V. A. (1975). Holographic miniaturization of scientific and technical documents. *Optics & Spectroscopy*, **38**, 571–3.

Vagin, L. N. & Shtan'ko, A. E. (1974). Copying holograms by stamping on a thermoplastic. *Optics & Spectroscopy*, **36**, 597–8.

van Deelen, W. & Nisenson, P. (1969). Mirror blank testing by real time holographic interferometry. *Applied Optics*, **8**, 951–5.

Vander Lugt, A. (1964). Signal detection by complex spatial filtering. *IEEE Transactions on Information Theory*, **IT-10**, 139–45.

Vander Lugt, A. (1973). Design relationships for holographic memories. *Applied Optics*, **12**, 1675–85.

Vander Lugt, A., Rotz, F. B. & Klooster Jr., A. (1965). Character reading by optical spatial filtering. In *Optical & Electro-Optical Information Processing*, eds. J. T. Tippett, D. A. Berkowitz, L. C. Clapp, C. M. Koester & A. Vanderburgh Jr., pp. 125–41. Cambridge: Massachusetts Institute of Technology Press.

van Heerden, P. J. (1963a). A new optical method of storing and retrieving information. *Applied Optics*, **2**, 387–92.

van Heerden P. J. (1963b). Theory of optical information storage in solids. *Applied Optics*, **2**, 393–400.

Vanin, V. A. (1978). Hologram copying. *Soviet Journal of Quantum Electronics*, **8**, 809–18.

Vanin, V. A. (1979). Influence of the polarization of the object and reference waves on the hologram quality. *Soviet Journal of Quantum Electronics*, **9**, 774–6.

van Ligten, R. (1966). Influence of photographic film on wavefront reconstruction. II. 'Cylindrical' wavefronts. *Journal of the Optical Society of America*, **56**, 1009–14.

van Ligten, R. F. & Osterberg, H. (1966). Holographic microscopy. *Nature*, **211**, 282–3.

van Renesse, R. L. (1980). Scattering properties of fine-grained bleached emulsions. *Photographic Science & Engineering*, **24**, 114–19.

Vasara, A., Turunen, J. & Friberg, A. T. (1989). Realization of general nondiffracting beams with computer-generated holograms. *Journal of the Optical Society of America. A*, **6**, 1748–54.

Velzel, C. H. F. (1973). Non-linear holographic image formation. *Optica Acta*, **20**, 585–606.

Verboven, P. E. & Lagasse, P. E. (1986). Aberration coefficients of curved optical elements. *Applied Optics*, **25**, 4150–4.

Vest, C. M. (1979). *Holographic interferometry*. New York: John Wiley.

Vest, C. M. (1981). *Holographic NDE: Status & Future*. National Bureau of Standards Report NBS-GCR-81-318. Springfield: National Technical Information Service.

Vikhagen, E. (1991). TV holography: spatial resolution and signal resolution in deformation analysis. *Applied Optics*, **30**, 420–5.

Vikram, C. S. (1974a). Quadruple-exposure holographic interferometry for analysis of superposition of ramp motion and sinusoidal vibration. *Optics Communication*, **10**, 290–1.

Vikram, C. S. (1974b). Stroboscopic holographic interferometry of vibration simultaneously in two sinusoidal modes. *Optics Communications*, **11**, 360–4.

Vikram, C. S. (1975). Holographic interferometry of superposition of two motions. *Optik*, **43**, 65–70.

Vikram, C. S. (1976). Holographic interferometry of superposition of motions with different time functions. *Optik*, **45**, 55–64.

Vikram, C. S. (1977). Mechanical vibrations: mapping their phase with hologram interferometry. *Applied Optics*, **16**, 1140–1.

Vikram, C. S. (1992). *Particle Field Holography*. Cambridge: Cambridge University Press.

Vikram, C. S. & Billet, M. L. (1983). Gaussian beam effects in far-field in-line holography. *Applied Optics*, **22**, 2830–5.

Vikram, C. S. & Billet, M. L. (1984). Optimizing image-to-background irradiance ratio in far-field in-line holography. *Applied Optics*, **23**, 1995–8.

Vikram, C. S. & Billet, M. L. (1988). Some salient features of in-line Fraunhofer holography with divergent beams. *Optik*, **78**, 80–3.

Vilkomerson, D. H. R. & Bostwick, D. (1967). Some effects of emulsion shrinkage on a hologram's image space. *Applied Optics*, **6**, 1270–2.

Vlasov, N. G., Ryabova, R. V. & Semenov, S. P. (1977). Leith holograms reconstructed in white light. *Zhurnal Nauchnoi i Prikladnoi Fotografii i Kinematografii*, **22**, 384–5.

von Bally, G., ed. (1979). *Holography in Medicine and Biology*. Berlin: Springer-Verlag.

Walker, S. J. & Jahns, J. (1990). Array generation with multilevel phase gratings. *Journal of the Optical Society of America. A*, **7**, 1202–8.

Walles, S. (1969). Visibility and localization of fringes in holographic interferometry of diffusely reflecting surfaces. *Arkiv för Fysik*, **40**, 299–403.

Walkup, J. F. (1980). Space-variant coherent optical processing. *Optical Engineering*, **19**, 339–46.

Wang, M. R., Sonek, G. J., Chen, R. T. & Jannson, T. (1992). Large fanout optical interconnects using thick holographic gratings and substrate wave propagation. *Applied Optics*, **31**, 236–48.

Wang, X., Magnusson, R. & Haji-Sheikh, A. (1993). Real-time interferometry with photorefractive holograms. *Applied Optics*, **32**, 1983–6.

Ward, A. A. & Solymar, L. (1989). Diffraction efficiency limitations of holograms recorded in silver-halide emulsions. *Applied Optics*, **28**, 1850–5.

Waters, J. P. (1968). Three-dimensional Fourier transform method for synthesizing binary holograms. *Journal of the Optical Society of America*, **58**, 1284–8.

Waters, J. P. (1972). Object motion compensation by speckle reference beam interferometry. *Applied Optics*, **11**, 630–6.

Watrasiewicz, B. M. & Spicer, P. (1968). Vibration analysis by stroboscopic holography. *Nature*, **217**, 1142–3.

Watt, D. W., Gross, T. S. & Henning, S. D. (1991). Three-illumination beam phase-shifted holographic interferometry study of thermally induced displacements on a printed wiring board. *Applied Optics*, **30**, 1617–23.

Weaver, C. S. & Goodman, J. W. (1966). A technique for optically convolving two functions. *Applied Optics*, **5**, 1248–9.

Weber, A. M., Smothers, W. K., Trout, T. J. & Mickish, D. J. (1990). Hologram recording in Du Pont's new photopolymer materials. In *Practical Holography IV*, Proceedings of the SPIE, vol. 1212, ed. S. A. Benton, pp. 30–9. Bellingham: SPIE.

Webster, J. M., Tozer, B. A. & Davis, C. R. (1979). Holography of large volumes using holographic scatter plates. *Optics & Laser Technology*, **11**, 157–9.

Weingärtner, I. & Rosenbruch, K. J. (1980*a*). Incoherent polychromatic imaging with holographic optical elements and systems. *Optik*, 57, 103–22.

Weingärtner, I. & Rosenbruch, K. J. (1980*b*). Incoherent polychromatic imaging with holographic optical elements and systems. II. *Optik*, **57**, 161–71.

Weingärtner, I. & Rosenbruch, K. J. (1982). Chromatic correction of two- and three-element holographic imaging systems. *Optica Acta*, **29**, 519–29.

Weissbach, S., Wyrowski, F. & Bryngdahl, O. (1989). Digital phase holograms: coding and quantization with an error diffusion concept. *Optics Communications*, **72**, 37–41.

Welford, W. T. (1973). Aplanatic hologram lenses on spherical surfaces. *Optics Communications*, **9**, 268–9.

Wilson, A. D. (1970). Characteristic functions for time-average holography. *Journal of the Optical Society of America*, **60**, 1068–71.

Wilson, A. D. (1971). Computed time-average holographic interferometric fringes of a circular plate vibrating simultaneously in two rationally or irrationally related modes. *Journal of the Optical Society of America*, **61**, 924–9.

Wilson, A. D. & Strope, D. H. (1970). Time-average holographic interferometry of a circular plate vibrating simultaneously in two rationally related modes. *Journal of the Optical Society of America*, **60**, 1162–5.

Wilson, S. J. & Hutley, M. C. (1982). The optical properties of 'moth eye' antireflection surfaces. *Optica Acta*, **29**, 993–1009.

Windischbauer, G., Keck, F. G., Cabaj, A., Langschwert, H., Ranninger, G. & Tomiser, J. (1973). Polarization holography – a critical review. *Optik*, **37**, 385–90.

Witherow, W. K. (1979). A high resolution particle sizing system. *Optical Engineering*, **18**, 249–55.

Withrington, Q. J. (1978). Optical design for a holographic vision helmet-mounted display. In *Computer-Aided Optical Design*, Proceedings of the SPIE, vol. 147, ed. R. E. Fischer, pp. 161–70. Bellingham: SPIE.

Wolfke, M. (1920). Uber der Möglichkeit der optischen Abbildung vom Molekulargittern. *Physikalische Zeitschrift*, **21**, 495–7.

Wood, V. E., Cressmann, P. J., Holman, R. L. & Verber, C. M. (1989). Photorefractive effects in waveguides. In *Photorefractive Materials & Their Applications*, eds. P. Günter & J.-P. Huignard, vol. 2, pp. 45–100. New York: Springer-Verlag.

Worthington Jr., H. R. (1966). Production of holograms with incoherent illumination. *Journal of the Optical Society of America*, **56**, 1397–8.

Woznicki, J. (1980). Geometry of recording and color sensitivity for evanescent wave holography using a Gaussian beam. *Applied Optics*, **19**, 631–7.

Wu, S., Song, Q., Mayers, A., Gregory, D. A. & Yu, F. T. S. (1990). Reconfigurable interconnections using photorefractive holograms. *Applied Optics*, **29**, 1118–25.

Wuerker, R. F. & Heflinger, L. O. (1970). Pulsed laser holography. In *The Engineering Uses of Holography*, ed. E. R. Robertson & J. M. Harvey, pp. 99–114. Cambridge: Cambridge University Press.

Wüthrich, A. & Lukosz, W. (1975). Holography with evanescent waves. II. Measurements of the diffraction efficiencies. *Optik*, **42**, 315–34.

Wüthrich, A. & Lukosz, W. (1980). Holography with guided optical waves. I. Experimental techniques and results. *Applied Physics*, **21**, 55–64.

Wyant, J. C. (1977). Image blur for rainbow holograms. *Optics Letters*, **1**, 130–2.

Wyant, J. C. & Bennett, V. P. (1972). Using computer-generated holograms to test aspheric wavefronts. *Applied Optics*, **11**, 2833–9.

Wyant, J. C. & O'Neill, P. K. (1974). Computer generated hologram: null lens test of aspheric wavefronts. *Applied Optics*, **13**, 2762–5.

Wyant, J. C., Oreb, B. F. & Hariharan, P. (1984). Testing aspherics using two-wavelength holography: application of digital electronic techniques. *Applied Optics*, **23**, 4020–3.

Xu, S., Mendes, G., Hart, S. & Dainty, J. C. (1989). Pinhole hologram and its applications. *Optics Letters*, **14**, 107–9.

Yamazaki, K., Ichikawa, T., Aritake, H., Yamagishi, F., Ikeda, H. & Inagaki, T. (1990). New holographic technology for a compact POS scanner. *Applied Optics*, **29**, 1666–70.

Yan-Song, C., Yu-Tang, W. & Bi-Zhen, D. (1978). A new method of colour holography. *Acta Physica Sinica*, **27**, 723–8.

Yariv, A. & Koch, T. L. (1982). One-way coherent imaging through a distorting medium using four-wave mixing. *Optics Letters*, **7**, 113–15.

Yaroslovskii, L. P. & Merzlyakov, N. S. (1980). *Methods of Digital Holography*. New York: Consultants Bureau, Plenum Publishing Co.

Yatagai, T., Nakadate, S., Idesawa, M. & Saito, H. (1982). Automatic fringe analysis using digital image processing techniques. *Optical Engineering*, **21**, 432–5.

Yatagai, T. & Saito, H. (1978). Interferometer testing with computer-generated holograms: aberration balancing method and error analysis. *Applied Optics*, **17**, 558–65.

Yatagai, T. & Saito, H. (1979). Dual computer-generated holograms for testing aspherical surfaces. *Optica Acta*, **26**, 985–93.

Young, M. & Hicks, A. (1974). Holographic ruby laser with long coherence and precise timing. *Applied Optics*, **13**, 2486–8.

Yu, F. T. S. & Chen, H. (1978). Rainbow holographic interferometry. *Optics Communications*, **25**, 173–5.

Yu, F. T. S., Ruterbusch, P. H. & Zhuang, S. L. (1980). High-resolution rainbow holographic process. *Optics Letters*, **5**, 443–5.

Yu, F. T. S., Tai, A. & Chen, H. (1978). Archival storage of color films by rainbow holographic technique. *Optics Communications*, **27**, 307–10.

Yu, F. T. S., Tai, A. & Chen, H. (1979). Multiwavelength rainbow holographic interferometry. *Applied Optics*, **18**, 212–18.

Yu, F. T. S., Tai, A. & Chen, H. (1980). One-step rainbow holography: recent development and application. *Optical Engineering*, **19**, 666–78.

Yu, F. T. S. & Wang, E. Y. (1973). Speckle reduction in holography by means of random spatial sampling. *Applied Optics*, **12**, 1656–9.

Zager, S. A. & Weber, A. M. (1991). Display holograms in Du Pont's OmniDex films. In *Practical Holography V*, Proceedings of the SPIE, vol. 1461, ed. S. A. Benton, pp. 58–67. Bellingham: SPIE.

Zaidel, A. N., Ostrovskaya, G. V. & Ostrovskii, Yu. I. (1969). Plasma diagnostics by holography. *Soviet Physics – Technical Physics*, **13**, 1153–64.

Zambuto, M. & Lurie, M. (1970). Holographic measurement of general forms of motion. *Applied Optics*, **9**, 2066–72.

Zambuto, M. H. & Fischer, W. K. (1973). Shifted reference holographic interferometry. *Applied Optics*, **12**, 1651–5.

Zarrabi, K., Oreb, B. F. & Hariharan, P. (1990). Laser holographic interferometry and finite element analysis: a comparative assessment with reference to pressure vessels. *Non-destructive Testing Australia*, **27**, 64–8.

Zelenka, J. S. & Varner, J. R. (1968). New method for generating depth contours holographically. *Applied Optics*, **7**, 2107–10.

Zelenka, J. S. & Varner, J. R. (1969). Multiple-index holographic contouring. *Applied Optics*, **8**, 1431–4.

Zemtsova, E. G. & Lyakhovskaya, L. V. (1976). A study of a method of copying three-dimensional holograms. *Soviet Journal of Optical Technology*, **43**, 744–6.

Zetsche, C. (1982). Simplified realization of the holographic inverse filter: a new method. *Applied Optics*, **21**, 1077–9.

Zhivkova, S. & Miteva, M. (1990). Holographic recording in photorefractive crystals with simultaneous electron/hole transport and two active centers. *Journal of Applied Physics*, **68**, 3099–103.

# Author index

389

398                              *Author index*

# Subject index

Abel transform, 255
aberrations, 31
  classification, 32
  correction, 201
  correction of, 221
  in microscopy, 197
  of holographic optical elements, 221
achromatic images, 9, 156
  by dispersion compensation, 156
  with computer-generated holograms, 176
  with image holograms, 22, 37, 156
  with rainbow holograms, 157
  with white-light recording, 189
aliasing, 6, 320
amplitude holograms, 45, 85
  thin, 45
  volume reflection, 57
  volume transmission, 54
amplitude transmittance, 11
  and H & D curve, 335
  complex, 15, 85
  *vs* exposure curve, 15, 85
  *vs* log-exposure curve, 88
analytic signal, 313
anamorphic imaging, 35
angular magnification, 28
argon-ion laser, 37, 75, 77, 79, 108
  for colour holography, 145
  for contouring, 286
  output wavelengths, 79
  use with photoresists, 113
associative storage, 225
astigmatism, 34
autocorrelation, 319

beam fanning, 242
beam polarization, 72
beam ratio, 72, 75
beam splitters, 72, 83
  holographic, 223

polarizing prism, 73
binary detour-phase hologram, 164
  generalized, 168
binary hologram, 164, 176
$Bi_{12}GeO_{20}$, 120, 256
$Bi_{12}SiO_{20}$, 120, 202, 294
  for contouring, 286
  for vibration analysis, 291
bleach techniques, 102
  conventional, 102
  light stability, 106
  noise, 103
  rehalogenating, 104
  reversal, 102
Borrmann effect, 65
Bragg condition, 48–52, 55, 57, 60, 65, 120
  effect of emulsion shrinkage, 66
  effect of changes in permittivity, 62
  in multiply-exposed holograms, 65
  in three dimensions, 234

C.I.E chromaticity diagram, 145
carrier frequency, 6, 14
  minimum, 16
characteristic function, 272, 274–280
  generalized, 279
  probabilistic interpretation, 279
cinematography, 142
coded reference waves
  for multicolour imaging, 147
  for multiplexing images, 230
  for polarization recording, 181
  for space-variant image processing, 234
coherence, 312
  length, 75–77, 79, 80, 315
  of laser light, 76
  requirements for holography, 75
  spatial, 75, 314
  temporal, 75, 76, 314
  time, 315